Gorrod/Damani

Biological Oxidation of Nitrogen in Organic Molecules

Biological Oxidation of Nitrogen in Organic Molecules

Chemistry, Toxicology and Pharmacology

Edited by
John W. Gorrod and L. A. Damani

ELLIS HORWOOD
international publishers in science and technology

Prof. Dr. J. W. Gorrod
Head of Pharmacy
Chelsea College
University of London
London
England

Dr. L. A. Damani
Lecturer in Pharmacy
University of Manchester
Manchester
England

Library of Congress Card No.

Deutsche Bibliothek Cataloguing-in-Publication Data

Biological oxidation of nitrogen in organic molecules/ed. by John W. Gorrod and L. A. Damani.–
Weinheim; Deerfield Beach, Fl.: VCH; Chichester: Horwood, 1985.
(Ellis Horwood health science series)
   ISBN 3-527-26299-7 (Weinheim);
   ISBN 0-89573-422-2 (Deerfield Beach, FL.)

NE: Gorrod, John W. [Hrsg.]

British Library Cataloguing-in-Publication Data

Biological oxidation of nitrogen in organic molecules:
chemistry, toxicology and pharmacology.
   – (Ellis Horwood series in health science)
   1. Organonitrogen compounds – Metabolism
   2. Oxidation, Physiological
   I. Gorrod, J. W. II. Damani, L. A.
   574.19'24 QP801. N 55
   ISBN 3-527-26299-7 (VCH Verlagsgesellschaft)
   ISBN 0-89573-422-2 (VCH Publishers)

Published jointly in 1985 by
Ellis Horwood Ltd., Chichester, England
and VCH Verlagsgesellschaft mbH, Weinheim, Federal Republic of Germany

Distribution:

VCH Verlagsgesellschaft, P.O. Box 1260/1280, D-6940 Weinheim
(Federal Republic of Germany)

USA and Canada: VCH Publishers, 303 N.W. 12th Avenue, Deerfield Beach,
FL 33442-1705 (USA)

Dr. Robert E. McMahon 1923–1980

This volume is respectfully dedicated to the memory of Bob McMahon, a pioneer of xenobiochemistry, who with his colleagues at Lilly Research discovered the N-hydroxylation of imines as a metabolic process.

TABLE OF CONTENTS

Part 3.   Aromatic Amine Oxidation and Formation

Part 4. Oxidation of Amides and Carbamates

PREFACE

The metabolism of organic nitrogen compounds by N-oxygenation was discovered during the present century; Ellinger in 1920 apparently isolated
N-acetylphenylhydroxylamine from the blood of cats dosed with acetanilide
(Ellinger, A. *Hoppe.Sey.A.*, **111**, 86, 1920). Lintzel in 1934 reported that
trimethylamine administered to humans was largely excreted as
trimethylamine N-oxide (Lintzel, W. *Biochem.Z.*, **273**, 243, 1934). Since that
time, N-oxidation has been recognized as a fairly general route of metabolism
of nitrogenous xenobiotics. Interest in this metabolic pathway (i.e.
N-oxidation) was partly stimulated because of the discovery of the role of
products of such reactions in certain pharmacological and toxicological processes, although it is now recognized that N-oxidation does not always lead to
toxication. Much of the early work on this metabolic pathway was reviewed in
an edited book in 1978 (Gorrod, J. W., editor, *Biological Oxidation of Nitrogen*, Elsevier/North Holland, 1978). There have been many major
advances since that time, not only in the recognition of a vast array of
N-oxidation products in addition to hydroxylamines and N-oxides, but also in
the understanding of the physiological, pharmacological and toxicological
significance of these products. It therefore seemed appropriate to attempt to
collate this new data not only as an information dissemination exercise, but to
see if concepts were beginning to emerge from the detailed studies being
carried out in different laboratories. To this end, we invited our colleagues
involved in active research in this area, not only to provide critical reviews on
work published during the period 1978 to 1984, but also to make available
manuscripts on current work on N-oxidation. The response from the leading
authorities in this field are embodied in this book.

This book starts with an introductory chapter by the editors outlining the
different types of nitrogenous compounds, and their expected N-oxidation

products. This is followed by a section on the analysis of N-oxygenated products (hydroxylamines and N-oxides). The remainder of the texts have been arranged depending upon either the type of amino functionality involved (e.g. sections on azaheteroaromatic N-oxygenation, aromatic amine oxidation, oxidation of amides and carbamates, etc.), or on the enzyme system mediating the reaction (e.g. section on the flavin-containing monooxygenase).

The most exciting and novel findings in the last five years are those relating to the role of prostaglandin H synthetases and peroxidases in N-oxidation; data pertaining to these processes are given in two separate sections. The interaction of N-oxidized compounds with cellular constituents to initiate a tissue lesion still attracts a lot of attention. In addition to the pharmacological and toxicological aspects highlighted throughout, those reviews exclusively on the role of N-oxidation in toxicology are grouped together in the final section of this book. We would like to thank all contributors for submitting their manuscripts in time and for the way they have accepted our editorial decisions. Finally we hope that this compilation of data focuses attention on aspects of N-oxidation that still require more investigation and stimulates more research in these areas.

J. W. Gorrod and L. A. Damani
January 1985

INTRODUCTION

The biological activity of numerous xenobiotics is dependent on the presence of nitrogen atoms. It is therefore surprising that the metabolic fate of nitrogen functional groups was not studied, and the effect of such biotransformations on the pharmacological and toxicological properties of drug molecules not fully realized, until many years after those reactions involving oxidations at carbon. Nitrogen occurs in a wide variety of drugs, natural products, environmental chemicals, pesticides, food additives, etc., and there is now a considerable interest in metabolic N-oxidation of such nitrogenous chemicals.

The terms 'N-oxidation' and 'N-oxygenation' are often used interchangeably, but it must be stressed that oxidation at nitrogen can occur without the presence of oxygen in the product. The term N-oxygenation should be used to describe reactions that result in formation of N–O bonds. Much of the early work on metabolic N-oxygenation, on which this overview is based, has been extensively reviewed and readers new to this field should refer to the following articles for more detailed accounts: Bickel (1969); Weisburger & Weisburger (1973); Gorrod (1973); Coutts & Beckett (1977); Ziegler (1980); Damani (1982); Hlavica (1982).

The enzymatic N-oxygenation reactions occurring at organic nitrogen centres may be subdivided into N-hydroxylation and N-oxide formation, leading to formation of N-hydroxy compounds and amine N-oxides, respectively. N-Hydroxylation may be considered as the substitution of one of the hydrogens of an amino group by a hydroxyl group, although mechanistically the reaction may not be as simple. Primary and secondary aliphatic or aromatic amines afford N-hydroxy compounds referred to as hydroxylamines (see Table 1). The corresponding N-oxygenated metabolites of amides are hydroxamic acids. Imino functions are oxidized to oximes or to nitrones, whereas tertiary amines afford N-oxides. In many cases the primary N-oxygenated metabolites

Table 1

Common nitrogen functionalities in xenobiotics and their expected
N-oxygenated metabolites

Functional group	*N*-Oxygenated product(s)
Primary and Secondary aliphatic amines	*Hydroxylamines*
$R-CH_2-\ddot{N}H_2$	$R-CH_2-\ddot{N}HOH$
$R-CH_2-\ddot{N}HR'$	$R-CH_2-\ddot{N}(OH)R'$
Primary and Secondary aromatic amines	*Hydroxylamines*
$Ar-\ddot{N}H_2$	$Ar-\ddot{N}HOH$
$Ar-\ddot{N}HR'$	$Ar-\ddot{N}(OH)R'$
Aliphatic and Aromatic Amides	*Hydroxamic acids*
$R-\overset{\underset{\|}{O}}{C}-NH_2$	$R-\overset{\underset{\|}{O}}{C}-NHOH$
$R-\overset{\underset{\|}{O}}{C}-NHR'$	$R-\overset{\underset{\|}{O}}{C}-N(OH)R'$
Aromatic (and Aliphatic) Hydroxylamines	*Nitroso & Nitro*
$Ar-\ddot{N}HOH$	$Ar-\ddot{N}=O$ & $Ar-\overset{+}{N}\overset{\nearrow O}{\searrow O^-}$
Imines	*Oximes & Nitrones*
$R-CH=\ddot{N}H$	$R-CH=\ddot{N}-OH$
$R-CH=\ddot{N}-R'$	$R-CH=\underset{\underset{O^-}{\|}}{N^+}-R'$
Tertiary amines	*N-oxides*
$R_3N:$	$R_3N^{\pm}O^-$
$Ar-\underset{\underset{R_2}{\|}}{\overset{\overset{R_1}{\|}}{N}}:$	$Ar-\underset{\underset{R_2}{\|}}{\overset{\overset{R_1}{\|}}{N}}{}^{\pm}O^-$

are unstable and may undergo enzymic and/or non-enzymic conversion to other products. For example, hydroxylamines are further metabolized to nitroso and nitro compounds (see Table 1), whereas N-oxides may undergo reduction to the parent tertiary amine. In addition, some nitrogenous chemicals and/or their metabolites may afford reactive species such as nitroxide radicals and nitrenium ions. Specific examples of drugs and other xenobiotics that undergo the reactions outlined above will appear in subsequent sections of the book. The pharmacological and toxicological implications of such reactions are emphasized wherever appropriate, but those with a specific interest in postenzymatic chemistry and reactivity of N-oxygenated metabolites should refer to the last section on 'toxicological implications'.

REFERENCES

Bickel, M. H. (1969), *Pharmacol. Revs.*, **21**, 325.

Coutts, R. T. & Beckett, A. H. (1977), *Drug Metab. Revs.*, **6**, 51.

Damani, L. A. (1982), in *Metabolic Basis of Detoxication* (eds Jakoby, W. B., Bend, J. R. & Caldwell, J.) Adacemic Press, New York, p. 127.

Gorrod, J. W. (1973), *Chem. Biol. Interact.*, **7**, 289.

Hlavica, P. (1982), *CRC Critical Revs. in Biochem.*, **12**, 39.

Weisburger, J. H. & Weisburger, E. K. (1973), *Pharmacol. Revs.*, **25**, 1.

Ziegler, D. M. (1980), in *Enzymatic Basis of Detoxication* (ed. Jakoby, W. B.) Academic Press, New York, p. 201.

Part 1

ANALYSIS OF *N*-OXYGENATED COMPOUNDS

Chapter 1

APPROACHES TO THE ANALYSIS OF ARYLHYDROXYLAMINES WITH CONSIDERATION OF THEIR CHEMICAL REACTIVITY

L. A. Sternson, Department of Pharmaceutical Chemistry, University of Kansas, Lawrence, Kansas 66045, USA

1. The analysis of *N*-hydroxy metabolites of arylamines, which have been implicated as proximal carcinogens, is complicated by their chemical instability.
2. In aqueous solution, the arylhydroxylamine (AH) is oxidized to the corresponding nitroso, nitro, azo, azoxy compounds and nitrosophenol; at acidic pH, oxidation is minimized but rapid conversion to aminophenol takes place.
3. The mechanistic pathway of these reactions is described as are analytical methods for the AHs which consider contamination from these degradation products.
4. Methodology is based on a combination of liquid–liquid extraction, reversed phase h.p.l.c with and without chemical derivatization. Derivatization results in conversion of the AH to a stable *N*-hydroxyurea. Both the AH and the derivatized products can be detected spectrophotometrically or amperometrically.

INTRODUCTION

The mutagenicity and carcinogenicity of aromatic amines and nitro compounds has been related to their metabolic conversion to arylhydroxylamines
(AH) in the liver (Kadlubar *et al.*, 1976). The hydroxylamines are themselves
highly reactive species which are subject to conversion to proximal carcinogens (postulated to be various electrophilic intermediates) through
enzyme or chemically mediated processes (Miller & Miller, 1969). The need
for analytical methodology capable of monitoring the formation and subsequent turnover of AHs in biological media at metabolically achieved levels
has existed for many years. Unfortunately, success has been limited by (1) the
chemical lability of AHs resulting in spontaneous conversion to multiple products, (2) the need for high resolution separation techniques capable of isolating the AH from the biological matrix, as well as from its chemical and
biochemical degradation products, under conditions and in a time frame that
minimizes significant degradation during the analysis sequence, and (3) the
low steady-state levels of AHs that are achieved under biologically relevant
conditions necessitating the availability of exquisitely sensitive detection systems. Thus, progress in analytical development has been limited by both the
chemical reactivity of the analyte and technological inadequacies.

CHEMICAL REACTIVITY OF ARYLHYDROXYLAMINES

Before one can attempt to understand the metabolic activation of AHs, an
appreciation of their chemical reactivity in aqueous solution (in the absence of
biological material) at physiologically relevant pH must be realized. Accordingly, we initiated an investigation of the chemistry of phenylhydroxylamine
(PHA, in aqueous buffers), which, although not a potent carcinogen, was
chosen because structurally it is the simplest representative of this class of
compounds (Becker & Sternson, 1981). In the absence of O_2, PHA was stable in
aqueous solution, but following the introduction of O_2, PHA was converted
to nitrosobenzene, azoxybenzene (AzB) and nitrobenzene. At pH $\leq$ 5.8, a
fourth product, *p*-nitrosophenol (PNP) was also found, and at pH $\leq$ 2, PHA
rearranges to the corresponding aminophenol (PAP, a reaction not requiring
O_2). The loss of PHA from O_2-saturated aqueous solution (pH 5–7.6; $[O_2]$,
2.2 mM) followed first order kinetic behaviour with a first order dependency
on the partial pressure of O_2. The h.p.l.c procedure developed to monitor the
reaction was capable of monitoring substrate loss as well as product appear-

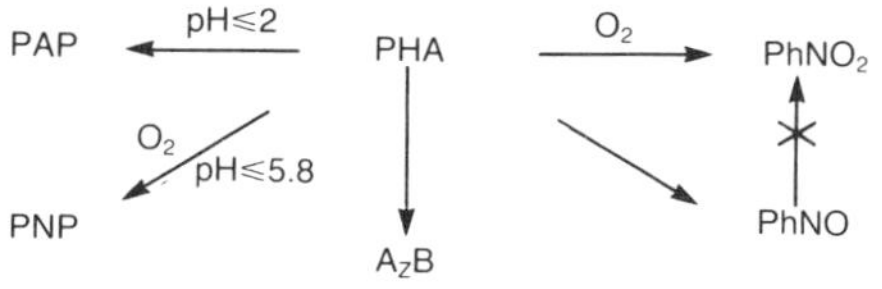

Scheme 1 – Chemical degradation of PHA in aqueous buffer solution

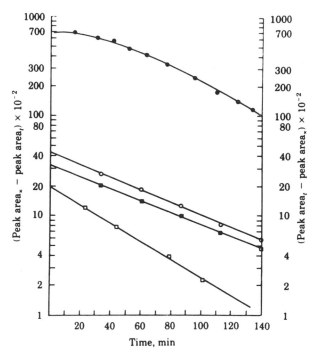

Fig. 1 – Distribution profile. PHA (○) disappearance (calculated as peak area$_t$–peak area$_\infty$, where peak area$_\infty$ = 0) and AzB (●), PNP (■), and PhNO$_2$ (□) appearance (calculated as peak area$_\infty$ – peak area$_t$) 0.01 M cacodylate buffer (pH 5.3; μ = 0.5, NaClO$_4$) at 25 °C.

ance. A representative plot (shown in Fig. 1) of substrate disappearance and product formation in pH 5.3 (μ = 0.5, NaClO$_4$) cacodylate buffer (25 °C) shows the apparent first order loss of PHA, first order formation of PNP and PhNO$_2$ (no lag phase) and production of AzB following a lag phase representing the time needed for appreciable quantities of PhNO to be generated which subsequently condense with PHA in a bimolecular process. The rate of PHA degradation was dependent on buffer concentration in both cacodylate and phosphate buffers. Plots of $k_{obs}/[\mathrm{O_2}]$ vs total buffer concentration were linear, with non-zero intercepts, suggesting that PHA disappearance proceeds through both buffer-catalysed (k_B) and buffer independent (k_{OX}) pathways, as described by equation (1),

$$k_{obs} = (k_B B_T + k_{OX})[\mathrm{O_2}], \tag{1}$$

where B_T is total buffer concentration. Plots of k_{OX} vs $1/a_H$ were linear but had non-zero intercepts, suggesting that k_{OX} is a composite of a hydroxide-dependent (k_{HO}) and a pH-independent (k_O) term as described by equation (2),

$$k_{OX} = k_{HO} K_W a_H^{-1} + k_O, \tag{2}$$

where a_H is the activity of $\mathrm{H_3O^+}$, as measured with a glass electrode. k_{HO} was

Table 1
Kinetic parameter for PHA oxidation

Buffer	$k_o \times 10^1$, $M^{-1} sec^{-1}$	$k_{HO} \times 10^{-1}$, $M^{-2} sec^{-1}$	k_{ga} $M^{-2} sec^{-1}$	k_{gb} $M^{-2} sec^{-1}$
Phosphate	1.5	2.3	2.3	2.3×10^5
Cacodylate	0.6	2.3	3.5	4.1

PHA was incubated in the appropriate O_2-saturated buffer (five buffer concentrations at each pHj), and substrate disappearance was monitored by HPLC. k_o and k_{HO} were determined graphically from equations (1) and (2). k_{ga} and k_{gb} were determined graphically. Values represent the average of triplicate determinations.

calculated from the slope of plots of k_{OX} vs a_H^{-1} and was identical for cacody-late and phosphate buffers, lending credibility to comparisons of these two systems (Table 1). The inconsistency of k_O values between the two buffer systems (a difference of a factor of 2.5) probably reflects the inaccuracy inherent in calculation or a minor kinetic salt effect (a factor of 2.5 in k_O corresponds to an energy difference of <0.55 kcal mol^{-1}).

The buffer-dependent term ($k_B B_T$, see equation 1) was, in both buffer systems, assumed to consist of general acid (k_{ga}) and a general base (k_{gb})-catalysed component, as described by equation (3),

$$k_B = k_{ga}[\text{HX}] + k_{gb}[\text{X}^-], \tag{3}$$

where HX and X$^-$ are the acid and conjugate-base forms of the buffer, respectively.

From equations (1) and (3), an expression (for k_{obs}) for PHA degradation in cacodylate buffer is generated (equation 4),

$$k_{obs}/[\text{O}_2] = k_{OX} + (k_{ga}[\text{HA}] + k_{gb}[\text{A}^-]), \tag{4}$$

where HA and A$^-$ are cacodylic acid and cacodylate, respectively. Rearranging equation (4) and expressing HA and A$^-$ in terms of total cacodylate (C_T) concentration gives equation (5),

$$k_{obs} = \left[\left(k_{ga}\frac{a_H}{K_a + a_H} + k_{gb}\frac{K_a}{K_a + a_H}\right)C_T + k_{OX}\right][\text{O}_2], \tag{5}$$

where K_a is the dissociation constant for cacodylic acid ($10^{-6.2}$).

k_B is given by the slope of plots of $k_{obs}/[\text{O}_2]$ vs C_T, and, from it, k_{gb} and k_{ga} can be determined graphically. This treatment of the kinetic data supports the hypothesis that buffer catalysis by cacodylate involves both general acid and general-base terms (or kinetically equivalent expressions). Similar terms were generated in phosphate buffer where $H_2PO_4^-$ and PO_4^{3-} (or kinetically equi-valent terms) were found to be the general acid and base species, respectively.

The product distribution showed no simple correlation with the degree to which the kinetic terms of equations (4) and (5) contribute to the rate of PHA degradation, suggesting that rates and products are determined in different processes, and that therefore, there must be at least one intermediate in the reaction sequence.

ROUTES TO FORMATION OF PRODUCTS

Several observations suggest that $PhNO_2$ and PNP form independently of AzB and PhNO: (1) both PhNO and AzB are stable in O_2-saturated buffers, (2) no simple relationship exists between the kinetic terms of equation (4) and the measured yields of $PhNO_2$ and PNP, and (3) the production of $PhNO_2$ and PNP proceeds with apparent first-order kinetic behaviour. A lag phase would be anticipated if PhNO, AzB, or some other compound was a precursor to PNP or $PhNO_2$ formed in the rate-determining step. Although this lag phase would not be observed if the final oxidation step was fast relative to the rate-determining step, under these conditions, a proportionality would exist between the concentration of PNP or $PhNO_2$ and that of PhNO or AzB. Neither of these situations was observed, however; thus, it appears that PNP and $PhNO_2$ form directly for PHA. Their production is, however, dependent on the presence of O_2.

Bamberger (1894, 1900) has reported that the autoxidation of PHA generates hydrogen peroxide, and the involvement of H_2O_2 as an oxidizing agent in producing $PhNO_2$ or PNP was investigated. PhNO is quantitatively converted to $PhNO_2$ by H_2O_2, and the rate of disappearance of PhNO is equal to the rate of $PhNO_2$ production. However, the rate constant for this reaction ($k_{H_2O_2}$ = 2.8×10^{-5} M^{-1} sec^{-1}) is too small to account for the observed yield of $PhNO_2$. The possibility that $PhNO_2$ and/or PNP is produced by H_2O_2 oxidation of PHA was investigated; but similarly ruled out based on kinetic evidence (Becker & Sternson, 1981).

Superoxide anion (O_2^-) was investigated as an alternative oxidizing agent for the production of $PhNO_2$ and PNP. It reacted instantaneously in buffer (pH 6.4–7.4) with PHA or PhNO to form $PhNO_2$ in quantitative yield. To clarify its role, the effect of superoxide dismutase (SOD), an enzyme that catalyses disproportionation of O_2^- to O_2 and H_2O_2 (McCord & Fridovich, 1969), on the kinetics of PHA disappearance and $PhNO_2$ formation was studied. SOD (1 mg/25 ml of buffer) failed to retard $PhNO_2$ formation, suggesting that superoxide does not participate in the reaction sequence.

By elimination of these simple oxidizing species, it appears that a reactive intermediate, perhaps one forming between O_2 and PHA, is responsible for generating $PhNO_2$ and PNP. The direct involvement of O_2 in the formation of $PhNO_2$ was determined by carrying out the oxidation of PHA in phosphate buffer in an environment supplemented with ^{18}O-labelled O_2. The $PhNO_2$ formed in this experiment was enriched to an extent of $82 \pm 4\%$ with one atom of ^{18}O per molecule, as determined by isotope-ratio mass spectrometry

(M^+, 123 and 125). Doubly-labelled $PhNO_2$ (M^+, 127) was not formed. A similar experiment with $^{16}O_2$ in buffer enriched with $H_2^{18}O$ resulted in formation of $PhNO_2$ with no ^{18}O incorporation. Thus, $PhNO_2$ appears to form by direct involvement of O_2 with PHA to give of a reactive oxygenated intermediate that subsequently collapses to yield product. A possible mechanism

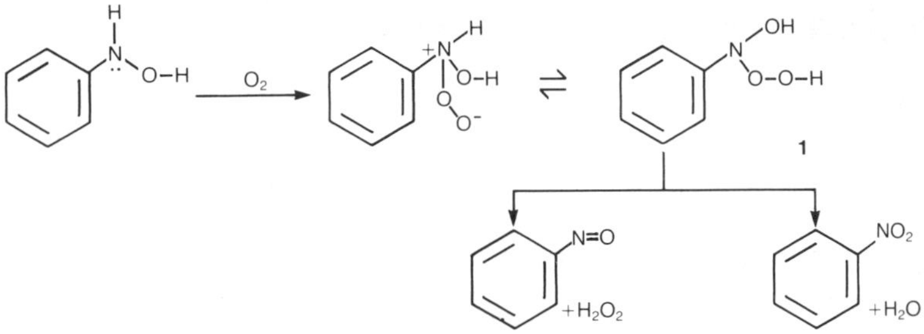

for $PhNO_2$ formation is shown below, in which elimination of water from the oxygenated intermediate, **1**, produces $PhNO_2$. The PhNO formed in these reactions contained no ^{18}O label suggesting that its formation either does not involve an oxygenated intermediate (but rather dehydrogenation of PHA) or that the oxygenated intermediate eliminates the elements of H_2O_2 to form PhNO. **1** is also consistent with the O_2-dependent formation of PNP. Attack of water (or HO^-) at the para position of this intermediate concomitant with (or subsequent to) buffer/solvent-assisted expulsion of HO^- would give **2**, which, on elimination of the elements of water, gives PNP.

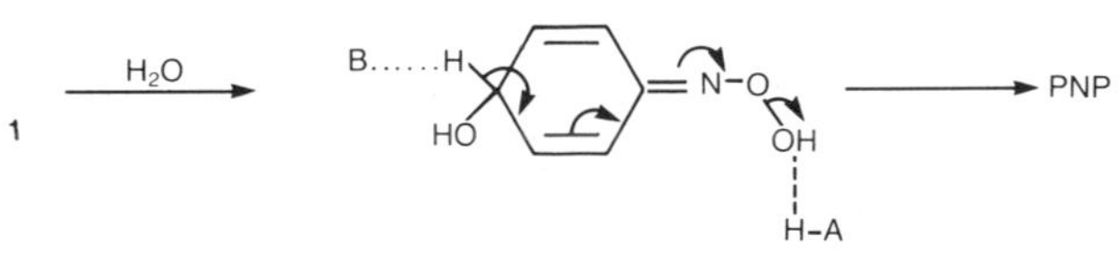

Rearrangement of AHs to aminophenols only occurred in strongly acidic (pH $\leq$ 2) solutions (Sternson & Chandrasakar, 1984). Both intra-(Yukawa, 1950) and intermolecular (Meller *et al.*, 1951) pathways have been postulated for this reaction. To help elucidate the course of reaction, two model compounds 3-(2-hydroxy)ethoxyphenylhydroxylamine, **3**, and 3-ethoxyphenylhydroxylamine, **4**, were synthesized and subjected to acidic media under anaerobic conditions. If rearrangement were to proceed *via* an S_N2

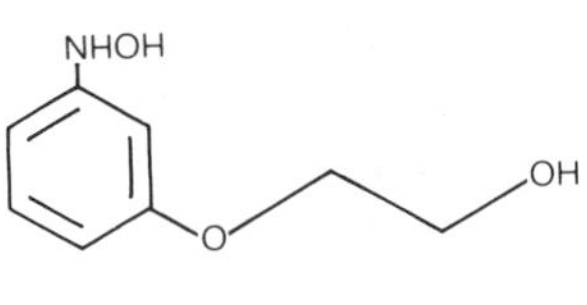

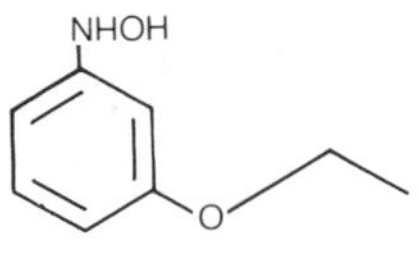

mechanism, the hydroxyl group in **3** may be capable of providing intra-molecular catalytic rate enhancement of AH degradation while **4** is not capable of participating in such anchimerically assisted rearrangements. The rates of disappearance of the two substrates were not significantly different nor was their degradation pH dependent over the range (pH 0.5–2.0) studied. Whereas **4** was converted to the corresponding aminophenol (**5**), 1 degraded to 6-aminobenzo-1,4-dioxin, **6** (Sternson & Chandrasakar, 1984).

5   6   7

These results are consistent with a unimolecular mechanism for AH rearrangement in which nitrenium ion (**7**) formation from the hydroxyl-ammonium ion is rate determining followed by rapid attack by available nucleophilic species and support the recent findings of Sone *et al.* (1981) and Kohnstan *et al.* (1984).

HPLC ANALYSIS OF ARYLHYDROXYLAMINES AND THEIR DEGRADATION PRODUCTS

The component complexity of solutions containing AHs (both of chemical and biological origin) necessitates a high efficiency separation step as part of the analysis sequence. The physical and chemical properties of AHs suggests the use of a mild technique such as reversed phase partition chromatography (r.p.–h.p.l.c.) to achieve the separation. We have carried out such chromatography on AHs themselves, as well as on products formed following their pre-column derivation with various acylating agents. The initial discussion will focus on chromatography of PHA and its degradation products without inclusion of a pre-column derivatization step.

Analytes could conveniently be separated on a RP-18 column eluted with a binary mobile phase consisting of alcohol and aqueous buffer (Sternson & DeWitte, 1977a,b; Sternson *et al.*, 1983a). To minimize on column degradation of AHs, mobile phases should be deoxygenated. This is most effectively accomplished with helium, which minimizes 'out-gassing' in the detector. We also attempted to protect PHA from autoxidation by the addition of bisulphite (HSO_3^-) to the mobile phase as an antioxidant. However, HSO_3^- significantly accelerated PHA degradation (at pH 5–7) converting PHA to aniline, *o*- and *p*-aminophenol and *o*- and *p*-aminobenzenesulphonate. This reaction sequence was investigated (Sternson *et al.*, 1983b) further and shown to exhibit a first order dependency on HSO_3^- (equation 6).

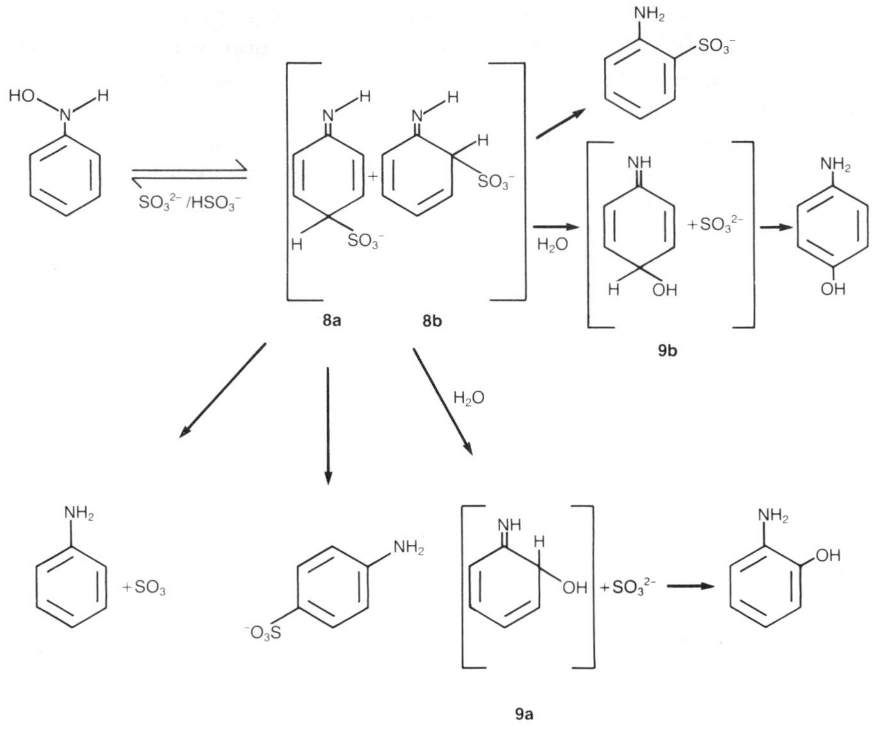

Scheme 2 – (Bi)sulphite initiated degradation by PHA

$$k_{\text{obs}} = \overset{\circ}{\overrightarrow{k_o}} + K_s S_T = k' C_{\text{HSO}_3} + k'' C_{\text{SO}_3} \tag{6}$$

where k_s is the sulphite dependent term, S_T is total sulphite concentration and the spontaneous rate constant, k_o, is equal to zero. All products thus result from intermediates formed from nucleophilic attack of both sulphite ($k'' = 103$ $\text{M}^{-1}\,\text{s}^{-1}$) and bisulphite ($k' = 40.5$ $\text{M}^{-1}\,\text{s}^{-1}$) on PHA with subsequent covalent addition–elimination processes leading to products (Scheme 2). Bisulphite and sulphite undergo similar reversible covalent addition–elimination reactions with uracils and 5-halouracils (Rork & Pitman, 1974, 1975) and with thiamine (Zoltewicz & Kauffman, 1977).

USE OF METALS IONS IN MOBILE PHASE

Solutions containing AHs and also *o*-aminophenol (OAP) were not resolvable on reverse phase systems irrespective of the mobile phase composition tried Sternson & DeWitte, 1977a,b). In all cases, these two analytes coeluted, not allowing their quantitative determination. Resolution was achieved by the addition of Ni^{2+} (in the form of nickel acetate, ~50 mM) to the mobile phase. Separation was achieved by virtue of the fact that Ni^{2+} selectively complexes with *o*-aminophenol forming a hydrophilic chelate (whereas PHA does not) which elutes more rapidly than the uncomplexed hydroxylamine. In the above example, although the capacity factor, k'_{OAP}, decreased with addition of Ni^{2+}, k'_{PhNHOH}, k'_{PNP}, $k'_{p\text{-}HOPhNH_2}$, k'_{PhNH_2}, k'_{PhNO} and k'_{PhNO} were unchanged. Since only certain functionalities with particular orientations bind to specific metal ions, such secondary equilibria can provide unique opportunities to achieve selective separations. The effectiveness of metal ion addition in modifying retention behaviour is determined by the magnitude of the complexation constant,

which in turn is a function of mobile phase pH, organic modifier concentration and metal ion chosen (Sternson et al., 1983a). In the presence of Ni^{2+}, the retention of all analytes tested (except OAP) increased slightly ($\leq 10\%$) with pH in comparison with the response observed in the absence of Ni^{2+}. However, o-aminophenol exhibits different behaviour (Fig. 2), in that retention increased up to pH 5.8 and then decreased. The increased retention would be expected to result from deprotonation of the anilinium ion to yield the less polar amine. However, as this basic form increases in concentration, it can chelate with Ni^{2+} to form a charged hydrophilic species, which elutes more rapidly. Prior to deprotonation, the nitrogen electron pair is unavailable for interaction with Ni^{2+}. Thus, it is to be expected that maximum chelation should be observed when the amine is in the free base form.

The effect of addition of nickel ion (up to 0.07 M) to the mobile phase was greatly dependent on the volume fraction of organic modifier in the mobile phase (Fig. 3). The addition of nickel ion only led to measurable changes in retention with mobile phases containing $\leq 23\%$ methanol (or $\leq 20\%$ acetonitrile), the effect of the metal ion being more pronounced as the fraction of organic modifier decreased. With no organic modifier in the mobile phase, addition of increasing amounts of Ni^{2+} resulted in progressively decreased retention. The slopes of plots of k' vs Ni^{2+} concentration decreased as organic modifier concentration in the mobile phase increased up to 30% MeOH (or 20% acetonitrile). This behaviour is to be anticipated since complexation constants for such chelates decrease rapidly as the organic composition of the solvent mixture is increased.

The influence of ions other than Ni^{2+} on analyte retention was also examined. Zn^{2+} and Cd^{2+} produced minimal effects on retention, whereas Hg^{2+} was a less selective complexing agent, interacting with not only OAP but also with aniline, p-aminophenol and PHA (which are incapable of chelation) as evidenced by decreases in their respective k' with increasing concentration of Hg^{2+}. The degree of which Hg^{2+} facilitated elution varied, reflecting differences in the extent to which the solutes complex with this ion.

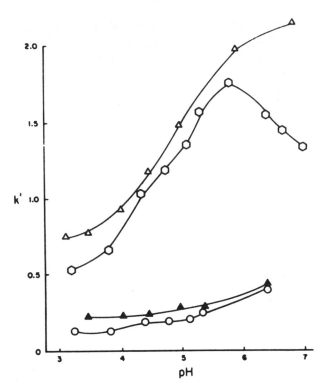

Fig. 2 – Retention (as capacity factor, k') of o- and p-aminophenol with methanol–water (15:85 v/v) as eluent in the presence (50 mM) and absence of nickel acetate, determined as a function of pH; the ionic strength remained constant at 0.47 M. o-Aminophenol: no Ni^{2+}, $\triangle$; with Ni^{2+}, $\bigcirc$. p-aminophenol: no Ni^{2+}, $\blacktriangle$, with Ni^{2+}, $\bigcirc$.

Thus, a variety of parameters can be utilized to control the chromatographic behaviour of AHs and their chemical and biodegradation products. Arylhydroxylamines and their reaction products are conveniently detected in h.p.l.c column effluent by spectrophotometry (although modest absorptivities limit sensitivity), although certain functionalities can alternatively be monitored amperometrically (Sternson & DeWitte, 1977b). In particular, AHs and aminophenols are electroxidizable (at carbon anodes), whereas aromatic nitro, nitroso and azoxy compounds are reduced at mercury cathodes. Unfortunately, the poor behaviour of many of these analytes at electrode surfaces limits the achievable detectability.

PRE-COLUMN DERIVATIZATION OF ARYLHYDROXYLAMINES

To improve sensitivity to a level sufficient to monitor AHs in biologically relevant situations and to protect them from autoxidation during analysis, pre-column derivatization schemes have been developed for AHs. Taking

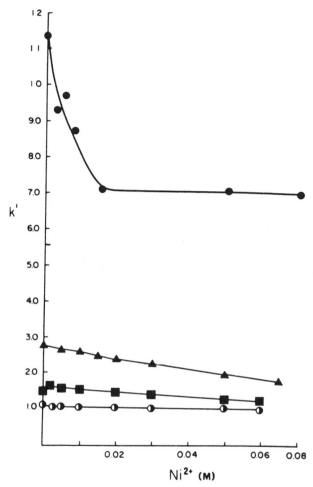

Fig. 3 – Retention (as capacity factor, k') of o-aminophenol on a reversed-phase column, with aqueous eluents containing 0% (●), 15% (▲), 23% (■) and 30% (◑) methanol, determined as a function of nickel acetate concentration in the eluent.

advantage of their nucleophilicity, AHs were acylated with isocyanates to form stable hydroxyureas (Sternson *et al.*, 1978). Whereas isocyanates directed acylation exclusively *at nitrogen*, other common acylating agents (anhydrides; acyl halides) gave predominantly O-acylation resulting in rapid degradation of the derivatives via Bamberger-type rearrangements (Bamberger, 1894). The N-acylation products were stable. Using methyl isocyanate as a model reagent, stability of PHA was enhanced from $t_{1/2} < 2$ min to $t_{1/2} \sim 24$ h for the corresponding hydroxyurea. The derivative did have significantly greater absorptivity than the AH but it was still insufficient to monitor these species at the levels anticipated ($\sim 10^{-8}$–10^{-9} M) in biochemical experiments.

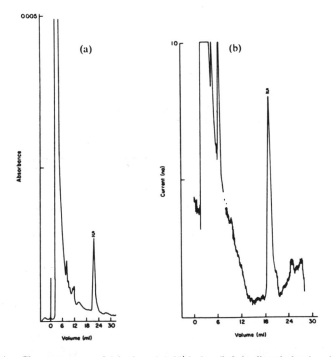

Accordingly, a new isocyanate, *p*-*N*,*N*-dimethylaminophenyl isocyanate, **10**, was designed to offer improved sensitivity (Musson & Sternson, 1980). Although **10** is electrochemically inactive (at carbon anodes at potentials ≤ 0.9 V vs Ag/AgCl), it reacts with AHs to form the corresponding hydroxyurea, **11**, which is readily oxidized at $E_p \sim 0.3$ V. Thus, **10** is an *electrogenic*

Fig. 4 – Chromatogram of 1-hydroxyl-1-(5'-indanyl)-3-(*p*-dimethylaminophenyl)-urea (**11**) obtained by reaction of hydroxyaminoindan extracted from liver microsomal homogenates with *p*-dimethylaminophenylisocyanate (**10**). Separation carried out on an RP-18 column with methanol–water–acetic acid (33:49:1) mobile phase and flow rate = 1.5 ml/min. Analyte detected (a) spectrophotometrically at 254 nm and (b) amperometrically at +0.5 V vs Ag/AgCl reference electrode.

reagent, not responding at the detector but forming a derivative that is responsive. The minimal background signal and electroinactivity of the corresponding urea (the major breakdown product of 10), at the applied potential optimal for analysis, allows for low level amperometric detection of the AHs (10^{-9} M). Additionally, the high molar absorptivity of the derivative (37,200 at λ = 254 nm) allows AH monitoring by spectrophotometric transducers at levels approaching 10^{-7} M. A sample chromatogram showing both the UV and amperometric tracing of the derivative formed from hydroxylaminoindan following chromatographic separation on a reversed phase column is shown in Fig. 4. This analysis illustrates AH analyses from a liver microsomal preparation, involving initial extraction with dichloromethane, derivatization followed by an h.p.l.c separation step.

In summary, the development of analytical methodology for monitoring AHs in biological systems is a formidable problem intensified by (1) their profound reactivity resulting in transformation to multiple products, (2) the low steady-state levels achieved in biological systems, and (3) the multiplicity of components in mixtures containing AHs and their chemical and biodegradation products. Efforts need to continue in describing the chemistry of AHs and utilizing this information in the design of improved analytical methodology. Such objectives will be further aided by the achievement of new improved chromatographic approaches and technology as well as more sensitive, universal detection devices.

REFERENCES

Bamberger, E. (1894), *Chem. Ber.*, **27**, 1548.
Bamberger, E. (1900), *Chem. Ber.*, **33**, 113.
Becker, A. R. & Sternson, L. A. (1981), *Proc. Natl. Acad. Sci.*, **78**, 2003.
Kadlubar, F. F., Miller, J. A. & Miller, E. C. (1976), *Cancer Res.*, **36**, 1196.
Kohnstam, G., Petch, W. A. & Williams, D. L. H. (1984), *J. Chem. Soc. Perkin Trans.*, 423.
McCord, J. M. & Fridovich, I. (1969), *J. Biol. Chem.*, **244**, 6049.
Meller, H. E., Hughes, E. D. & Ingold, C. K. (1951), *Nature*, **168**, 909.
Miller, J. A. & Miller, E. C. (1969), *Progr. Exp. Tumor Res.*, **11**, 273.
Musson, D. & Sternson, L. A. (1980), *J. Chromatogr.*, **188**, 159.
Rork, G. S. & Pitman, I. H. (1974), *J. Amer. Chem. Soc.*, **96**, 4654; (1975) **97**, 5559; **97**, 5566.
Sone, T., Tokuda, Y., Sakai, T., Shinkai, S. & Manabe, O. (1981), *J. Chem. Soc. Perkin Trans.*, 298.
Sternson, L. A. & DeWitte, W. J. (1977a), *J. Chromatogr.*, **137**, 305.
Sternson, L. A. & DeWitte, W. J. (1977b), *J. Chromatogr.*, **138**, 229.
Sternson, L. A., DeWitte, W. J. & Steven, J. (1978), *J. Chromatogr.*, **153**, 481.

Sternson, L. A., Dixit, A. S. & Riley, C. M. (1983a), *J. Pharm. Biomed. Analysis*, **1**, 105.
Sternson, L. A., Dixit, A. S. & Becker, A. R. (1983b), *J. Org. Chem.*, **48**, 57.
Sternson, L. A. & Chandrasakar, R. C. (1984), *J. Org. Chem.*, **49**, 4295.
Yukawa, Y. (1950). *J. Chem. Soc. Jap.*, **71**, 547.
Zoltewicz, J. A. & Kauffman, G. M. (1977), *J. Amer. Chem. Soc.*, **99**, 3134.

Chapter 2

COLORIMETRIC AND CHROMATOGRAPHIC ASSAY OF TERTIARY AMINE *N*-OXIDES

L. A. Damani, Department of Pharmacy, University of Manchester, Manchester, M13 9PL, UK.

1. *N*-Oxygenation of tertiary amines (i.e. *N*-oxide formation) is now recognized as being an important metabolic pathway for many xenobiotics.
2. Sensitive colorimetric methods are available for the assay of some classes of amine *N*-oxides; such methods are, however, often non-specific.
3. Sensitive and specific chromatographic assay methods based on gas liquid chromatography (g.l.c.) and high performance liquid chromatography (h.p.l.c.) are gaining in popularity.
4. This review outlines the various assay methods available, and attempts to demonstrate the advantages of chromatographic techniques by comparing the chromatographic and colorimetric assays of *N,N*-dimethylaniline *N*-oxide.

INTRODUCTION

N-Oxides may be formed as metabolites of drugs and other foreign compounds containing a tertiary aliphatic, tertiary alicyclic, *N,N*-dialkylaryl, or heteroaromatic amino functionality (Bickel, 1969; Gorrod, 1973; Ziegler, 1980; Damani, 1982; Hlavica, 1982). *N*-Oxygenation of tertiary amines (i.e. *N*-oxide formation) involves the nitrogen lone pair of electrons in the bonding orbital linking the nitrogen and oxygen atoms (equation 1). Most *N*-oxide

$$R_3N: \longrightarrow R_3N^+\!\!-\!O^- \tag{1}$$

metabolites are extremely water soluble and do not partition into commonly used organic solvents. In addition, most *N*-oxides are thermolabile, whereas others are unstable to extremes of pH. These factors play an important part in the choice of an assay method for quantitation of *N*-oxide metabolites. The problems in the isolation and unequivocal identification of *N*-oxides have been discussed in a previous review (Patterson *et al.*, 1978). This review examines the various colorimetric and chromatographic methods used for *N*-oxide assay, with emphasis on recent developments in the use of h.p.l.c. for the analysis of thermolabile, non-volatile *N*-oxides.

COLORIMETRIC METHODS FOR *N*-OXIDE ASSAY

Quantitation of drug metabolites by colorimetry has been popular in the past, and methods based on formation of coloured complexes with chemical reagents are available for many metabolic reactions, e.g. the determination of formaldehyde by the Nash reagent as a measure of the amount of oxidative demethylation (Nash, 1953). Ideally the reagent should be specific to the metabolite under investigation, but in practice, colorimetric methods tend to be non-specific and are being replaced by more sensitive chromatographic methods.

The colorimetric methods available for *N*-oxides are compound specific, i.e. a general method applicable to all types of *N*-oxides is not available. For example, the colorimetric method described by Ziegler & Pettit (1964) is restricted to the *N*-oxides of *N*,*N*-dialkylanilines. This method is outlined in Scheme 1. Biological samples (e.g. tissue incubates) containing *N*,*N*-dialkylaniline *N*-oxides are deproteinated by the addition of perchloric acid and centrifugation. The clear supernatant is adjusted to pH 9.4, and unreacted substrate and *N*-dealkylated metabolites removed by exhaustive extraction with diethylether. The remaining supernatant (containing the *N*-oxide) is titrated to pH 2.5 with trichloracetic acid, a solution of sodium nitrite added and the reaction heated at 60 °C for 5 minutes. The *N*-oxide is reduced to the parent tertiary aniline under the conditions employed, converted to its *p*-nitroso derivative by the nitrous acid produced *in situ*, and the colour measured in a spectrophotometer at 420 nm. This method has been widely used for studies on the microsomal *N*-oxidation of the model substrate *N*,*N*-dimethylaniline (see Patterson, *et al.*, 1978 and references cited therein). This method was found unsuitable for hepatocyte studies (Damani *et al.*, 1984) in that quantitation was obfuscated by interfering materials in the media used for hepatocyte culture. It is possible to modify the Ziegler and Pettit method, by incorporating extraction and back-extraction steps further to clean up samples prior to the colour reaction. A comparison of such a modified colorimetric method with g.l.c. and h.p.l.c. methods for the assay of

Scheme 1
Colorimetric assay of *N,N*-Dimethylaniline *N*-oxide
(Ziegler & Pettit, 1964)

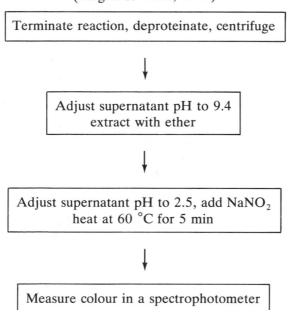

| Terminate reaction, deproteinate, centrifuge |

↓

| Adjust supernatant pH to 9.4 extract with ether |

↓

| Adjust supernatant pH to 2.5, add NaNO$_2$ heat at 60 °C for 5 min |

↓

| Measure colour in a spectrophotometer |

N,N-dimethylaniline *N*-oxide is given in Table 1, and a discussion ensues in the chromatography section.

Mitchell & Ziegler (1969) have described a more general colorimetric method for measuring low levels of *N,N*-dimethylalkyl- and *N*-methyl-heterocyclic *N*-oxides. This method involves the extraction of the *N*-oxide from the biological media and its reaction with SO$_2$; at 0–5 °C in acidic aqueous solutions, *N*-oxides react with SO$_2$ to yield addition complexes

Table 1
Comparison of colorimetric and chromatographic methods
for measuring *N,N*-dimethylaniline *N*-oxide

Method	Limit of detection (nmol per 2-ml sample)	S.D.[*]	C.V. (%)
Colorimetric	50	3.7	8.2
G.l.c.	5	1.4	28.5
H.p.l.c.	1	0.95	9.5

[*]Mean of six separate determinations. Data from Damani *et al.*, (1985, in preparation). Experimental detail of the three methods published in abstract form (Damani *et al.*, 1984).

that quickly decompose to yield formaldehyde and the secondary amine (Lecher & Hardy, 1948). The formaldehyde produced is readily measured by the Nash method. This method was slightly modified by Fok & Ziegler (1970) for use with N-oxides that are not readily extracted from biological materials. The modifications permitted the measurement of N-oxides in the presence of microsomal protein, i.e. without prior extraction. The N-oxide–SO_2 adducts are formed and hydrolysed in the aqueous reaction media, and neither protein nor lipid interfere with these processes. An important point to bear in mind is that microsomes also catalyse the oxidative N-demethylation of tertiary amines, a process that results in the production of formaldehyde. Therefore, an additional aliquot of the biological sample must be taken to measure this formaldehyde by the Nash method. The difference between the two measurements, i.e. Nash assays directly and after the SO_2 reaction, represents the amount of formaldehyde generated from the N-oxide. A large number of tertiary amines of pharmacological interest fall into this group, i.e. they are N,N-dimethylalkyamines or N-methylheterocyclic amines. The method of Fok & Ziegler (1970) is therefore of use for a large number of N-oxide metabolites, e.g. chlorpromazine N-oxide, nicotine N-oxide, atropine N-oxide.

Mattocks & White (1970) have described a very sensitive colorimetric method for measuring the N-oxides of unsaturated pyrrolizidine alkaloids (e.g. retrorsine) which depends on a modified Ehrlich reagent. Pyrrolizidine alkaloids are metabolized *in vitro* and *in vivo* by the following routes: hydrolysis of the ester groups, N-oxidation and dehydrogenation to dihydropyrrolizine esters, or pyrrole derivatives (Mattocks, 1968). Since many components in this complex mixture are likely to interfere with the colorimetric assay of the N-oxide, the estimation of N-oxide involves several steps. A preliminary chloroform extraction is carried out to remove unreacted substrate. The pyrrolic metabolites, which would give a colour with the Ehrlich reagent, are removed by extracting with ether containing iodine. This was much better than ether on its own, probably through charge-transfer complex formation aiding extraction. The N-oxides (e.g. retrorsine N-oxide), which remain in the aqueous phase are estimated using sodium nitroprusside as the ferrous complex; this apparently converts the N-oxide to the pyrrole derivative, which is then estimated by coupling it with 4-dimethylaminobenzaldehyde (Ehrlich reagent) to give a colour which is measured spectrophotometrically.

The colorimetric methods described above are limited in their use, in some cases because of their low sensitivity, or their lack of specificity, and because they are usually multistep methods that are quite tedious for routine analysis of a large number of samples. There are several other classes of N-oxides (e.g. azaheteroaromatic N-oxides) for which there are no colorimetric methods. These methods are therefore losing their popularity to chromatographic methods (see below).

CHROMATOGRAPHIC METHODS IN *N*-OXIDE ASSAY

Gas liquid chromatographic analysis of *N*-oxides is achieved in one of two ways; either their direct g.l.c. (for *N*-oxides that are volatile and thermostable), or after their reduction to the parent tertiary amine. Heteroaromatic *N*-oxides (e.g. pyridine *N*-oxide) are usually much more stable than tertiary aliphatic or alicyclic amine *N*-oxides, and when such *N*-oxides are also volatile or can be readily converted to volatile derivatives, their direct g.l.c. should be possible. Such direct g.l.c. assay procedures have been reported for pyridine *N*-oxide itself, and for several of its alkyl and halogen substituted derivatives (Gorrod & Damani, 1978), and these methods have successfully been used for estimating *N*-oxide formation in microsomal incubates (Gorrod & Damani, 1979a,b) and in urine samples (Gorrod & Damani, 1980). Most other classes of *N*-oxide are thermolabile and breakdown or rearrange during g.l.c. Biological samples are examined for the presence of such *N*-oxides by the method of Beckett *et al.* (1971) (see Scheme 2). This involves exhaustive extraction of the parent amine using an organic solvent in which the *N*-oxide is insoluble. The *N*-oxide, which is left in the aqueous phase, can be assayed after its reduction to the tertiary amine using titanous chloride. The method was originally used for estimating formation of nicotine 1'-*N*-oxide, but has been used for the *N*-oxides of many other tertiary amines, e.g. *N,N*-dialkylanilines (Gorrod *et al.*, 1975).

High performance liquid chromatography (h.p.l.c.) has found increased application in the analysis of *N*-oxides in the last decade. The earliest reported use of h.p.l.c. in *N*-oxide assay was given by Mrochek *et al.* (1976); they quantified nicotinamide *N*-oxide in human urine using an anion exchange h.p.l.c. column. This was followed by a report on the h.p.l.c. separation of nicotine, *cis* and *trans* isomers of nicotine 1'-*N*-oxide, *cis* and *trans* isomers of nicotine-1,1'-di-*N*-oxide, and nicotine 1-*N*-oxide (Brandänge *et al.*, 1977). Since these studies, h.p.l.c. has been routinely used for the assay of a large number of other thermolabile and non-volatile *N*-oxides. Since detection in h.p.l.c. is normally by u.v. monitors, this methd is only applicable to compounds with a suitable chromophore. Highly sensitive h.p.l.c. assays have been developed for azaheteroaromatic *N*-oxides, since they usually have a very high molar absorptivity. Amongst these are *N*-oxides of metyrapone (Damani *et al.*, 1979), of rosoxacin (Kullberg *et al.*, 1979), of pyridine, nicotinamide and isonicotinamide (Blaauboer & Paine, 1980) and pinacidil (Eilertsen *et al.*, 1982).

Although *N,N*-dialkylaniline *N*-oxides can be chromatographed on normal or reversed phase h.p.l.c. columns, development of sensitive direct assay procedures for such *N*-oxides is not practical because of their very low molar absorptivities. These *N*-oxides are best assayed after reduction to the parent tertiary amines, which have much higher molar extinction coefficients. The titanous chloride reduction technique (see Scheme 2) used with g.l.c., can be

Scheme 2
G.l.c. analysis of *N*-oxides using the
titanous chloride reduction technique
(Beckett, Gorrod & Jenner, 1971)

Exhaustive ether extraction of parent amine and basic
metabolites from biological fluids at alkaline pH

↓

Examination of ether extracts for estimation of
parent amine and basic metabolites by g.l.c.
using suitable internal or external standards

↓

Reduction of *N*-oxide (in aqueous phase), at acid
pH using titanous chloride

↓

Basification and ether extraction of 'reduced *N*-oxide'

↓

Quatitative determination of 'reduced *N*-oxide' in
ether extracts by g.l.c. using a suitable internal
or external standard

modified for use with h.p.l.c. A comparison of the colorimetric, g.l.c. and
h.p.l.c. assays of *N*,*N*-dimethylaniline *N*-oxide has recently been carried out
(see Table 1). The colorimetric method utilized (a slightly modified Ziegler
and Pettit method) was the least sensitive, whereas the h.p.l.c. method was the
most sensitive. The h.p.l.c. assay was of particular use in studies with hepatocyte monolayers, where the amount of *N*,*N*-dimethylaniline *N*-oxide produced is often very low. Both the chromatographic methods have the additional advantage of being able to monitor the *N*-dealkylated metabolites. In
the present reviewers' experience, such h.p.l.c. methods are far more reproducible and reliable than colorimetric methods, and more sensitive than g.l.c.
with flame ionization detection. The full potential of h.p.l.c. has not been fully
realized in the analysis of *N*-oxidation products, and it is likely that more
detailed studies will be forthcoming.

REFERENCES

Beckett, A. H., Gorrod, J. W. & Jenner, P. (1971), *J. Pharm. Pharmacol.*, **23**, 55S.

Bickel, M. H. (1969), *Pharmacol. Rev.*, **21**, 325.

Blaauboer, B. J. & Paine, A. J. (1980), *Xenobiotica*, **10**, 655.

Brandänge, S., Lindblom, L. & Samuelsson, D. (1977), *Acta Chem. Scand.*, Ser B., **31**, 907.

Damani, L. A. (1982), in *Metabolic Basis of Detoxication* (eds. Jakoby, W. B., Bend, J. R. & Caldwell, J.) Academic Press, New York, p. 127.

Damani, L. A., Crooks, P. A. & Gorrod, J. W. (1979), *J. Pharm. Pharmacol.*, **31**, 94P.

Damani, L. A., Hoodi, A. A., Sherratt, A. J. & Waghela, D. (1984), *J. Pharm. Pharmacol.*, **36**, 30P.

Damani, L. A., Hoodi, A. A., Sherratt, A. J. & Waghela, D. (1985), in preparation.

Eilertsen, E., Magnussen, M. P., Petersen, H. J., Rastrup-Anderson, N., Sørensen, H. & Arrigoni-Martelli, E. (1982), *Xenobiotica*, **12**, 187.

Fok, A. K. & Ziegler, D. M. (1970), *Biochem. Biophys. Res. Commun.*, **41**, 534.

Gorrod, J. W. (1973), *Chem. Biol. Interact.*, **7**, 289.

Gorrod, J. W. & Damani, L. A. (1978), *J. Chromatogr.*, **155**, 349.

Gorrod, J. W. & Damani, L. A. (1979a), *Xenobiotica*, **9**, 209.

Gorrod, J. W. & Damani, L. A. (1979b), *Xenobiotica*, **9**, 219.

Gorrod, J. W. & Damani, L. A. (1980), *European J. Drug Metab. Pharmacokinet.*, **5**, 53.

Gorrod, J. W., Temple, D. J. & Beckett, A. H. (1975), *Xenobiotica*, **5**, 453.

Hlavica, P. (1982), *CRC Critical Rev. Biochem.*, **12**, 39.

Kullberg, M. P., Koss, R., O'Neil, S. & Edelson, J. (1979), *J. Chromatogr.*, **173**, 155.

Lecher, H. Z. & Hardy, W. B. (1948), *J. Am. Chem. Soc.*, **70**, 3789.

Mattocks, A. R. (1968), *Nature*, **217**, 723.

Mattocks, A. R. & White, I. N. H. (1970), *Anal. Biochem.*, **38**, 529.

Mitchell, C. H. & Ziegler, D. M (1969), *Anal. Biochem.*, **28**, 261.

Mrochek, J. E., Jolley, R. L., Young, D. S. & Turner, W. J. (1976), *Clin. Chem.*, **22**, 1821.

Nash, T. (1953), *Biochem. J.*, **55**, 416.

Patterson, L. H., Damani, L. A., Smith, M. R. & Gorrod, J. W. (1978), in *Biological Oxidation of Nitrogen in Organic Molecules* (ed. Gorrod, J. W.) Elsevier, Amsterdam, p. 213.

Ziegler, D. M. (1980), in *Enzymatic Basis of Detoxication* (ed. Jakoby, W. B.) Academic Press, p. 201.

Ziegler, D. M. & Pettit, F. H. (1964), *Biochem. Biophys. Res. Commun.*, **15**, 188.

Part 2

SUBSTRATES AND MOLECULAR MECHANISMS OF THE FLAVIN-CONTAINING MONOOXYGENASE

Chapter 3

MOLECULAR BASIS FOR N-OXYGENATION OF *sec*- AND *tert*-AMINES

D. M. Ziegler, Clayton Foundation Biochemical Institute and Department of Chemistry, University of Texas at Austin, 78712, USA

1. The nature of products formed upon oxidation of amines is a function of reaction mechanism. Oxidants that generate intermediate amine cation radicals lead to oxygenation of the alpha carbon. N-Oxygenation of most secondary and tertiary amines usually requires direct two-electron oxidation of the heteroatom by an ionic mechanism.
2. Since the flavin-containing monooxygenase is the only mammalian enzyme known capable of oxygenating amines by an ionic mechanism, the N-oxygenation of the vast majority of alkaloids and medicinal amines depend almost exclusively on the activity and substrate specificity of this enzyme.
3. Some of the known structural features of potential amine substrates that facilitate or prevent binding to the catalytic site of the flavin-containing monooxygenase are reviewed.
4. The available evidence suggests that in contrast to N-demethylation, which frequently generates more-reactive intermediates, N-oxygenation of tertiary amines produces less-reactive metabolites and N-oxygenation appears to be a major route for detoxication of a significant number of medicinal amines.

INTRODUCTION

Nitrogen, after carbon and hydrogen, is one of the more abundant elements present in drugs and other foreign compounds. While this element appears in a variety of different molecular configurations, it is most common as an amine, and because of its prevalence, the metabolism of the amine functional group has been studied extensively. Oxidation is usually the first step in the metabolic transformation of amines and the nature of products formed (Lindeke & Cho, 1982; Damani, 1982; Coutts & Beckett, 1977; Hlavica, 1982) as well as the biochemical basis (Ziegler, 1980; Wislocki *et al.*, 1980; Gorrod, 1978) for the reactions are described in detail in the reviews cited.

This report will summarize our current knowledge regarding the molecular basis and contribution of *N*-oxygenation to the metabolism of secondary and tertiary aliphatic amines.

OXIDATION PRODUCTS AS A FUNCTION OF REACTION MECHANISM

It is generally accepted that the oxidation of xenobiotic amines by oxygen is catalysed by NADPH-dependent monooxygenases, mostly concentrated in liver microsomes. Tertiary amines are *N*-dealkylated or *N*-oxygenated and the nature of the products formed is dependent on the structure of the amine as well as on the catalytic mechanism of the monooxygenases involved.

There is almost universal agreement that the oxidative *N*-dealkylation of aliphatic amines is catalysed by specific isozymes of the cytochrome P-450 family of monooxygenases. These enzymes catalyse oxygenation of a variety of xenobiotics by sequential one-electron oxidations. All definitive studies on the catalytic mechanism of cytochrome P-450 monooxygenases indicate that the insertion of oxygen is always preceded by electron (or hydrogen) abstraction from substrate (White & Coon, 1980; Guengerich & MacDonald, 1984). Regardless of whether the initial oxidative attack involves the abstraction of an electron from nitrogen or a hydrogen atom from carbon, oxygen addition generally occurs on the alpha carbon[†] for the reasons summarized below.

Abstraction of an electron from nitrogen generates the amine cation radical (equation 1).

$$-\overset{\cdot\cdot}{N}-CH_2- \quad \xrightarrow{\;e^-\;} \quad -\overset{+}{N}-CH_2- \qquad (1)$$

If there is a hydrogen on the alpha carbon, the loss of a proton and concomitant electron transfer from carbon to nitrogen generates the carbon centred radical as illustrated in equation (2).

[†]The recent excellent review by Guengerich & MacDonald (1984) summarizes the biochemical and chemical evidence for this conclusion. After a detailed analysis of reactions catalysed by cytochrome P-450s they point out that all the reactions are consistent with a general mechanism involving stepwise electron abstraction followed by oxygen transfer (rebound) from the oxygenated enzyme intermediate.

$$-\overset{+}{\underset{|}{N}}-CH_2- \quad \xrightarrow{\quad H^+ \quad} \quad -\overset{\cdot\cdot}{\underset{|}{N}}-\overset{\cdot}{C}H- \tag{2}$$

In oxygenation of amines by a radical mechanism, oxygen (at the oxidation state of the hydroxyl radical) adds to the electron-deficient atom. Whether oxygenation occurs on the nitrogen or carbon depends solely on the rate of reaction (2). For most alicyclic and monocyclic amines the rate of proton loss and intramolecular electron transfer from carbon to nitrogen is no less than 10^4/sec and essentially irreversible (Chow, 1980). Therefore, the oxidation of aliphatic amines by a radical mechanism inevitably leads to hydroxylation of carbon. Subsequent decomposition of the intermediate carbinolamine yields an aldehyde and the dealkylated amine, which are the predominant products obtained upon the oxidation of *sec*- and *tert*-amines via the cytochrome P-450 system (Lindeke & Cho, 1982).

While amines can be N-*oxidized* by a radical mechanism, only a few can be N-*oxygenated* by this route. N-Oxygenations catalysed by cytochrome P-450, or any other enzyme that generates intermediate radicals, are largely restricted to those few amines that upon abstraction of an electron from nitrogen do not readily rearrange to the carbon centred radical. These would include amines that lack alpha carbon hydrogens (e.g. phentermine) and bridged amines that cannot violate Bredt's rule (e.g. nortropines) and the cytochrome P-450 catalysed N-oxygenation of both phentermine (Cho *et al.*, 1974) and norcocaine (Rauckman *et al.*, 1982) have been documented. However, the reported N-oxygenation of N,N-dimethylaniline by cytochrome P-448 (see Hlavica, 1982, for a review of this work) is difficult to reconcile with the mechanism of the enzyme. It is quite likely that the N-oxidation of this amine in the presence of the purified reconstituted cytochrome P-448 system is largely non-enzymic. This system generates copious quantities of peroxide *in vitro* and it is without question that hydrogen peroxide or lipid hydroperoxides can N-oxygenate tertiary amines. While the peroxides are generated enzymically, the subsequent formation of N-oxide by this route is non-enzymic and of little or no physiological significance.

The vast majority of medicinal amines or naturally occurring alkaloids are, however, readily N-oxygenated by the flavin-containing monooxygenase which catalyses the direct insertion of oxygen without the formation of intermediate radicals. The catalytic mechanism of this monooxygenase is known in great detail (Poulsen & Ziegler 1979; Beaty & Ballou 1981a, b) and oxygen transfer from the peroxyflavin intermediate is similar to the oxidation of amines by peroxides (Ball & Bruice, 1980) (equation 3).

$$-\overset{\cdot\cdot}{\underset{|}{N}}-CH_2- + ROOH + H^+ \quad \longrightarrow \quad -\overset{\overset{\displaystyle OH^+}{|}}{\underset{|}{N}}-CH_2- + ROH \tag{3}$$

The oxidation of amines by this mechanism is an ionic two-electron oxidation of the heteroatom without the formation of intermediate radicals. Therefore, unlike the oxidation of amines by radical mechanisms, this route will yield *N*-oxygenated products regardless of the nature of the substituents on the amino nitrogen. Since the flavin-containing monooxygenase is the only mammalian enzyme known that can catalyse two-electron oxidation of amines by an ionic mechanism, the *N*-oxygenation of the vast majority of *sec*- and *tert*-amines *in vivo* depends almost exclusively on the activity and substrate specificity of the flavin-containing monooxygenase. The substrate specificity of the enzyme from hog liver has been reviewed in detail (Ziegler, 1980, 1984) and only some of the more general aspects will be considered in the following section.

SUBSTRATE SPECIFICITY OF THE FLAVIN-CONTAINING MONOOXYGENASE

The flavin-containing monooxygenase has been purified to homogeneity from hog (Ziegler & Mitchell, 1972), rat (Kimura *et al.*, 1983) and mouse (Sabourin *et al.*, 1984) liver, and most recently, from rabbit lung microsomes (personal communication, Williams & Masters). A detailed analysis of the substrate specificities of this enzyme in these species has not been carried out but a recent report by Sabourin *et al.* (1984) indicates that the hog and mouse liver enzymes are immunologically and mechanistically similar, although some quantitative differences in kinetic constants for amine substrates are observed.

On the other hand, the rabbit lung monooxygenase is immunologically distinct from the liver enzymes and may be a different gene product. Although the composition and catalytic mechanism of the rabbit lung and liver enzymes are similar, preliminary studies indicate that access of amines to the catalytic site of the lung enzyme is much more restricted. Some of the better amine substrates for the liver enzyme show no detectable substrate activity with the purified lung monooxygenase. If these preliminary studies can be fully substantiated it will be the first demonstration that isozymes of the flavin-containing monooxygenase (with overlapping but distinct substrate specificities) may be present in different organs.

However, there have been no studies that show significant quantitative differences in the liver monooxygenase from different species, therefore, the following summary of the properties of the hog liver enzyme may, in general, apply to the flavin-containing monooxygenase in this organ from other mammals.

The monooxygenase from hog liver microsomes catalyses *N*-oxygenation of a variety of different lipophilic alicyclic and cyclic aliphatic amines. A few representative examples are listed in Table 1. Any lipophilic *sec*- or *tert*-amine is a potential substrate and drugs from virtually all therapeutic categories are *N*-oxygenated by this enzyme. However, known amine substrates for the monooxygenase are most prevalent among antipsychotic, antihistaminic and narcotic drugs (Ziegler, 1980). The number of drugs from other categories

Table 1

The flavin-containing monooxygenase: amine substrates

EXAMPLES	REACTION

t-AMINES
   Chlorpromazine
   Guanethidine
   Brompheniramine
   Morphine
   Nicotine

$$R-N\binom{R}{R} \longrightarrow \overset{\overset{\displaystyle OH^+}{|}}{R\text{-}N\text{-}R} \atop R$$

sec-AMINES
   Propranolol
   Desipramine
   Methamphetamine
   Desmethyltrifluperazine

$$\overset{\overset{\displaystyle H}{|}}{R\text{-}N\text{-}R} \longrightarrow \overset{\overset{\displaystyle OH}{|}}{R\text{-}N\text{-}R}$$

N-HYDROXYLAMINES
   N-Methyl-N-hydroxybenzylamine
   N-Benzyl-N-hydroxyamphetamine
   N-hydroxydesmethyltrifluperazine

$$\overset{\overset{\displaystyle OH}{|}}{R\text{-}N\text{-}R} \longrightarrow \overset{\overset{\displaystyle O}{|}}{R\text{-}N\text{=}R}$$

The oxidation of amine functional groups listed are all NADPH- and oxygen-dependent and the enzymic nature of these oxygenations has been unambiguously demonstrated (Ziegler, 1980).

tested have been quite limited, but because of the rather broad substrate specificity of this enzyme, it is likely that most medicinal *sec-* and *tert-*amines would show substrate activity.

While substrate structural elements that determine the magnitude of kinetic constants are not well defined, some molecular features that prevent substrate activity are known (Ziegler *et al.*, 1980). For example, the flavin-containing monooxygenase does not catalyse oxygenation of aromatic heterocyclic amines or other amines where the electrons on nitrogen are delocalized by conjugation with an aromatic system. The lack of substrate activity of aromatic heterocyclic amines is consistent with the known mechanism of the enzyme. Activation of substrate does not occur prior to oxygen transfer and the peroxyflavin is apparently not a sufficiently strong oxidant to oxygenate these less nucleophilic amines.

Access to the catalytic site is markedly affected by the number and position of ionic groups on the amine. Diamines where both nitrogens are protonated at physiological pH are not substrates. Studies with both nitrogen- and sulphur-containing nucleophiles (Ziegler *et al.*, 1980) indicate that it is the dication rather than the diamine character that prevents access to the catalytic site. The presence of one or more anionic groups on the molecule also prevents binding of amines.

From the nature and number of ionic groups and the relative nucleophilicity of the amino nitrogen, it is frequently possible to predict whether a specific amine will be *N*-oxygenated by the flavin-containing monooxygenase. In general, neutral amines or those bearing a single positive charge at pH 7–8 are *N*-oxygenated and because of the unusual catalytic mechanism of the monooxygenase, V_{max} is essentially the same for all substrates (Poulsen & Ziegler, 1979; Sabourin *et al.*, 1984).

ROLE OF *N*-OXYGENATION

Almost without exception, standard texts and reviews on the metabolism of xenobiotics state that *N*-oxygenation is a minor route for the metabolism of amines. However, this conclusion is largely based on the nature of metabolic end-products excreted by whole animals. While the identities of metabolic end-products excreted in urine or bile may be useful for delineating routes for disposition, they are of little value in determining the nature or rates of intermediate metabolic reactions. Pathways, as well as the contribution of different routes, are more accurately assessed from studies with tissue homogenates or purified preparations.

Figure 1 illustrates the major known pathways for the oxidative metabolism of *sec*- and *tert*-amines. All the steps listed are based on enzyme catalysed reactions that have been described for various *N*-methyl and *N,N*-dimethyl-alkyl- and arylamines. Although the relative velocities of the different steps are dependent on the nature of the drug, it is evident that the intermediary metabolism of *sec*- and *tert*-amines can be quite complex. Steady state concentrations of the parent amine and of *N*-oxygenated or *N*-demethylated metabolites depend on the rates of reduction and/or remethylation as well as on the rates of metabolite formation by oxidative pathways. Enzymic methylation of a number of medicinal amines has been demonstrated (Narasimhachari & Lin, 1974, 1976) and various routes for the reduction of *N*-oxygenated amines are known (Bickel, 1971; Kato *et al.*, 1978; Suguira *et al.*, 1976; Kadlubar *et al.*, 1973). While these reactions undoubtedly limit disposition of amine metabolites generated by oxidation, their contribution to half-life or to overall metabolism of medicinal amines has received relatively little attention. More detailed work on mechanisms and properties of *N*-methylases as well as the *N*-oxide and *N*-hydroxy amine reductases are necessary for a more complete understanding of the intermediary metabolism of amine drugs.

The relative contribution of *N*-oxygenation and *N*-demethylation to the oxidative metabolism of amine drugs has received considerably more attention. Partitioning between these two routes depends largely on the substrate specificities and tissue concentrations of the flavin-containing and the cyto-

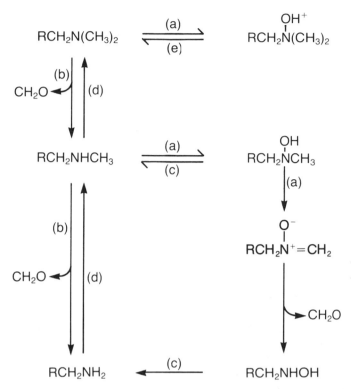

The metabolic reactions listed are catalysed by the following enzymes: (a) the flavin-containing monooxygenase; (b) cytochrome P-450 dependent monooxygenase; (c) microsomal *N*-hydroxyamine reductase; (d) *S*-adenosylmethionine-dependent *N*-methylases, and (e) microsomal *N*-oxide reductase. The hydrolysis of the intermediate nitrone formed upon enzymic *N*-oxygenation of the *N*-hydroxyamine is apparently non-enzymic. The experimental evidence for the steps listed are described in the references given in the text.

Fig. 1 – Oxidative metabolism of tertiary alkylamines by hepatic microsomes.

chrome P-450 dependent monooxygenases, respectively. While both enzymes are present in liver from all species examined, their relative concentrations based on activity measurements *in vitro* must be carefully evaluated. These monooxygenases differ considerably in stability and the extreme thermal lability of the flavin-containing monooxygenase above 35 °C in the absence of NADPH (Uehleke, 1973) was not generally recognized in many of the older studies. Furthermore, the almost universal practice of preincubating microsomes in the absence of NADPH selectively destroys the flavin-containing monooxygenase without affecting activities of the far more stable cytochrome P-450 system (Uehleke, 1973).

Inactivation of the flavin-containing monooxygenase during tissue collection and processing is also a major problem in some species. For instance Dannan & Guengerich (1982) have shown with an immunochemical method that in liver from adult female hogs the concentration of this monooxygenase is rather constant at about 45 μM. The observed eightfold variations in activity in the samples from different animals has apparently been due to enzyme inactivation during tissue collection and processing. Although enzyme inactivation usually presents less problems with tissues from laboratory animals, the large differences in *N*-oxygenase activities of liver microsomes from the same species reported from different laboratories (see Ziegler, 1980, for review) are undoubtedly due to inadvertent enzyme inactivation. As a result, the concentration of the flavin-containing monooxygenase in liver of species other than the hog is not accurately known. However, the recent report of Sabourin *et al.* (1984) suggests that the level in normal mouse liver is almost one-half that of the hog. This conclusion is based on the finding that the mouse enzyme purified to homogeneity is concentrated 120-fold vs 60-fold in the hog. A twofold variation in enzyme concentration for species as diverse as the mouse and hog is surprisingly small and species differences in the hepatic concentrations of the flavin-containing monooxygenase may be far less than assumed from activity measurements.

N-Oxygenation vs *N*-dealkylation also depends upon kinetic constants for enzymic oxidation of specific amines. While the majority of the trialkyl medicinal amines are substrates for both cytochrome P-450 and the flavin-containing monooxygenases, V_{max} for cytochrome P-450 catalysed dealkylations decreases with increasing base strength of the substrate (McMahon & Easton, 1961). With few exceptions basic medicinal amines are relatively poor substrates for this monooxygenase. On the other hand, basicity has less effect on the rate of *N*-oxygenation. The more basic amines (e.g. the tricyclic antipsychotic, antihistaminic and narcotic drugs) exhibit K_m values in the μM range for the hog liver flavin-containing monooxygenase (Ziegler, 1980). Furthermore, at saturation, all substrates for this enzyme are oxygenated at the same velocity so that measurements on one substrate accurately predict V_{max} for all other substrates.

The enzyme is also fully active in intact cells. McManus *et al.* (1983) have shown that the rate of guanethidine *N*-oxygenation in rat hepatocytes is essentially equal to the rate predicted from measurements with isolated microsomes. They point out that guanethidine *N*-oxide is one of the few tertiary amine oxides that is not readily reduced or subject to further metabolism and its rate of formation may be a useful probe for assessing activity of the flavin-containing monooxygenase *in vivo*. In any event, the high catalytic capacity of this enzyme suggests that *N*-oxygenation is a major pathway for the metabolism of medicinal trialkylamines.

The apparent preferential *N*-oxygenation of such a large number of medicinal amines may not be fortuitous. Trialkylamine oxides are polar, relatively non-toxic compounds and *N*-oxygenation is a true route for detoxication of medicinal trialkylamines. On the other hand recent work (Krieter *et al.*, 1984) shows that glutathione depletion (and possible hepatotoxicity) of aminopyrine is due to its rapid *N*-demethylation. Aminopyrine, in contrast to more basic amines, is metabolized exclusively by *N*-demethylation and the rate of formaldehyde formation from this drug exceeds the catalytic capacity of the formaldehyde dehydrogenase in rat livers. In perfused rat liver, the excess formaldehyde is excreted as the glutathione conjugate, which depletes the liver of this important nucleophile and opens up the tissue to potential damage from formaldehyde.

From the chemical and pharmacological properties of the products formed, it would appear that N-oxygenation of tertiary N-methylamines is a route for detoxication, whereas N-demethylation produces more toxic metabolites. Selective, rapid N-oxygenation of medicinal trialkylamines may have been an important (but unrecognized) factor that led to their selection for therapeutic use during toxicity screening of potential drugs.

REFERENCES

Ball, S. & Bruice, T. C. (1980), *J. Am. Chem. Soc.*, **102**, 6498.
Beaty, N. B. & Ballou, D. P. (1981a), *J. Biol. Chem.*, **256**, 4611.
Beaty, N. B. & Ballou, D. P. (1981b), *J. Biol. Chem.*, **256**, 4619.
Bickel, M. H. (1971), *Arch. Biochem. Biophys.*, **148**, 54.
Cho, A. K., Lindeke, B. & Sum, C. Y. (1974), *Drug Metab. Dispos.*, **2,** 1.
Chow, Y. L. (1980), in *Reactive Intermediates*, (ed. R. A. Abramovitch), Vol. 1, Plenum Press, New York, p. 151.
Coutts, R. T. & Beckett, A. H. (1977), *Drug Metab. Rev.*, **6,** 51.
Damani, L. A. (1982) in *Metabolic Basis of Detoxication*, (eds. Jakoby, W. B., Bend, J. R. & Caldwell, J.) Academic Press, New York, p. 127.
Dannan, G. A. & Guengerich, F. P. (1982), *Mol. Pharmacol.*, **22**, 787.
Gorrod, J. W. (1978), in *Mechanisms of Oxidizing Enzymes* (eds. Singer, T. P. & Ondarza, R. N.) Elsevier/North Holland Inc., New York, p. 189.
Guengerich, F. P. & McDonald, T. L. (1984), *Acct. Chem. Res.*, **17**, 9.
Hlavica, P. (1982), *CRC Critical Revs. Biochem.*, **12**, 39.
Kadlubar, F. F., McKee, E. M. & Ziegler, D. M. (1973), *Arch. Biochem. Biophys.*, **156**, 46.
Kato, R., Iwasaki, K. & Noguchi, H. (1978), *Mol. Pharmacol.*, **14**, 654.
Kimura, T., Kodama, M. & Nagata, C. (1983), *Biochem. Biophys. Res. Comm.*, **110**, 640.
Krieter, P. A., Ziegler, D. M., Hill, K. E. & Burk, R. F. (1984), *Biochem. Pharmacol.*, in press.
Lindeke, B. & Cho, A. K. (1982), in *Metabolic Basis of Detoxication*, (eds. Jakoby, W. B., Bend, J. R. & Caldwell, J.) Academic Press, New York, p. 105.
McMahon, R. E. & Easton, N. R. (1961), *J. Med. Pharm. Chem.*, **4**, 437.
McManus, M. E., Grantham, P. H., Cone, J. L., Roller, P. P., Wirth, P. J. & Thorgeirsson, S. S. (1983), *Biochem. Biophys. Res. Comm.*, **112**, 437.
Narasimhachari, N. & Lin, R.-L. (1974), *Res. Comm. Chem. Path. Pharmacol.*, **8**, 341.
Narasimhachari, N. & Lin, R.-L. (1976), *Psychopharmacol. Comm.*, **2**, 27.

Poulsen, L. L. & Ziegler, D. M. (1979), *J. Biol. Chem.*, **254**, 6449.

Rauckman, E. J., Rosen, G. M. & Cavagnaro, J. (1982), *Mol. Pharmacol.*, **21**, 458.

Sabourin, P. J., Smyser, B. P. & Hodgson, E. (1984), *Int. J. Biochem.*, **16**, 713.

Sugiura, M., Iwasaki, K. & Kato, R. (1976), *Mol. Pharmacol.*, **12**, 322.

Uehleke, H. (1973), *Drug Metab. Dispos.*, **1**, 299.

White, R. E. & Coon, M. J. (1980), *Ann. Rev. Biochem.*, **49**, 315.

Wislocki, P. G., Miwa, G. T. & Lu, A. Y. H. (1980), in *Enzymatic Basis of Detoxication*, (ed. Jakoby, W. B.), Vol. 1, Academic Press, New York, p. 136.

Ziegler, D. M. (1980), in *Enzymatic Basis of Detoxication*, (ed. Jakoby, W. B.), Vol. 1, Academic Press, New York, p. 201.

Ziegler, D. M. (1984), in *Drug Metabolism and Drug Toxicity*, (eds. Mitchell, J. R. & Horning, M. G.), Raven Press, New York, p. 33.

Ziegler, D. M. & Mitchell, C. H. (1972), *Arch. Biochem. Biophys.*, **150**, 116.

Ziegler, D. M., Poulsen, L. L. & Duffel (1980), in *Microsomes, Drug Oxidations and Chemical Carcinogenesis*, (eds. Coon, M. J., Conney, A. H., Estabrook, R. W., Gelboin, H. V., Gillette, J. R. & O'Brien, P. J.), Vol. II, Academic Press, New York, p. 637.

Chapter 4

STRUCTURE ACTIVITY RELATIONSHIPS IN THE METABOLIC *N*-OXIDATION OF PARGYLINE AND RELATED AMINES

Afaf M. Weli and **B. Lindeke**, Department of Organic Pharmaceutical Chemistry, Biomedical Centre, University of Uppsala, Box 574, S-751 23 Uppsala, Sweden

1. The mtabolism of pargyline (*N*-benzyl-*N*-methylpropargylamine, PARG), and two congeners, *N*-α-methylbenzyl-*N*-methylpropargyl-amine (MBMP) and *N*-benzyl-*N*-methyl-1-methylpropargylamine (BMMP) was investigated in liver microsomes from control and PB-pretreated rats.

2. The metabolism of all three substrates, which was very rapid, comprised three different *N*-dealkylation reactions and *N*-oxidation. *N*-Oxidation of PARG was a reaction of utmost importance and was not sensitive to phenobarbital induction. Considerably less *N*-oxide was recovered in the incubations of MBMP or BMMP.

3. Hydrogen peroxide, together with cytochrome P-450, had the capacity to promote all the metabolic conversions seen in the NADPH-dependent microsomal metabolism. This oxidation was inhibited by cyanide and DPEA.

4. A marked enantioselectivity was noted in the hydrogen peroxide promoted *N*-demethylation and *N*-depropargylation of MBMP while no enantioselectivity was seen in any of the NADPH-dependent monoxygenations.

INTRODUCTION

Pargyline (*N*-benzyl-*N*-methylpropargylamine, PARG) and related tertiary α-acetylenic amines appear among enzyme inhibitors which have shown a potential in the treatment of hypertension and mental depression. Although, PARG is no longer used clinically, due to undesirable side effects (Fowler *et al.*, 1981), it is still frequently being made use of in experimental pharmacology as a tool to examine, for example, MAO-active centres (Parkinson & Callingham, 1980). In addition this drug is an interesting substrate for mechanistic investigations of drug metabolizing enzymes.

In the course of studies on metabolic functionalization of α-acetylenic amines, the biotransformation of PARG has been investigated, and we have previously reported (Hallström *et al.*, 1981) that *N*-demethylation and *N*-oxidation are important metabolic routes. The development of a sensitive analytical assay (Weli *et al.*, 1982) for the simultaneous quantitation of PARG and its four primary metabolites, *N*-benzylpropargylamine (BPA), *N*-methylpropargylamine (MPA), *N*-benzylmethylamine (BMA) and pargyline *N*-oxide (PNO) (Fig. 1), has now made possible more detailed studies on the metabolism of PARG (Weli & Lindeke, 1985).

This contribution summarizes some recent findings on the metabolism of PARG and two α-methylsubstituted congeners, *N*-benzyl-*N*-methyl-(1-methylpropargyl)amine (BMMP) and *N*-methyl-*N*-(1-phenylethyl)-propargylamine (MBMP) (Fig. 1). The study consists of *in vitro* investigations of the NADPH-dependent metabolism in rat liver microsomes, as well as the conversion of these substrates by peroxidative mechanisms.

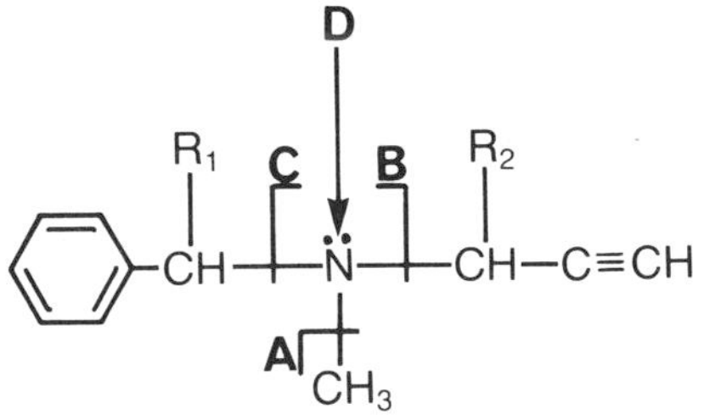

Fig. 1 – General structure for PARG ($R_1 = R_2 = H$), MBMP ($R_1 = CH_3$, $R_2 = H$) and BMMP ($R_1 = H$, $R_2 = CH_3$). A, B, and C indicate the bonds broken in the different *N*-dealkylation reactions yielding BPA, BMA and MPA from PARG and the analogous secondary amines from MBMP and BMMP. D indicates the position for *N*-oxide formation i.e. PNO and congeners.

MATERIALS AND METHODS

Details of most experimental procedures are found in Weli *et al.* (1982) and Weli & Lindeke (1985). The analytical assay used for the quantitation of PARG, BPA, MPA, BMA and PNO is described in Weli *et al.* (1982).

Analogous methodology was used for the determination of BMMP and MBMP and their respective metabolites. The preparation of BMMP and MBMP, as racemates and as resolved enantiomers, and their corresponding N-oxides will be published elsewhere. Standard incubations were carried out in 25-ml Erlenmeyer flasks at 37 °C. Microsomes from untreated or phenobarbital pretreated rats were used. An NADPH-regenerating system was added to the substrate and the incubations were started by addition of the microsomes. Hydrogen peroxide-promoted oxidation was studied using 25 or 100 mM H_2O_2 omitting the NADPH-regenerating system. Other agents and enzyme inhibitors were sometimes introduced at concentrations as stated.

RESULTS

Pargyline is rapidly metabolized in rat liver microsomes. At an initial concentration of 100 μM most of the substrate is consumed within the first 5 minutes in microsomes from untreated rats and, if microsomes from PB-pretreated rats are used, essentially all the substrate is consumed within the first minute (Table 1) (Weli & Lindeke, 1985). N-Oxidation (D in Fig. 1) is a reaction of

Table 1
Substrate consumption and metabolite formation in liver microsomes from control and PB-pretreated rats

Substrate[a]	Substrate consumed[a]	Metabolites formed[b]			
		A	B	C	D
Control					
PARG	30.5 ± 3.7	3.5 ± 0.5	6.8 ± 0.4	1.0 ± 0.1	10.9 ± 1.2
MBMP(+)	59.3 ± 0.8	15.9 ± 0.2	8.3 ± 0.4	n.d.[c]	n.d.
MBMP(−)	55.0 ± 0.4	14.0 ± 0.4	7.7 ± 0.6	n.d.	n.d.
BMMP(+)	48.0 ± 0.4	13.1 ± 0.1	3.5 ± 0.3	1.1 ± 0.1	n.d.
BMMP(−)	42.1 ± 0.2	14.1 ± 0.1	2.4 ± 0.1	1.3 ± 0.1	n.d.
PB-pretreated					
PARG	38.7 ± 1.4	5.4 ± 0.7	9.5 ± 0.6	2.4 ± 0.1	5.7 ± 0.9
MBMP(+)	38.2 ± 1.0	0.9 ± 0.1	6.2 ± 0.1	n.d.	n.d.
MBMP−)	38.2 ± 1.2	1.6 ± 0.3	6.5 ± 0.2	n.d.	n.d.
BMMP(+)	41.0 ± 1.1	< 0.2	4.4 ± 0.1	n.d.	n.d.
BMMP(−)	41.0 ± 0.8	< 0.2	6.3 ± 0.3	n.d.	n.d.

[a] The substrate concentration was 100 μM which corresponds to 40–70 nmole (mg protein)$^{-1}$.
[b] Mean values of at least three experiments (nmole mg prot^{-1} ± SEM) after 5 min of incubation. A,B,C and D refer to metabolites formed in the various reactions depicted in Fig. 1.
[c] *n.d.* = not detected.

Table 2

H_2O_2-cytochrome P-450 promoted oxidation of pargyline

Components added[a]	Metabolites formed[b]			
	A	B	C	D
Control				
+NADPH generating system	2.5 ± 0.5	3.4 ± 1.0	2.0 ± 1.0	7.0 ± 0.8
+ NaN$_3$(5 mM)	5.2 ± 0.6	5.6 ± 0.4	2.8 ± 0.4	6.4 ± 1.5
+ NaN$_3$(5 mM) + H$_2$O$_2$(100 mM)	40.3 ± 1.0	50.2 ± 1.6	17.1 ± 0.7	11.5 ± 2.6
+ NaN$_3$(5 mM) + H$_2$O$_2$(25 mM)	14.8 ± 2.6	17.4 ± 3.5	11.9 ± 3.5	14.1 ± 3.3
+ NaN$_3$(5 mM) + H$_2$O$_2$(25 mM) + NaCN(5 mM)	8.6 ± 1.5	7.2 ± 1.3	4.8 ± 1.8	4.6 ± 1.3
+ NaN$_3$(5 mM) + H$_2$O$_2$(25 mM) + DPEA(100 μM)	1.3 ± 0.1	2.5 ± 0.2	n.d.[c]	5.5 ± 1.0
PB-pretreated				
+ NADPH generating system	13.5 ± 0.7	13.6 ± 0.9	10.6 ± 1.5	7.3 ± 0.6
+ NaN$_3$(5 mM)	11.6 ± 0.6	13.1 ± 0.7	7.0 ± 0.9	6.9 ± 0.7
+ NaN$_3$(5 mM) + H$_2$O$_2$(100 mM)	49.6 ± 4.6	49.6 ± 4.6	12.1 ± 2.1	11.8 ± 2.6

[a] The substrate concentration was 1 mM.
[b] Mean values of at least three experiments expressed as nmole (mg prot)$^{-1}$ min^{-1}. A,B,C, and D refer to metabolites formed in the various reactions depected in Fig. 1.
[c] n.d. = not detected.

Table 3
Enantioselectivity of H_2O_2-cytochrome P-450 promoted oxidation

Substrate[a] (1 mM)	A[b]	B[b]	C[b]	D[b]
MBMP(+)	70.0 ± 9.7	42.9 ± 9.0	4.0 ± 0.5	5.8 ± 1.0
(−)	20.7 ± 6.2	12.6 ± 2.8	3.6 ± 0.3	4.9 ± 0.7
BMMP(+)	22.1 ± 3.4	4.2 ± 0.2	12.5 ± 2.0	4.3 ± 0.3
(−)	29.4 ± 6.5	5.8 ± 0.4	12.4 ± 1.8	4.2 ± 1.0

[a] H_2O_2(25 mM) and NaN_3(5 mM).
[b] Expressed as nmole (mg protein)$^{-1}$ min^{-1}. A,B,C, and D refers to metabolites formed in the various reactions depicted in Fig. 1.

utmost importance. Depending on animal pretreatment, the second most important reaction is N-demethylation (A in Fig. 1), or N-depropargylation (B in Fig. 1) (Weli & Lindeke, 1985). N-Oxidation of PARG is not sensitive to PB-induction and its inhibition by different cytochrome P-450 inhibitors is marginal (Weli & Lindeke, 1985).

The metabolism of the α-substituted congeners of PARG—MBMP and BMMP (Table 1)—appears to be as extensive as that of PARG. However, at a substrate concentration of 100 μM essentially no N-oxide could be retrieved from the incubation of either substrate. Also, N-dealkylation at the site of methyl substitution in MBMP and BMMP, C and B in Fig. 1, respectively, is decreased as compared to what was seen in PARG. At the 100 μM concentration no enantioselectivity could be noted in any of the NADPH-dependent metabolic reactions in either MBMP or BMMP.

It should be noted that the amounts of metabolites formed in 5 minutes, due to N-demethylation (A in Fig. 1) and given in Table 1, greatly underestimate the importance of this reaction. This is because the N-demethylated metabolites are rapidly further metabolized (Weli & Lindeke, 1985).

Hydrogen peroxide when substituted for the NADPH-generating system (Tables 2 and 3) had the capacity to promote all the metabolic conversions depicted in Fig. 1. However, in this case N-demethylation (A in Fig. 1) and N-depropargylation (B in Fig. 1) greatly exceed N-oxidation (D in Fig. 1). The H_2O_2-cytochrome P-450 promoted oxidation of PARG occurred also under nitrogen (data not shown) and was inhibited by cyanide and the cytochrome P-450 inhibitor DPEA (2,4-dichloro-6-phenylphenoxyethylamine). A marked enantioselectivity (Table 3) was noted in the H_2O_2-dependent N-demethylation (A in Fig. 1) and N-depropargylation (B in Fig. 1) of MBMP, while, as for the NADPH-dependent metabolism no enantioselectivity was seen in the 'N-debenzylation' (C in Fig. 1) and N-oxidation (D in Fig. 1) or in the H_2O_2-dependent conversion of BMMP.

DISCUSSION

The results summarized in this contribution show that α-acetylenic amines related to PARG are excellent substrates for membrane-bound microsomal enzymes. The *N*-dealkylation reactions, of which *N*-demethylation and *N*-depropargylation are considerably more important than *N*-debenzylation, appear to be catalysed by cytochrome P-450 enzymes, while *N*-oxide formation is essentially cytochrome P-450 independent (Weli & Lindeke, 1985). The latter reaction is probably promoted by the NADPH-dependent flavin-containing monooxygenase originally described by Ziegler & Mitchell (1972). Introduction of a methyl group α to the nitrogen in PARG significantly decreases *N*-dealkylation at that position which is consistent with that *N*-dealkylation reactions yielding ketones are less effective than those generating aldehydes (Lindeke & Cho, 1982).

That PNO is an abundant metabolite of PARG, while no *N*-oxides are retrieved in the incubations of MBMP or BMMP, could be due to the fact that the introduction of a methyl group on either side α to the nitrogen in PARG prevents *N*-oxidation. However, all three *N*-oxides are chemically unstable. Pargyline *N*-oxide undergoes what is recognized as the Meisenheimer rearrangement (Hallström *et al.*, 1980, 1981), while the *N*-oxides of MBMP and BMMP, which contain β-hydrogens, undergo Cope elimination (Cope *et al.*, 1953). As the Cope elimination in the *N*-oxides of MBMP and BMMP appears to occur under milder conditions than the Meisenheimer rearrangement in PNO (Weli & Lindeke, unpublished results) this could be an alternative reason why no *N*-oxides of MBMP and BMMP were detected.

That hydrogen peroxide–cytochrome P-450 promoted oxidation of PARG generated all four metabolites found in the NADPH-dependent microsomal metabolism is somewhat interesting. To our knowledge, PARG is the first drug for which H_2O_2-cytochrome P-450 promoted *N*-dealkylation reactions other than *N*-demethylation has been reported. Even more intriguing is the enantioselectivity seen in the hydrogen peroxide-dependent *N*-demethylation and *N*-depropargylation of MBMP. That these reactions are stereoselective in the peroxidative system, but not when cytochrome P-450 acts in the presence of a NADPH-generating system, clearly shows that these reactions occur by different reaction mechanisms (Estabrook *et al.*, 1984) and probably with different rate-limiting steps.

ACKNOWLEDGEMENTS

The work discussed in this contribution was supported by the Swedish Medical Research Council (grant No. 5645-05A).

REFERENCES

Cope, A. C., Pike, R. A. & Spencer, C. F. (1953), *J. Am. Chem. Soc.*, **75**, 3212.

Estabrook, R. W., Martin-Wixstrom, C., Saeki, Y., Renneberg, R., Hildebrandt, A. & Weingloer, J. (1984), *Xenobiotica*, **14**, 87.

Fowler, C. J., Norqvist, A., Oreland, L., Saramies, E. & Wiberg, A. (1981), *Comp. Biochem. Physiol.*, **68c**, 145.

Hallström, G., Lindeke, B., Khutier, A.-H. & Al-Iraqi, M. A. (1980), *Tetrahedron Lett.*, **21**, 667.

Hallström, G., Lindeke, B., Khutier, A.-H. & Al-Iraqi, M. A. (1981), *Chem.-Biol. Interactions*, **34**, 185.

Lindeke, B. & Cho, A. K. (1982), in *Metabolic Basis of Detoxication*, eds, Jacoby, W. B., Bend, J. R. & Caldwell, J. Academic Press, New York, p. 105.

Parkinson, D. & Callingham, B. A. (1980), *J. Pharm. Pharmacol.*, **32**, 49.

Weli, A. M., Ahnfelt, N.-O. & Lindeke, B. (1982), *J. Pharm. Pharmacol.*, **34**, 771.

Weli, A. M. & Lindeke, B. (1985), *Biochem. Pharmacol.* in press.

Ziegler, D. M. & Mitchell, C. H. (1972), *Arch. Biochem. Biophys.*, **150**, 116.

Chapter 5

METABOLIC *N*-OXIDATION OF ALICYCLIC AMINES

H. Oelschläger and **M. Al Shaik**, Institut für Pharmazeutische Chemie der Johann Wolfgang Goethe-Universität, D-6000 Frankfurt am Main, FR Germany

1. Alicyclic amines are very often the basic moiety of drugs.
2. Metabolic attack on such tertiary amines results in the formation of an *N*-oxide, normally accompanied by oxidation of the alicyclic ring carbons.
3. The subsequent ring opening, with formation of polar compounds, seems to be more frequent than has been supposed in earlier investigations.

INTRODUCTION

Drugs which are very important in different fields of therapeutic management are frequently bases, e.g. nearly all psychotherapeutics, analgesics, local anaesthetics, adrenergics and some chemotherapeutics are amines. In West European countries, drugs which are primary or secondary amines occur less frequently than drugs containing tertiary amino functions. Among these tertiary amine drugs alicyclic amines, ranging from three-membered to eight-membered rings, prevail over simple aliphatic tertiary amines. As research chemists, who are not only interested in the metabolism of drugs, but also engaged in the syntheses of new compounds with pharmacological activity, we examined the literature to investigate the biotransformation reactions of representative drugs containing different alicyclic amines. After reviewing the metabolic pathways of about one-hundred compounds, it becomes apparent that *N*-oxidation seemed to play a minor role in comparison with oxidative reactions occurring with the sp^2- or sp^3-hybrid C-atoms of the more lipophilic

Table 1

Comparison of the dipole moments

Amine N-oxide	μ (D)	Tertiary amine	μ (D)
$(CH_3)_3N{\rightarrow}O$	5.02	$(CH_3)_3N$	0.65
$C_6H_5(CH_3)_2N{\rightarrow}O$	4.79	$C_6H_5(CH_3)_2N$	1.58
$C_5H_5N{\rightarrow}O$	4.24	C_5H_5N	2.20

From Ochiai (1967).

part of the molecule. This may be due to the closer relationship of the lipophilic moiety of the compounds to the lipophilic parts of the oxidizing enzymes. The sole products of N-oxidation of tertiary amines are N-oxides, which themselves are sometimes reduced by enzymes back to the parent compounds. Therefore the amount of N-oxide found in urine sometimes does not agree with the amount of N-oxide in blood. Amine oxides possess a higher dipole moment than the corresponding tertiary amines (see Table 1) and are thus usually more water soluble, thereby aiding their excretion. Exceptions have, however, been observed.

N-Oxides, in spite of their frequently high water solubility, are very often targets for further metabolic attack as we have shown, with several N-oxides in the group of fomocaine derivatives (Oelschläger & Ewald, this volume). In several cases the alicyclic ring carbons are attacked, with subsequent formation of oxo-derivatives which have lost their basic properties. Metabolites have not always been found in the analysis of biological fluids, because changes in pK_a have not been taken into consideration. The change of basicity is of great importance in the further distribution of these metabolites in tissues. Oxo-derivatives frequently undergo ring opening reactions catalysed by hydrolases, with the subsequent formation of compounds with high polarity which may form zwitterions in equilibrium with the ionized compounds.

These general considerations may be illustrated by a survey of the metabolic pathways of the basic moiety of drugs containing different types of alicyclic amines. Such data have largely been obtained by investigation of the excretion products found in urine and faeces. In some cases, results have been obtained by incubating tissue preparations of different animal species. The following survey is classified on the basis of the number of hetero atoms and ring size of the molecule.

COMPOUNDS CONTAINING AZIRIDINE

Aziridine rarely occurs in drugs, with the exception of certain cancer chemotherapeutic agents, such as thiotepa. After ring opening, the resulting electrophilic carbon atoms cause interstrand linking of two DNA strands. The aziridine ring is analogous to an epoxide, which undergoes hydrolytic ring opening as one of the major pathways (Fig. 1). Studies with 2,3-iminocholestane as a model compound have shown that this molecule is converted by enzymic hydrolysis to the amino alcohol (Watabe *et al.*, 1971) by rabbit liver microsomes.

Another type of metabolic degradation of aziridines has been demonstrated by Hata *et al.* (1976). *N*-Methyl-*β*-naphthylaziridine was converted to *β*-vinylnaphthalene to an extent of 7–8% after incubation with rat liver microsomes. The authors postulated the intermediate formation of the corresponding aziridine *N*-oxide. After incubation of *N-p*-anisylaziridine with rat liver microsomes, the authors detected *p*-nitrosoanisol which was converted to *p*-nitroanisol and *p*-anisidine.

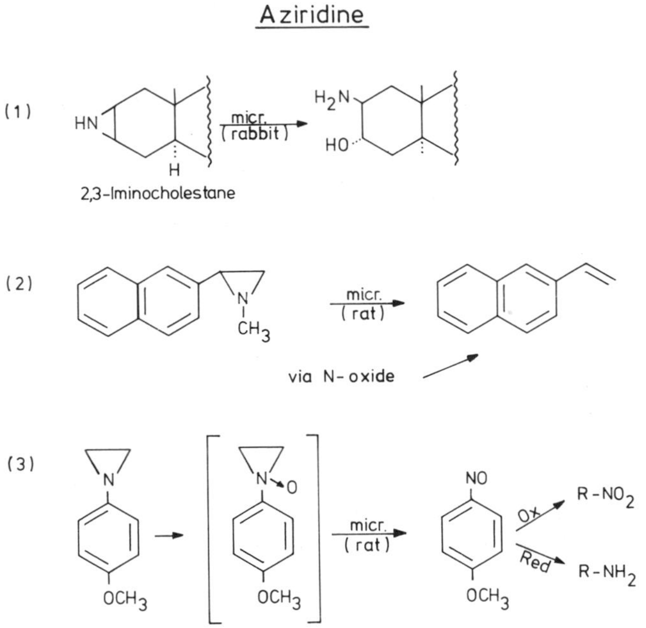

Aziridine

Fig. 1 – Metabolism of the aziridine ring.

COMPOUNDS CONTAINING AZETIDINE

The metabolism of azetidine derivatives has not yet been investigated, since these four-membered alicyclic amines have not been used widely in compounds with pharmacological activity. Bishop *et al.* (1968) described a series of compounds containing an azetidine moiety, which have analgesic activity, but as yet no metabolic studies have been reported.

COMPOUNDS CONTAINING PYRROLIDINE

Pyrrolidine is very often a constituent of drugs, for instance it is part of the neuroleptic sulpirid. Its chemical difference from other common psychotropic drugs is obvious. After i.p. injection of [^{14}C]sulpirid to rats, several metabolites were formed, but no *N*-oxide was identified in urine (Dross, 1978) (Fig. 2). Two 5-oxopyrrolidine derivatives were found, possibly formed due to an initial *α-C*-oxidation. Pyrrolidine also forms the basic moiety of the

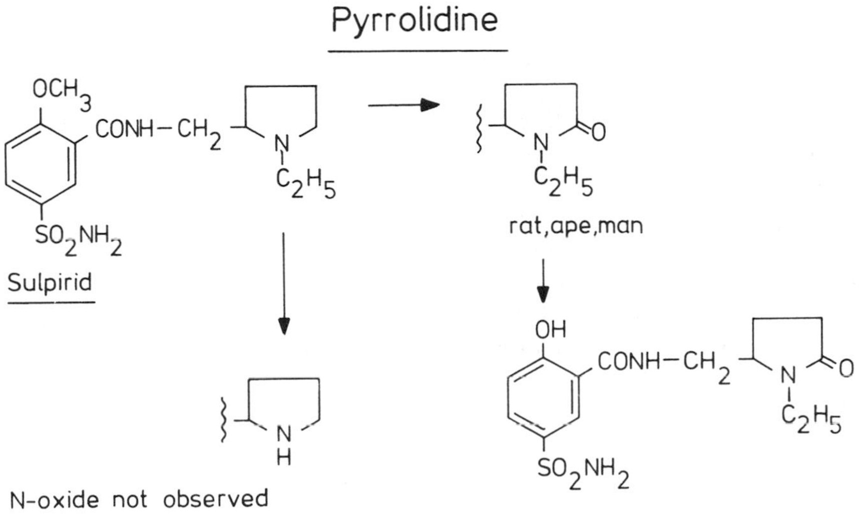

Pyrrolidine

Sulpirid

rat, ape, man

N-oxide not observed

Fig. 2 – Metabolism of the pyrrolidine moiety of sulpirid.

Pyrrolidine

Prolintane

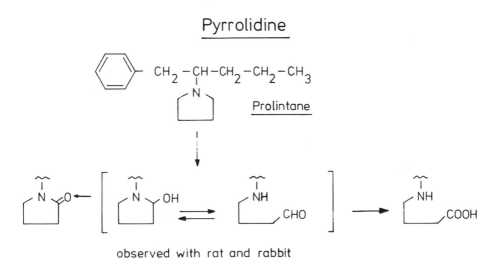

observed with rat and rabbit

Fig. 3 – Metabolism of the pyrrolidine moiety of prolintane.

psychomotor stimulant drug prolintane. The formation of the 5-oxo- derivative of prolintane was observed in rats and rabbits. Furthermore, an equilibrium exists between the 5-hydroxy derivative and its aminoaldehyde, which is oxidized to the corresponding amino acid (Fig. 3). This metabolic product has depressive properties on the psychomotor system (Yoshihara & Yoshimura, 1974).

COMPOUNDS CONTAINING PIPERIDINE

Piperidine is one of the most common alicyclic amines encountered in drugs. The important analgesic pethidine (mepiridine), for example, contains this amine, which after *N*-dealkylation undergoes an oxidation process, forming the hydroxylamine derivative (Yeh *et al.*, 1981). The parent compound is converted to the *N*-oxide to a small extent; no 6-oxo-derivative has been reported to date (Fig. 4).

On the other hand, the local anaesthetic mepivacaine does not form an *N*-oxide, but surprisingly does form the 6-oxo-derivative of the parent compound as well as that of the *N*-demethylated metabolite (Meffin *et al.*, 1973). The appearance of different amounts of the enantiomeric 6-oxo-compounds due to the chiral C-atom 2 in the urine of the volunteers has not been investigated as far as we know (Fig. 5).

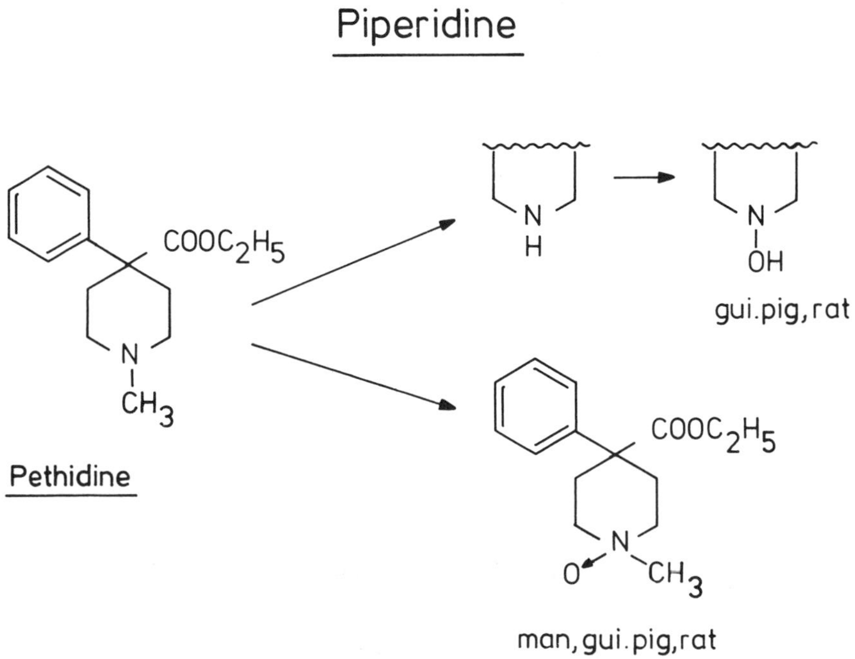

Piperidine

Fig. 4 – Metabolism of the piperidine moiety of pethidine.

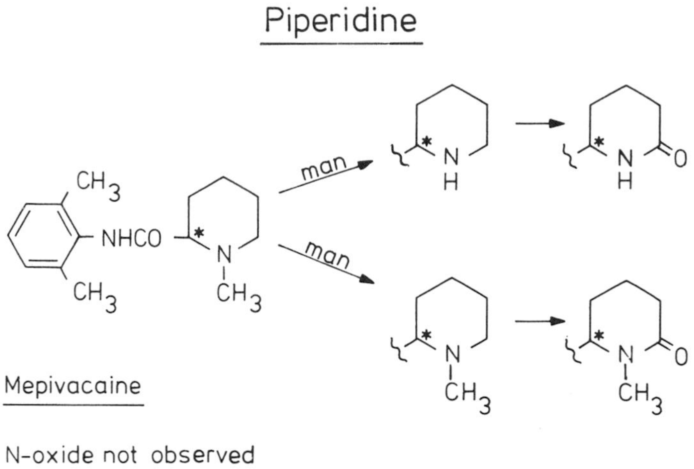

Piperidine

Mepivacaine

N-oxide not observed

Fig. 5 – Metabolism of the piperidine moiety of mepivacaine.

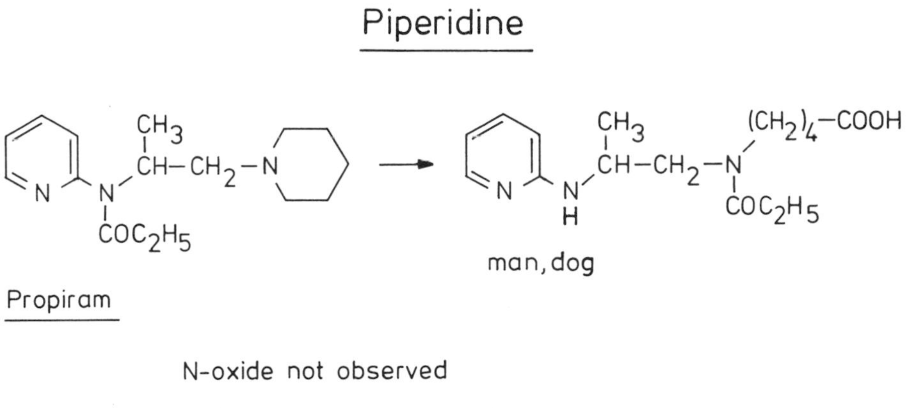

Piperidine

man, dog

Propiram

N-oxide not observed

Fig. 6 – Metabolism of the piperidine moiety of propiram.

A rare biotransformation reaction has been shown to occur in man and in dogs after administration of the analgesic propiram. By a conventional α-C-oxidative process, the piperidine ring is cleaved to give an aminovaleric acid, followed by migration of the propionyl group to the secondary N-atom generated from the piperidine ring (Fig. 6). The structure of this main metabolite was confirmed by synthesis (Duhm *et al.*, 1974).

COMPOUNDS CONTAINING HEXAMETHYLENIMINE

Ethoheptazin and meptazinol are closely related analgesics, which are extensively metabolized in the liver (Fig. 7). Their common characteristic is the N-methylhexamethylenimine moiety, which is oxidized. Whilst ethoheptazin

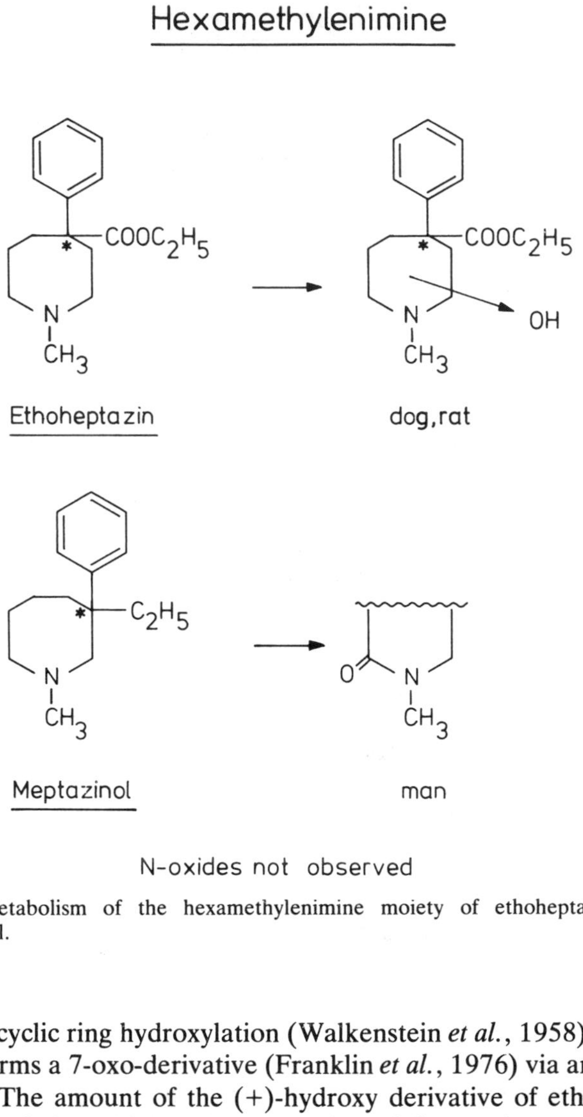

Hexamethylenimine

Fig. 7 – Metabolism of the hexamethylenimine moiety of ethoheptazin and meptazinol.

undergoes alicyclic ring hydroxylation (Walkenstein *et al.*, 1958), meptazinol specifically forms a 7-oxo-derivative (Franklin *et al.*, 1976) via an α-C- oxidative process. The amount of the (+)-hydroxy derivative of ethoheptazin in dog urine is 2–3 times higher than that of the (−)-enantiomer.

COMPOUNDS CONTAINING HEPTAMETHYLENIMINE

The antihypertensive agent guanethidin is converted primarily to its *N*-oxide while the guaninidine moiety does not undergo any metabolic transformation. The heptamethylenimine ring is cleaved by the conventional α-C-oxidative route to an acid (Fig. 8). The corresponding aminoaldehyde was not detected. Both metabolites had less than one tenth of the activity of guanethidin as an antihypertensive drug (Lukas, 1973).

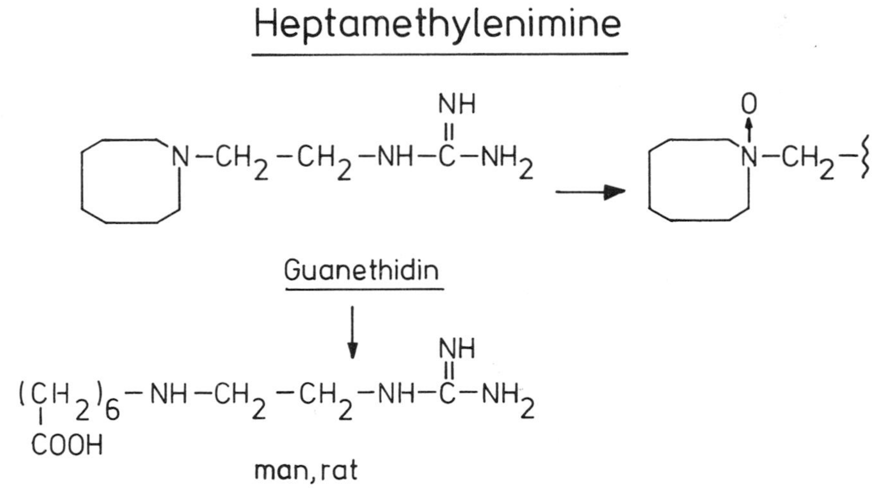

Heptamethylenimine

Guanethidin

man, rat

Fig. 8 – Metabolism of the heptamethylenimine moiety of guanethidin.

COMPOUNDS CONTAINING PIPERAZINE

A *N,N'*-disubstituted piperazine is the basic part of the neuroleptic fluphenazine. The drug forms only one *N*-oxide (Sofer & Ziegler, 1978), and undergoes a ring opening reaction connected with the loss of two carbon atoms (Gaertner *et al.*, 1975). The ethylenediamine derivatives formed accumulate after repeated administration of fluphenazine. The mechanism of the ring cleavage is under investigation in our laboratories (Fig. 9).

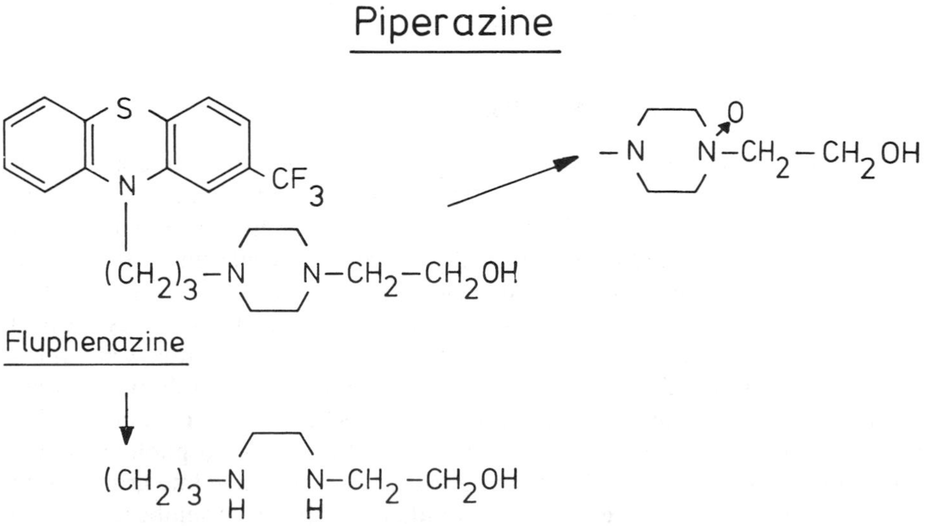

Piperazine

Fluphenazine

Fig. 9 – Metabolism of the piperazine moiety of fluphenazine.

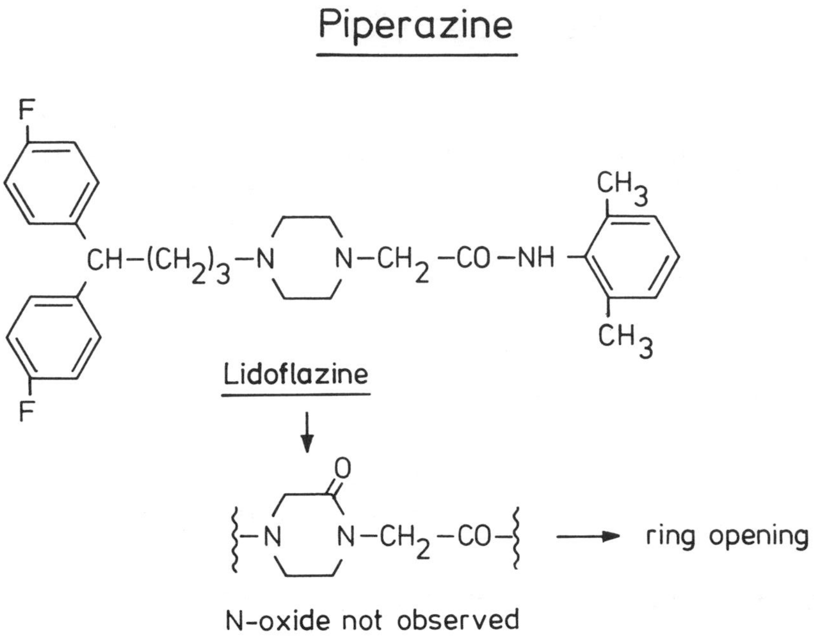

Piperazine

Lidoflazine

N-oxide not observed

Fig. 10 – Metabolism of the piperazine moiety of lidoflazine.

The Ca-blocking agent lidoflazine is not oxidized at either N-atom of the piperazine moiety, but an oxo-derivative has been observed, followed by cleavage of the piperazine ring (Fig. 10). Oxidative *N*-dealkylation and a subsequent ring opening reaction seem to be the dominant routes of biotransformation of this drug (Meuldermans *et al.*, 1977).

COMPOUNDS CONTAINING MORPHOLINE

Morpholine does not occur as frequently in drugs as piperidine or piperazine. Timolol is the L-enantiomer of 3-(3-tert-butylamino-2-hydroxypropoxy)-4-morpholino-1,2,5-thiadiazole. The major metabolic pathway in man includes the formation of highly oxidized morpholine ring-opened species in which the oxypropylamine side chain is unchanged (Fig. 11). The mechanism of the ring opening reaction has been proven by the mass-spectral data for the TMSi derivative of a monohydroxylated timolol found in human urine. This intermediate metabolite is transformed to the oxo-derivative (lactone), which is hydrolysed to the hydroxy carboxylic acid (Tocco *et al.*, 1980).

The anti-anginal agent molsidomine also contains a morpholine moiety which is connected by an N–N bond to the 1,2,3-oxadiazolidine group. Both heterocyclic rings show extensive degradation. The five-membered ring is cleaved first, followed by the morpholine group via formation of an inter-

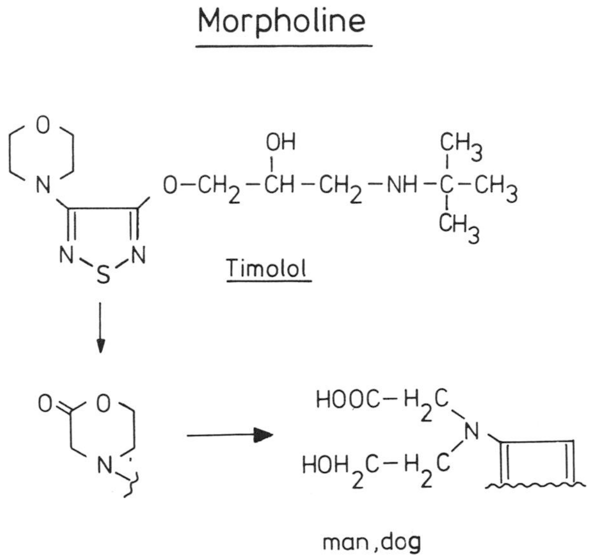

Morpholine

Timolol

man,dog

Fig. 11 – Metabolism of the morpholine moiety of timolol.

mediate lactone (Fig. 12). An *N*-oxide was not observed (Tanayama *et al.*, 1974).

Fomocaine is a potent local anaesthetic developed by Oelschläger and coworkers (1959, 1968) (Fig. 13). This compound is extensively metabolized after oral and i.v. administration to rats, beagles and man (Oelschläger *et al.*, 1975; Oelschläger & Ewald, this volume).

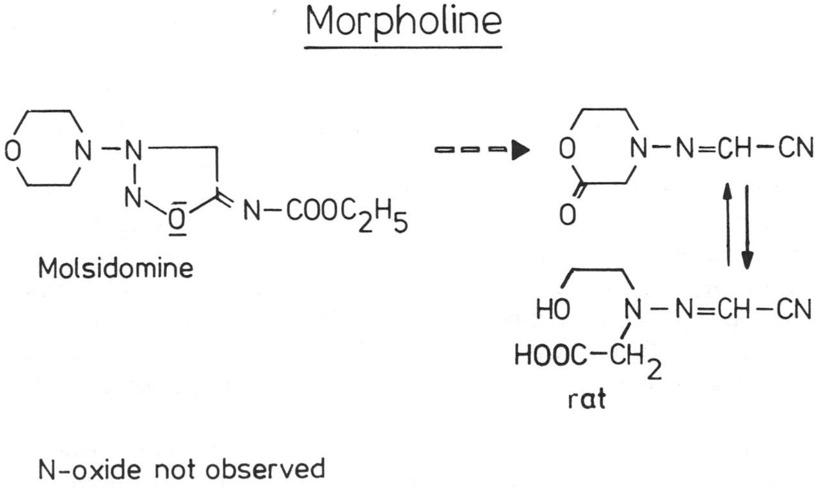

Morpholine

Molsidomine

rat

N-oxide not observed

Fig. 12 – Metabolism of the morpholine moiety of molsidomine.

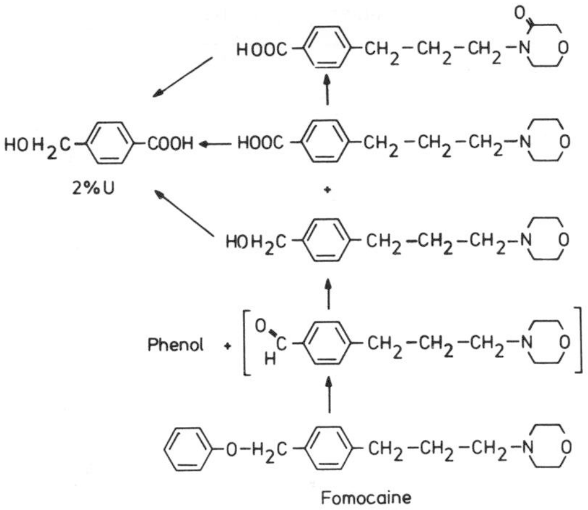

Fig. 13 – Metabolism of fomocaine in the rat.

Six derivatives of fomocaine with changed basic moieties and their corresponding *N*-oxides were synthesised (Table 2) and their metabolism investigated (Schatton, 1977).

Table 2

Synthesised derivatives of fomocaine

3-Diethylamino-1-(4-phenoxymethylphenyl)propane	O/S 4
3-Diethanolamino-1-(4-phenoxymethylphenyl)propane	O/S 5
3-Pyrrolidino-1-(4-phenoxymethylphenyl)propane	O/S 6
3-Piperidino-1-(4-phenoxymethylphenyl)propane	O/S 7
3-*N'*-Methylpiperazino-1-(4-phenoxymethylphenyl)propane	O/S 8
3-Hexamethylenimino-1-(4-phenoxymethylphenyl)propane	O/S 9

Most of the *N*-oxides are more soluble in water than the hydrochlorides of the corresponding amines. But we also observed exceptions, e.g. the *N*-oxide of the diethanolamine derivative (O/S 5) has poor water solubility, probably due to the formation of micelles ($c_K = 3.0 \times 10^{-7}$ mol). After i.v. injection of the *N*-oxides into rats we measured the amount of unchanged *N*-oxides in the 30-h urine by g.l.c. (Table 3). We did not even find traces of the *N*-oxides of the pyrrolidino (O/S 6), piperidino (O/S 7) and the hexamethylenimino derivative (O/S 9) but large amounts (36%) of the *N*-oxide of the diethanolamino derivative (O/S 5) and only a small amount of fomocaine *N*-oxide (7%).

The results demonstrate that *N*-oxides formed may be reduced or further metabolized to a variable extent. The main route of metabolism of these compounds in rats seems to be *p*-hydroxylation of the phenol moiety, as their

Table 3

G.l.c.-Determination of N-oxides after
i.v. administration to rats

N-Oxide of	N-Oxide found (%)
Fomocaine	7
O/S 5	36.0
O/S 6	0
O/S 7	0
O/S 8	45.0[+]
O/S 9	0

0 = not detectable by g.l.c.
[+] = N,N-dioxide was injected, which is not
formed by rats.

conjugates have been found in urine. Ring opening reactions could not be observed in any of these compounds.

3-(β-Morpholinoethoxy)-1H-indazol (O/Ma 35) is a very potent analgesic compound developed in our laboratory (Matthiesen, 1970; Möhrke, 1979) and is now under clinical trial. O/Ma 35 undergoes extensive biotransformation. In rat urine a metabolite was detected which is derived from the biotransformation of the morpholine moiety (Fig. 14); the structure of this

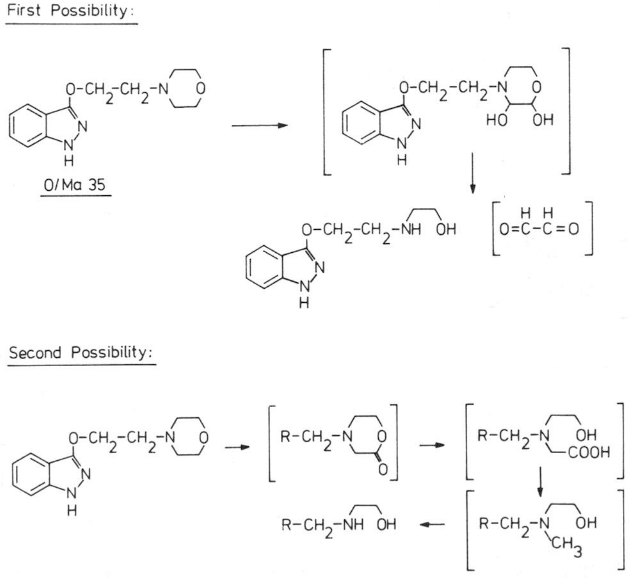

Fig. 14 – Two possible routes to the formation of an unexpected metabolite of O/Ma 35.

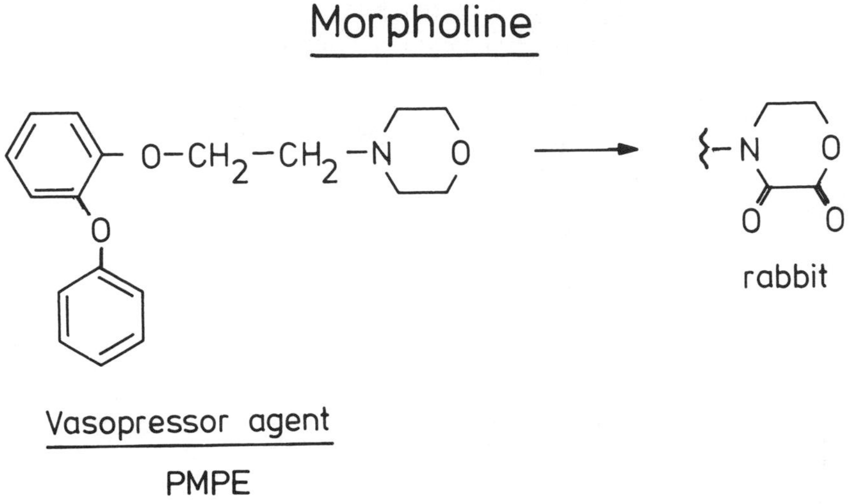
Morpholine
O-CH2-CH2-N O
N O
O O
O
rabbit
Vasopressor agent
PMPE

Fig. 15 – Metabolism of the morpholine moiety of PMPE.

metabolite was proven by synthesis. Schemes for the formation of this metabolite are proposed in Fig. 14.

Phenyl-*O*-(2-*N*-morpholinoethoxy)phenyl ether (PMPE) possesses a long-lasting vasopressor activity. Studies of its metabolism have shown the formation of a 2,3-dioxomorpholine for the first time (Fig. 15). Its structure was elucidated by spectral data (Tatsumi *et al.*, 1975). Thiomorpholine is the basic moiety of the antiparasitic compound 2-(5-nitro-2-furyl)-4-(morpho-

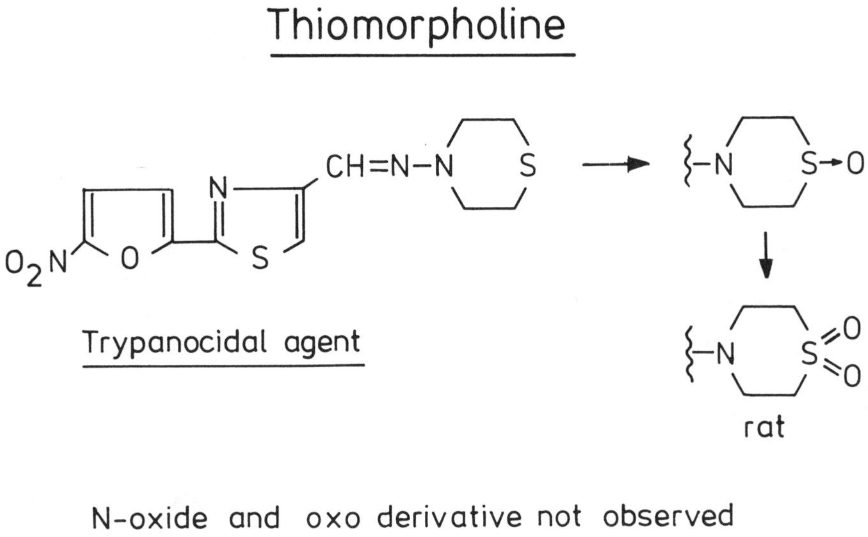
Thiomorpholine
CH=N-N S
N
O2N O S
Trypanocidal agent
N S→O
N S=O =O
rat
N-oxide and oxo derivative not observed

Fig. 16 – Metabolism of the thiomorpholino ring.

Metabolism of Nicotine

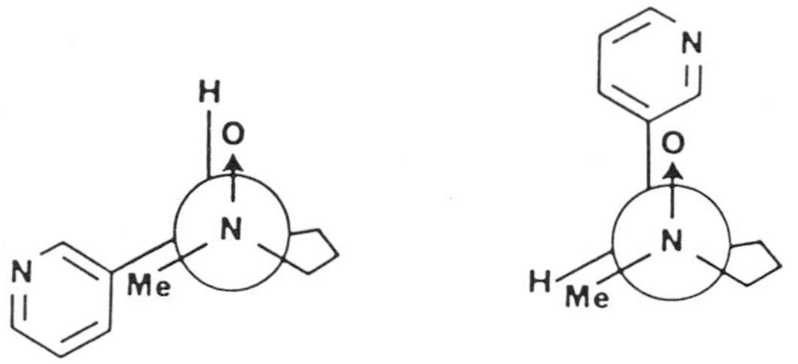

cis-derivative   trans-derivative

S-(-) nicotine ⟶ R,S-cis- nicotine-1'- N-oxide

R-(+) nicotine ⟶ S,R-trans-nicotine - 1'- N-oxide

predominantly

observed with liver 10000 xg supernatant preparations from rats, rabbits, mice, guinea-pigs
and hamsters

Fig. 17 – Metabolism of nicotine to stereoisomers of the 1'-*N*-oxide.

lino-iminomethyl)thiazole. Metabolic studies indicated that no *N*-oxide
was formed and that no ring opening occurred (Fig. 16). The S-atom,
instead of the N-atom, was oxidized via the sulphoxide to the corresponding
sulphone (Chatfield, 1976).

NICOTINE

The metabolism of nicotine is very interesting due to its stereochemi-
cal aspects. Such aspects have been investigated only with a few other
compounds. Nicotine-1'-*N*-oxide can exist as four diastereoisomers and examination of the isomeric ratio produced after incubation of the nicotine enantiomers revealed marked stereoselectivity (Fig. 17). Hepatic preparations from
rats, mice, hamsters, and guinea pigs produced more (1'*R*; 2'*S*)-*cis*- than (1'*S*;
2'*S*)-*trans*-*N*-oxide from (2'*S*)-(−)-nicotine, but more (1'*R*; 2'*R*)-*trans*- than
(1'*S*; 2'*R*)-*cis*-*N*-oxide from (2'*R*)-(+)-nicotine (Jenner *et al.*, 1973a,b). Both
enantiomers thus preferentially lead to the *N*-oxide with (1'*R*) configuration
of the nitrogen atom. A general and qualitative interpretation can nevertheless be attempted by postulating an enzymatic active site which is able to

distinguish between the two faces of the pyrrolidine ring, and which binds one of them with varying selectivity. The attacking oxygen must be postulated to enter from a constant spatial direction. The information on the stereoselective binding of the pyrrolidine ring is indicated by the observed product stereoselectivity of the reaction.

On the other hand, the position of the pyridyl ring above or below the average plane of the pyrrolidine ring does not markedly alter the stereoselective binding of nicotine in most cases. This conclusion is derived from the fact that the $(1'R)$ configuration is often generated independently of the configuration of $2'$-carbon. This lack of influence is not absolute, since the $(1'R)/(1'S)$ ratio is not invariable and displays a detectable substrate stereoselectivity (Jenner *et al.*, 1973b).

CONCLUSION

Summarizing, we have to face the fact that quantitative investigations concerning *N*-oxidation of alicyclic amines and the relationship between chemical structure and biotransformation have been very rare up till now. With respect to our own experiments *N*-oxidation is normally accompanied by oxidation of the alicyclic ring carbons. The subsequent ring opening seems to be more frequent than has been supposed in earlier investigations.

REFERENCES

Bishop, D. C., Cavalla, J. F., Lockhart, I. M., Wright, M., Winder, C. V., Wong, A. & Stephens, M. (1968), *J. Med. Chem.*, **11**, 466.
Chatfield, D. (1976), *Xenobiotica*, **6**, 509.
Dross, K. (1978), *Arzneim-Forsch. Drug Res.*, **28**, 824.
Duhm, B., Maul, W., Medenwald, H., Patzschke, K. & Wegner, L. A. (1974), *Arzneim.-Forsch./Drug Res.*, **24**, 632.
Franklin, R. A., Aldridge, A. & White, C. de B. (1976), *Brit. J. Clin. Pharmacol.*, **3**, 497.
Gaertner, H. J., Liomin, G., Villumsen, D., Bertele, R. & Breyer, U. (1975), *Drug Metab. Dispos.*, **3**, 437.
Hata, Y., Watanabe, M., Matsubaha, T. & Touchi, A. (1976), *J. Am. Chem. Soc.*, **98**, 6033.
Jenner, P., Gorrod, J. W. & Beckett, A. H. (1973a), *Xenobiotica*, **3**, 563.
Jenner, P., Gorrod, J. W. & Beckett, A. H. (1973b), *Xenobiotica*, **3**, 573.
Lukas, G. (1973), *Drug Metab. Rev.*, **2**, 101.
Matthiesen, U. (1970), PhD Thesis, University of Frankfurt a.M., FR Germany.
Meffin, P., v.Robertson, A., Thomas, J. & Winkler, J. (1973), *Xenobiotica*, **3**, 191.
Meuldermans, W., Lauwers, W., Knaeps, A. & Heykants, J. (1977), *Arzneim. Forsch./Drug. Res.*, **27**, 828.

Möhrke, W. (1979), PhD Thesis, University of Frankfurt a.M., FR Germany.

Ochiai, E. (1967), *Aromatic Amine Oxides*, Elsevier, New York, p. 6.

Oelschläger, H. (1959), *Arzneim. Forsch./Drug Res.*, **9**, 313.

Oelschläger, H., Nieschulz, O., Meyer, F. & Schulz, K. H. (1968), *Arzneim, Forsch/Drug Res.*, **18**, 729.

Oelschläger, H., Temple, D. & Temple, C. (1975), *Xenobiotica*, **5**, 309.

Schatton, W. (1977), PhD Thesis, University of Frankfurt a.M., FR Germany.

Sofer, S. & Ziegler, D. (1978), *Drug. Metab. Dispos.*, **6**, 232.

Tanayama, S., Nakai, Y., Fujita, T., Suzuoki, Z., Imashiro, Y. & Masuda, K. (1974), *Xenobiotica*, **4**, 175.

Tatsumi, K., Kitamura, S., Yoshimura, H., Tanaka, S., Hashimoto, K. & Igarashi, T. (1975), *Xenobiotica*, **5**, 377.

Tocco, D. J., Duncan, A. E. W., de Luna, F. A., Smith, J. L., Walkner, R. W. & Vandenheuvel, W. J. A. (1980), *Drug Metab. Dispos.*, **8**, 236.

Walkenstein, S., MacMullen, J. A., Knebel, C. & Seifter, J. (1958), *J. Am. Pharm. Assoc. Sci. Ed.*, **47**, 20.

Watabe, T., Kiyonaga, K. & Hara, S. (1971), *Biochem. Pharmacol.*, **20**, 1700.

Yeh, S. Y., Krebs, H. A. & Changchit, A. (1981), *J. Pharm. Sci.*, **70**, 867.

Yoshihara, S. & Yoshimura, H. (1974), *Xenobiotica*, **4**, 529.

Chapter 6

METABOLISM OF THE LOCAL ANAESTHETIC FOMOCAINE

H. Oelschläger and **H. W. Ewald**, Institut für Pharmazeutische Chemie der Johann Wolfgang Goethe-Universität, D-6000 Frankfurt am Main, FR Germany

1. The metabolism of the local anaesthetic fomocaine has been investigated by i.v. administration of [^{14}C]fomocaine to rats and beagles.
2. Contrary to earlier investigations, cleavage of the propyl chain was observed.
3. The formation of a 'lactam', due to oxidation of the morpholine moiety occurred.

INTRODUCTION

The local anaesthetic fomocaine was introduced into therapy in 1967, for use in dermatology as ointments, gels, suppositories, etc. Fomocaine is an accepted drug, officially in the German Extra Pharmacopoea (DAC) since

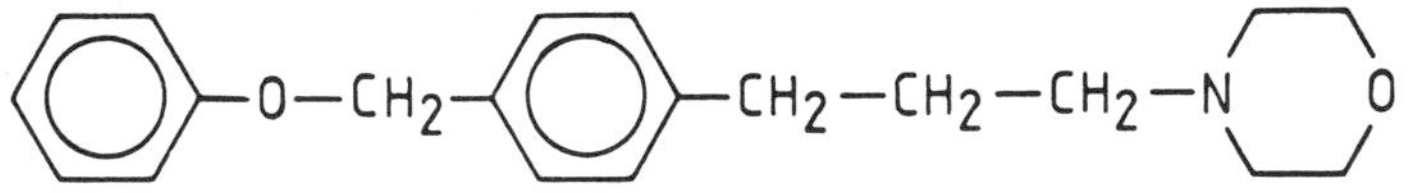

Fomocaine

1977. Its chemical structure as a basic phenolbenzylether is dissimilar to the common local anaesthetics of the ester type like procaine or of the amide type like lidocaine. The water soluble hydrochloride is relatively stable. Only at higher temperatures does a break of the ether bond, with subsequent formation of diphenylmethane derivatives occur to a small extent. The pharmacological properties of fomocaine are impressive: it has low toxicity, its topical anaesthetic activity is similar to tetracaine and the infiltration anaesthesia of fomocaine is 2–4 times stronger than procaine (Nieschulz *et al.*, 1958). No allergic responses have been noted since its clinical introduction (Oelschläger *et al.*, 1968) and the antihistaminic effect exceeds that of diphenhydramine (Oelschläger *et al.*, 1982). The low acute toxicity is due to the slow absorption from mucous membranes in comparison with tetracaine (Nieschulz *et al.*, 1958). Fomocaine is bound to plasma proteins to a great extent (91%), binding to human serum albumin (HSA), globulin fraction and erythrocytes (Nachev, 1982). In earlier investigations with unlabelled fomocaine, two groups of metabolites were observed (Oelschläger *et al.*, 1975; Sgoff, 1975; Jindrova, 1976; Oelschläger *et al.*, 1977). The first group includes metabolites with an intact fomocaine skeleton (fomocaine *N*-oxide, 2-hydroxyfomocaine, 4-hydroxyfomocaine and 4-hydroxyfomocaine *N*-oxide); the second group consists of metabolites formed following cleavage of the parent substance, i.e. 4-(γ-morpholinopropyl)benzoic acid, 4-(phenoxymethyl)phenylpropanoic acid.

Main metabolites in beagles and rats after oral administration of fomocaine were 4-hydroxyfomocaine, its *N*-oxide and fomocaine *N*-oxide (Oelschläger *et al.*, 1975). Only small amounts of the 4-(γ-morpholinopropyl)benzoic acid could be found, more being formed in the rat than in the beagle. All metabolites were less toxic than fomocaine itself and showed no local anaesthetic activity. About 20% of the administered dose could be recovered after 12 hours in urine and faeces, the rest of the administered dose could not be accounted for. The present communication describes further work on the metabolism of fomocaine.

EXPERIMENTAL

^{14}C-Labelled fomocaine was synthesised in a nine-step reaction sequence starting with 4-bromobenzaldehyde (Ewald, 1982). Due to the known metabolic pathways of fomocaine the benzylic C-atom was labelled. The yield of radioactivity was in the range of 7–8% over the last six steps, yielding fomocaine-HCl with a specific activity of 5 mCi/g.

In contrast to earlier investigations we (Oelschläger & Ewald, 1983) administered the labelled fomocaine i.v. to rats (40 mg/kg) and to beagles (5 mg/kg). The analytical procedure was liquid–liquid partitioning with assay of the radioactivity using a Beckman scintillation counter. In the rat we found 75% of the administered dose after 24 hours in urine and faeces (after 48 h, 82%, and after 168 h, 85%).

RESULTS AND DISCUSSION

The main route of metabolism in the rat is shown in Fig. 1. The phenolbenzyl ether bond of fomocaine is cleaved oxidatively to phenol and the unstable

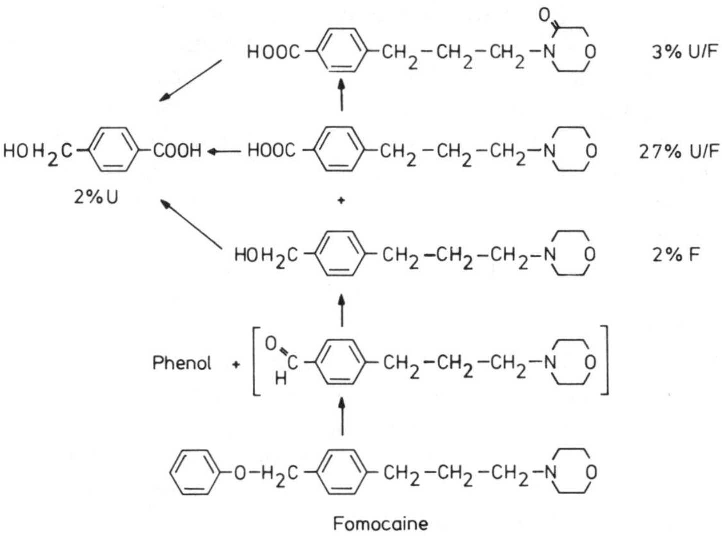

Fig. 1 – First route of [^{14}C]fomocaine metabolism in the rat (δ).

aldehyde, which is oxidized mainly to the corresponding benzoic acid (27%) and reduced to the corresponding benzylalcohol (2%). The acid is transformed to a small amount to the 'lactam' (3%). We also found 2% of 4-hydroxymethylbenzoic acid, which is probably the common end-product from several different metabolites. In the second route, fomocaine is oxidized to 4-hydroxyfomocaine (2%) and to the fomocaine *N*-oxide (0.9%). Contrary to earlier experiments with orally administered fomocaine, where we found small amounts of the ampholyte 4-(γ-morpholinopropyl)benzoic acid and 4-hydroxyfomocaine, 4-hydroxyfomocaine *N*-oxide and fomocaine *N*-oxide as main metabolites (Oelschläger *et al.*, 1975), the ampholyte is now the main metabolite; the other three metabolites are formed to a lesser extent. These differences may be explained partly by the difficult analysis of the ampholyte.

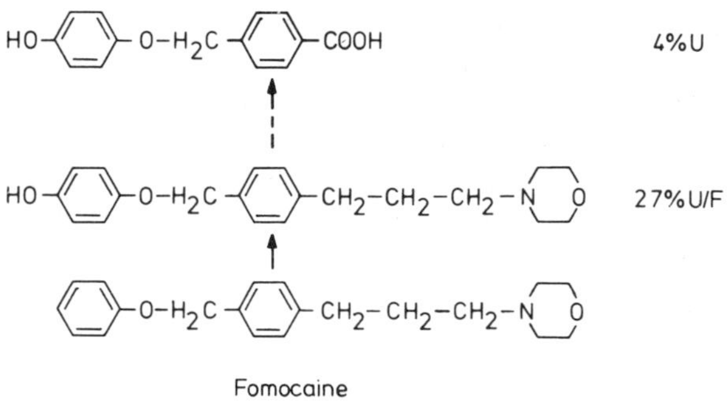

Fig. 2 – First route of [^{14}C]fomocaine metabolism in the beagle (δ).

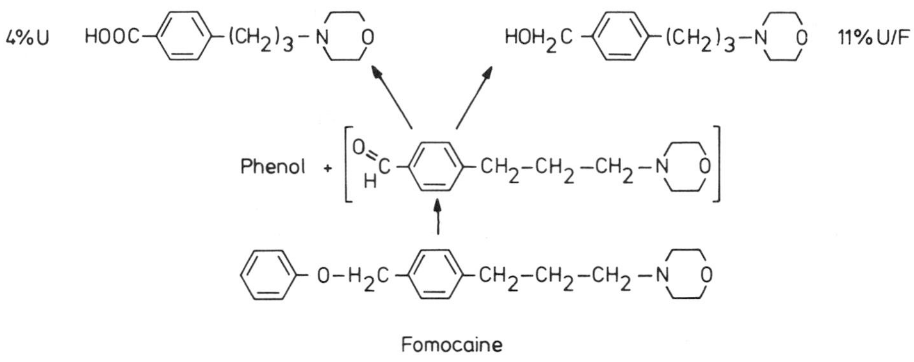

Fig. 3 – Second route of [^{14}C]fomocaine metabolism in the beagle (δ).

New metabolites now found in the rat are: 4-(γ-morpholinopropyl)benzyl alcohol, 4-hydroxymethylbenzoic acid and the 'lactam'.

In the beagle after 24 hours we found 38% of the administered dose in urine and faeces (after 48 h, 76%, and after 168 h, 94%). The main route of biotransformation in the dog is shown in Fig. 2. Fomocaine is oxidized to 4-hydroxyfomocaine (27%), which is further metabolized to 4-(4'-hydroxy)-phenoxymethylbenzoic acid (4%). The second route (Fig. 3) involves the oxidative cleavage of the phenol–benzyl ether bond and the aldehyde formed is either reduced to the benzyl alcohol (11%), or oxidized to the corresponding benzoic acid (4%).

Figure 4 shows that the C–N bond of fomocaine is cleaved oxidatively to morpholine and the unstable aldehyde, which could not be detected, due to its rapid further biotransformation. It is probable that the aldehyde is transformed to the propionic acid, which is β-oxidized to 4-(phenoxy-methyl)benzoic acid (4%). Moreover we found 3% of 4-hydroxymethyl-

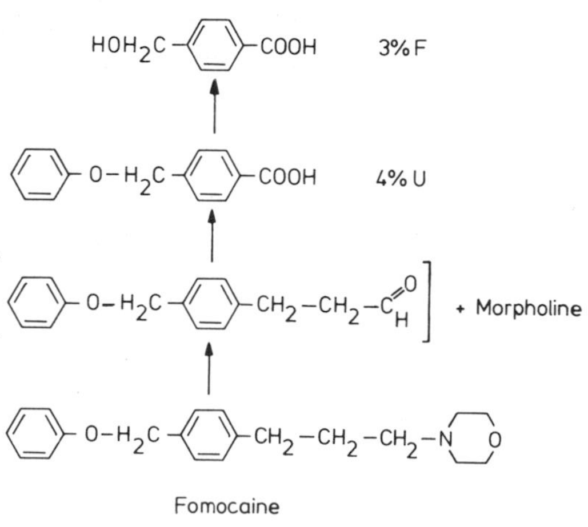

Fig. 4 – Third route of [^{14}C]fomocaine metabolism in the beagle (δ).

benzoic acid which would have been produced following dephenylation and reduction of the intermediate aldehyde.

In accordance with earlier investigations (Oelschläger *et al.*, 1975) 4-hydroxyfomocaine is the main metabolite (27%) and the ampholyte occurs to a small amount. We could not confirm, that 4-hydroxyfomocaine *N*-oxide and fomocaine *N*-oxide are main metabolites. The following metabolites were detected for the first time: 4-(γ-morpholinopropyl)benzyl alcohol, 4-phenoxymethylbenzoic acid, 4-hydroxymethylbenzoic acid and 4-((4'-hydroxy)phenoxymethyl)benzoic acid. The 'lactam' was not positively identified in the dog. All metabolites, found in urine of rat and beagle predominantly exist as conjugates, except 4-(γ-morpholinopropyl)benzoic acid.

In conclusion fomocaine metabolism has been fully elucidated and five new metabolites have been found. Furthermore, the cleavage of the three-carbon chain has been observed. The formation of the 'lactam', formed by oxidation of the morpholine moiety, has been found.

ACKNOWLEDGEMENTS

The authors wish to thank Hoechst AG, D-6000 Frankfurt am Main for generous support, especially Dr K. H. Bremer, Head of the Laboratory of Radiochemistry.

REFERENCES

Ewald, H. W. (1982), unpublished observations.
Jindrova, N. (1976), *Pharmazie*, **31**, 580.
Nachev, P. K. (1982), doctoral dissertation, University of Frankfurt a.M.
Nieschulz, O., Hoffmann, I. & Popendiker, K. (1958), *Arzneim.-Forsch./Drug Res.*, **8**, 539.
Oelschläger, H. & Ewald, H. W. (1983), unpublished observations.
Oelschläger, H., Nieschulz, O., Meyer, F. & Schulz, K. H. (1968), *Arzneim.-Forsch./Drug Res.*, **18**, 729.
Oelschläger, H., Temple, D. J. & Temple, C. F. (1975), *Xenobiotica*, **5**, 309.
Oelschläger, H., Temple, D. J. & Meier, J. I. (1977), *Arch.Pharm. (Weinheim)*, **310**, 579.
Oelschläger, H., Rothley, D. & Müller, M. (1982), *Arzneim.-Forsch./Drug Res.*, **32**, 72.
Sgoff, H. (1975), doctoral dissertation, University of Frankfurt a.M.

Chapter 7

MICROSOMAL *N*-OXIDATION OF SECONDARY AROMATIC AMINES

N. J. Gooderham[†] and **J. W. Gorrod**, Department of Pharmacy, Chelsea College, University of London, Manresa Road, London, SW3 6LX, UK

1. Nitrones are formed during microsomal *N*-oxidation of *N*-benzyl-4-substituted anilines. *N*-Benzyl-4-substituted phenylhydroxylamine, the putative product of *N*-oxidation was found, but only with the 4-chloro substrate.
2. Microsomal *N*-oxidation of *N*-benzylaniline differs quantitatively between species, the hamster being the most active.
3. Animal pretreatment and *in vitro* experiments with enzyme inhibitors indicated that *N*-benzylaniline *N*-oxidation is catalysed predominantly by microsomal flavin-containing monooxygenase, whereas *N*-oxidation of *N*-benzyl-4-chloroaniline may involve both the flavin-containing monooxygenase and cytochrome P-450.
4. We propose that the contribution of flavin-containing monooxygenase and cytochrome P-450 enzyme systems to *N*-oxidation of *N*-benzylanilines is determined by physico-chemical characteristics of the substrate and by the relative amounts of each enzyme in the system.

INTRODUCTION

Studies on metabolic *N*-oxidation of secondary aromatic amines were initiated in the early sixties by Manfred Kiese's group in Munich, Germany, as a

---

†Present Address. Department of Pharmacology, University of Minnesota Medical School, 3-260 Millard Hall, Minneapolis, Minnesota, 55455, USA.

result of their interest in the phenomenon of ferrihaemoglobin formation. These early studies on aromatic amines provided evidence that *N*-oxidation of such compounds generated metabolites capable of oxidizing haemoglobin (Kiese, 1966). However, a number of questions remained, in particular the identity of the primary product of *N*-oxidation, and the participation of the enzymes involved in these metabolic pathways. This communication attempts to address these questions in the light of more recent studies.

DETECTION OF THE PRODUCTS OF SECONDARY AROMATIC AMINE *N*-OXIDATION

Detection of *N*-oxygenated products of secondary aromatic amines has been difficult due to the inherent instability of these compounds. Most studies of metabolic *N*-oxidation of secondary aromatic amines have relied on indirect procedures based on the formation of a coloured product which infer rather than demonstrate identity. The method of Herr & Kiese (1959) has been used to quantify secondary arylhydroxylamines (Kiese & Uehleke, 1961; Kampffmeyer & Kiese, 1964, 1965; Appel *et al.*, 1965; Hlavica, 1970; Heinze *et al.*, 1970; Uehleke, 1971, 1973). During this procedure the putative *N*-hydroxy secondary aromatic amine is oxidized (with $Fe(CN)_6$) to the corresponding nitroso derivative which may be measured in a spectrophotometer. An alternative procedure developed by Das & Ziegler (1970) involves a modification of the method used to estimate tocopherols as reducing equivalents (Tsen, 1961). The hydroxylamine reduces Fe^{3+} to Fe^{2+} which yields a coloured product on reacting with bathophenanthroline. However, we have found that arylnitrones and certain aminophenols can participate in this reaction (Gorrod & Gooderham, 1985a).

Patterson *et al.* (1978) give a comprehensive review of chromatographic techniques used in determining *N*-oxygenated metabolites, but methods for *N*-oxygenated products of secondary aromatic amines are conspicuous by their absence. Kadlubar *et al.* (1976) report procedures for t.l.c. of *N*-hydroxy derivatives of *N*-methyl and *N*-ethyl-4-aminoazobenzene and also describe g.l.c. of the trimethylsilyl derivatives of these compounds. Recently we have reported t.l.c. (Gorrod & Gooderham, 1985a) and h.p.l.c. (Gooderham & Gorrod, 1984) techniques for detecting and estimating *N*-oxygenated metabolites of *N*-benzylanilines. Abbreviated versions of these are given in Table 1 and Fig. 1.

Development of chromatographic methods required authentic reference compounds. Examples of *N*-hydroxy secondary aromatic amines in the literature are scarce. Utzinger & Regenass (1954) and Renner (1963) prepared *N*-hydroxy derivatives of *N*-ethyl and *N*-butylanilines by reacting alkyl halide with phenylhydroxylamines, but report these compounds to be very unstable. Our attempts at reproducing this chemistry confirmed these observations (Gorrod & Gooderham, 1985b). Kadlubar *et al.* (1976) devised a Cope elimination procedure to synthesize *N*-hydroxy derivatives of *N*-methyl and *N*-ethyl-4-aminoazobenzene to facilitate the development of their analytical

Table 1

Thin layer chromatographic separation and response to visualizing reagents of *N*-benzylaniline and its metabolites

Compound	R_F Value (×100)				Chromogenic response				
	S1	S2	S3	S4	D1	D2	D3	D4	D5
N-Benzylaniline	78	80	73	86	—	—	blue/green	—	—
N-Benzyl-*N*-phenylhydroxylamine	60	52	49	70	black	red	blue/green	—	red
α,*N*-Diphenylnitrone	35	31	27	27	black	red	blue/green	purple	red
N-Benzyl-4-aminophenol	15	9	9	31	black	red	blue/green	green	orange
Benzanilide	53	42	39	51	—	—	—	—	—
Aniline	42	34	32	43	—	—	—	—	—

R_F Values determined using 20 × 20 cm glass plates coated with silica gel GF254 (0.25 mm thick).
S1 Chloroform;
S2 Petroleum ether (b.p. 40–60 °C):chloroform (25:75);
S3 Benzene:chloroform (25:75);
S4 Benzene:ethylacetate (90:10);
D1 Ammoniacal silver nitrate (Tollens reagent);
D2 Ferric chloride/bathophenanthroline (0.005%/0.1%);
D3 Ferric chloride/potassium ferricyanide (0.03%/0.03%);
D4 Sodium amminoprusside (0.05% in aqueous ethanol);
D5 Alkaline 2,3,5-triphenyltetrazolium (0.1% in 0.5 M sodium hydroxide/ethanol).

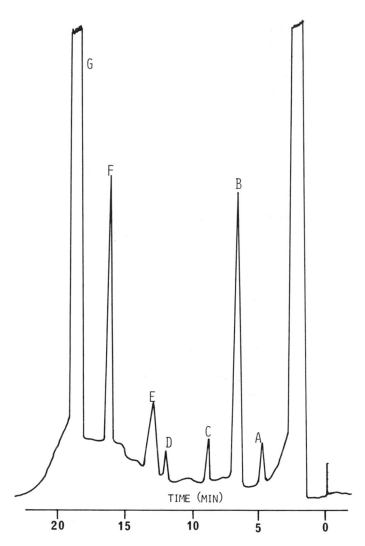

Fig. 1 – High-Performance Liquid Chromatography of the Products of
N-Benzylaniline Metabolism using Hamster Microsomal Preparations.

A = aniline;   B = benzaldehyde;   C = α,*N*-diphenylnitrone;   D = benzanilide;
E = *N*-benzyl-4-aminophenol; F = *N*-benzoyl-4-chloroaniline (internal standard);
G = *N*-benzylaniline.
Mobile phase: pump A, 0.005 M-acetic acid containing 0.005 M-heptane sulphonic
acid (pH of mixture = 3.5 ± 0.1); pump B, acetonitrile containing 0.005 M-acetic
acid + 0.005 M-heptane sulphonic acid.
Programme: solvent gradient, flow-rate 1.5 ml/min; time zero, pump B = 39% of
mobile phase, time 8 min, pump B = 39 to 60% of mobile phase in 5 min, time
17 min, pump B = 60 to 39% of mobile phase in 1 min.
Analytical column, 250 × 5 mm packed with Spherisorb 5ODS; detection, u.v.
254 nm; temperature, ambient.

systems. In establishing our t.l.c. and h.p.l.c. methodology we prepared samples of α,*N*-diphenylnitrones and *N*-hydroxy-*N*-benzylanilines by oxidizing *N*-benzyl-4-substituted anilines with 3-chloroperoxybenzoic acid to generate nitrones, which were then reduced with lithium aluminium hydride to the *N*-hydroxy-*N*-benzyl-4-substituted aniline (Gorrod & Gooderham, 1985b).

N-OXIDATION OF SECONDARY AROMATIC AMINES *IN VITRO*

Secondary aromatic amine *N*-oxidation has been observed using liver (Kiese & Uehleke, 1961; Kampffmeyer & Kiese, 1964; Appel *et al.*, 1965; Das & Ziegler, 1970), lung (Uehleke, 1971), kidney (Heinze *et al.*, 1970), bladder (Uehleke, 1966), intestinal mucosa (Uehleke, 1969) and corpus luteum (Heinze *et al.*, 1970) tissues. Our laboratory has examined the *in vitro* metabolism of *N*-benzylaniline by a variety of rat tissues. α-*N*-Diphenylnitrone was the only *N*-oxidized product detected. Using whole organ homogenates *N*-oxidation of *N*-benzylaniline was most active in the liver, kidney and lung (7.2, 4.0 and 9.1 nmole α,*N*-diphenylnitrone formed per g tissue per 10 minute incubation, respectively); there was negligible activity in the heart, brain, stomach, duodenum/jejunum, spleen and testes. We have established that the site for *N*-benzylaniline *N*-oxidation is the endoplasmic reticulum (Table 2) and the reaction requires NADPH or a regenerating system. In no metabolic experiment was the secondary hydroxylamine, *N*-benzyl-*N*-phenylhydroxylamine found, although we have previously shown that our analytical methods were capable of detecting it had it been present (Gooderham & Gorrod, 1984).

Table 2
In vitro N-oxidation of *N*-benzylaniline by various cell fractions from rat liver homogenate

Fraction	α-*N*-diphenylnitrone formed nmole/g liver/10 min
Whole liver homogenate	7.2
200 g pellet	8.1
Mitochondrial fraction	n.d.
9000 g supernatant	11.7
Microsomal fraction	27.9
Washed microsomal fraction	22.3
140,000 g supernatant	n.d.

n.d. not detected.
Incubation conditions: 37 °C in air for 10 min; 3 ml incubate contained, NADP (2 μmole), glucose-6-phosphate (10 μmole), glucose-6-phosphate dehydrogenase (1 unit), $MgCl_2$ (20 μmole), *N*-benzylaniline (5 μmole), tissue preparation (equivalent to 0.5 g original tissue) in phosphate buffer pH 7.4.

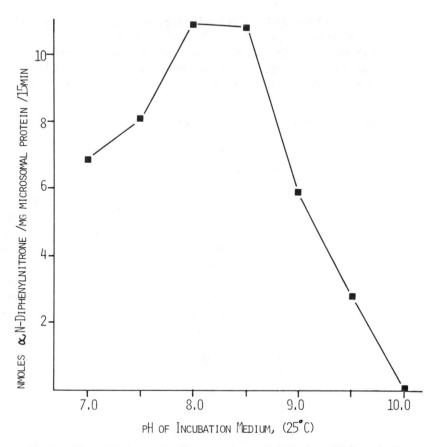

Fig. 2 – Effect of Hydrogen Ion Concentration on *N*-Oxidation of *N*-Benzylaniline by Hamster Hepatic Microsomes.

Incubation conditions were as described in Table 2, using washed microsomes (equivalent to 0.5 g original liver) as enzyme source and Tris/KCl buffer adjusted to the pH indicated.

Most *in vitro* metabolic experiments are carried out at pH 7.4, but this is not always optimal. Microsomal *N*-oxidation of primary aromatic amines is maximal at about pH 7.5 (Hlavica & Kiese, 1969; Hlavica, 1970) but tertiary aromatic amine *N*-oxidation is optimal at about pH 8.0–8.5 (Hlavica & Kiese, 1969; Das & Ziegler, 1970; Devereux & Fouts, 1974). These characteristics have been used to differentiate the enzymic processes involved in these reactions, but this approach must take into account product stability at different hydrogen ion concentrations (Gorrod, 1973). In our studies with *N*-benzylaniline, *N*-oxidation was optimal between pH 8.0–8.5 (Fig. 2), which is consistent with the pH optimum reported for *N*-oxidation of other secondary aromatic amines, e.g. *N*-ethylaniline, pH 8.5 (Hlavica & Kiese, 1969) and *N*-methyl-4-chloroaniline, pH 8.3 (Pan *et al.*, 1979).

The ability of various species to oxidize nitrogen in organic compounds differs. Appel *et al.* (1965) examined hepatic microsomal *N*-oxidation of aniline and *N*-ethylaniline using tissue obtained from guinea pig, rabbit and cat and found varied activities, the latter species being the least active. Similarly, the extent to which rabbit, guinea pig and dog microsomes *N*-oxidized *N*-ethylaniline differed widely (Kampffmeyer & Kiese, 1964, 1965). These species differences in the ability to *N*-oxidize may have significance when considering the extent and duration of a pharmacological or toxicological effect, e.g. ferrihaemoglobin formation (Kiese, 1966). Studies in our laboratory on *N*-benzylaniline *N*-oxidation by hepatic microsomes from a number of species also show considerable quantitative differences (Fig. 3). Clearly *N*-oxidation of *N*-benzylaniline to *α,N*-diphenylnitrone was much more active using hamster tissue than with microsomes from other species examined.

Aromatic ring substitution can have a pronounced effect on *N*-oxidation of aromatic amines. 4-Substitution of aniline with a chloro on methyl group increases *N*-oxidation three- to fourfold (Uehleke, 1973). Patterson & Gorrod (1978) showed that 4-substitution of *N*-ethyl-*N*-methylaniline with a chloro or phenyl group decreased *N*-oxide formation, whereas 4-substitution with a methyl group increased *N*-oxide production. Secondary aromatic amine *N*-oxidation is also affected; 4-substitution of *N*-benzylaniline alters metabolism both quantitatively and qualitatively (Figs 3 and 4). Hamster hepatic microsomal metabolism of *N*-benzyl-4-chloroaniline produced two *N*-oxidized metabolites, *α*-phenyl-*N*-(4-chloro)phenylnitrone and *N*-benzyl-*N*-(4-chloro)phenylhydroxylamine (Fig. 4). We believe this finding to be the first direct evidence for microsomal *N*-hydroxylation of a secondary aromatic amine. In the previous section it was mentioned that microsomal *N*-hydroxylation of secondary aromatic amines was mostly inferred and not proven in earlier studies. The exception to this was the report of Kadlubar *et al.* (1976) who used sensitive and specific methods for their examination of *N*-methyl-4-aminoazobenzene *N*-oxidation. They failed to detect the *N*-hydroxy secondary aromatic amine during microsomal metabolism because of its rapid decomposition. The presence of a secondary oxidation product, the hydroxylamine *N*-oxide or the nitrone, was proposed but not established. In other experiments with purified porcine microsomal flavin-containing monooxygenase, the *N*-hydroxy derivatives of *N*-methyl and *N*-ethyl-4-aminoazobenzene were detected (Kadlubar *et al.*, 1976). Our observations on the formation of both *N*-benzyl-*N*-(4-chloro)phenylhydroxylamine and *α*-phenyl-*N*-(4-chloro)phenylnitrone in microsomal incubates support the concept of the *N*-hydroxy compound being the primary metabolite, which upon further oxidation generates the nitrone. The role of a disubstituted hydroxylamine *N*-oxide intermediate in this type of reaction was first proposed by Poulsen *et al.* (1974) whilst investigating the properties of a disubstituted hydroxylamine oxidase. Evidence for the existence of hydroxylamine oxides and their rapid dehydration to nitrones has been documented (Klages *et al.*, 1963; Hoffman *et al.*, 1964).

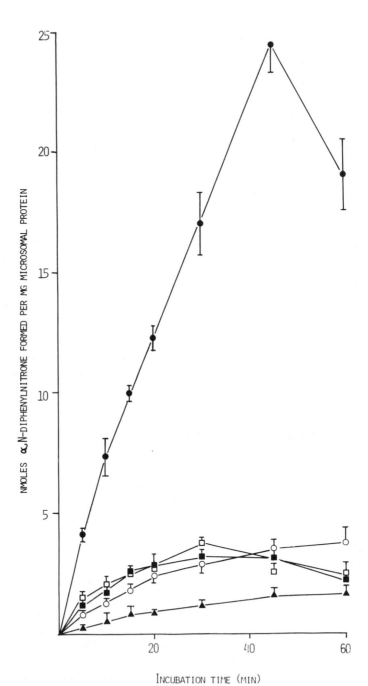

Fig. 3 – Effect of Incubation Time on the *N*-Oxidation of *N*-Benzylaniline using
Hepatic Microsomes from Male Animals of Various Species.

● hamster; ○ rat; ■ mouse; □ guinea pig; ▲ rabbit.

Incubation conditions as described in Table 2, using washed microsomes.

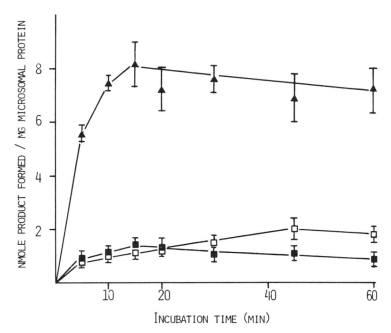

Fig. 4 – Effect of Incubation Time on the *N*-Oxidation of *N*-Benzyl-4-chloroaniline and *N*-Benzyl-4-toluidine using Hamster Hepatic Microsomes.

▲ *N*-benzyl-*N*-(4-chloro)phenylhydroxylamine
■ α-phenyl-*N*-(4-chloro)phenylnitrone
☐ α-phenyl-*N*-(4-tolyl)nitrone.

Incubation conditions as described in Table 2, using washed microsomes.

α-Phenyl-*N*-(4-tolyl)nitrone was the only *N*-oxidized product detected during hepatic microsomal oxidation of *N*-benzyl-4-toluidine (Fig. 4) and was formed in small amounts. This was not unexpected since *N*-benzyl-*N*-(4-tolyl)hydroxylamine was the least stable disubstituted hydroxylamine synthesized and it underwent spontaneous decomposition in aqueous media, forming predominantly the nitrone.

Kinetic analysis of *N*-benzyl-4-substituted aniline *N*-oxidation is summarized in Table 3. It may be seen that although the hamster has the highest V_{max} for *N*-benzylaniline *N*-oxidation it also has the highest K_m, whereas the rabbit has the lowest V_{max} and K_m. The K_m and V_{max} values for *N*-oxidation of the 4-substituted compounds (hamster) are all lower than the corresponding *N*-benzylaniline values.

The possibility of sex differences in *N*-benzylaniline *N*-oxidation has been examined for rats and mice. No differences were observed with mice, but male rats were considerably more active than females (1.71 ± 0.09 and 0.27 ± 0.02 nmoles α,*N*-diphenylnitrone formed per mg microsomal protein per 15 min incubation, respectively). This is consistent with the report from

Table 3

Apparent K_m and V_{max} values for the *N*-oxidation of *N*-benzyl-4-substituted anilines by hepatic washed microsomal preparations from various species

Substrate	Species	Metabolic product			
		α,-phenyl-*N*-(4-substituted) phenyl nitrone		*N*-benzyl-*N*-(4-substituted) phenylhydroxylamine	
		K_m [a]	V_{max} [b]	K_m [a]	V_{max} [b]
N-Benzylaniline	Hamster	5.62 ± 0.59	2.88 ± 0.61	—	—
	Guinea pig	2.20 ± 0.06	1.35 ± 0.11	—	—
	Mouse	2.03 ± 0.20	0.61 ± 0.04	—	—
	Rat	1.94 ± 0.08	0.16 ± 0.01	—	—
	Rabbit	0.42 ± 0.07	0.05 ± 0.01	—	—
N-Benzyl-4-chloroaniline	Hamster	0.42 ± 0.12	0.13 ± 0.01	0.30 ± 0.08	0.82 ± 0.05
N-Benzyl-4-toluidine	Hamster	0.86 ± 0.02	0.21 ± 0.01	—	—

[a] K_m values are mmolar.
[b] V_{max} values are nmole product formed per mg microsomal protein per min.

Kadlubar *et al.* (1976) for *N*-oxidation of *N*-methyl-4-aminoazobenzene who found male rats to be more active than female rats.

ENZYMES INVOLVED IN THE *N*-OXIDATION OF SECONDARY AROMATIC AMINES

Identification and estimation of metabolites formed in low amounts is difficult and one way to overcome this is to induce the enzymes involved in xenobiotic metabolism, e.g. with phenobarbitone or 3-methylcholanthrene. The effect of pretreating hamsters with inducing agents on their hepatic microsomal *N*-oxidation is summarized in Table 4. Formation of α,*N*-diphenylnitrone from *N*-benzylaniline is decreased by animal pretreatment implying that cytochrome P-450 enzymes are not involved in this reaction. Decreased *N*-oxidation may be due to dilution of the enzyme(s) responsible in the microsomes and/or an induction of proteins involved in further metabolism of α,*N*-diphenylnitrone. Similar arguments can be applied to the *N*-oxidation of *N*-benzyl-4-chloroaniline to *N*-benzyl-*N*-(4-chloro)phenylhydroxylamine. However, the increase in the production of α-phenyl-*N*-(4-chloro)phenylnitrone is an enigma and could indicate a role for cytochrome P-450.

Phenobarbitone pretreatment of rabbits failed to stimulate *N*-oxidation of *N*-methylaniline (Uehleke, 1973) or *N*-ethylaniline (Hlavica & Kiese, 1969; Lange, 1967, 1968). Similarly phenobarbitone and 3-methylcholanthrene pretreatments failed to induce *N*-oxidation of tertiary aromatic amines (Gorrod & Patterson, 1983). In contrast, phenobarbitone pretreatment of rabbits considerably stimulates liver microsomal *N*-oxidation of primary aromatic amines (Hlavica & Kiese, 1969; Smith & Gorrod, 1978).

The effect of including enzyme inhibitors in microsomal incubates on *N*-oxidation of *N*-benzylaniline is given in Table 5. Of the inhibitors of cytochrome P-450 used, only DPEA at the highest concentration depressed nitrone formation. *n*-Octylamine and carbon monoxide stimulate the reaction, but methimazole and 1-naphthylthiourea (inhibitors of the microsomal flavin-containing monooxygenase) inhibited *N*-benzylaniline *N*-oxidation. Collectively these studies and animal pretreatment experiments suggest *N*-oxidation of *N*-benzylaniline is catalysed by a cytochrome P-450 independent system and provide compelling evidence for an involvement of the flavin-containing monooxygenase. Similar studies on *N*-oxidation of *N*-benzyl-4-chloroaniline are given in Table 6. The sensitivity of disubstituted hydroxylamine formation to DPEA and high concentrations of *n*-octylamine suggests a cytochrome P-450 involvement. Formation of α-phenyl-*N*-(4-chloro)phenylnitrone was inhibited by SKF 525A, DPEA and high *n*-octylamine concentrations, indicating a cytochrome P-450 catalysed reaction, but this process was also inhibited by methimazole. Taking into account the effect of animal pretreatment on *N*-benzyl-4-chloroaniline *N*-oxidation, the overall evidence points to a cytochrome P-450 catalysed reaction for this substrate but a role for the flavin-containing monooxygenase cannot be eliminated.

Table 4

Effect of pretreating hamsters with inducers of cytochrome P-450 on the microsomal *N*-oxidation of *N*-benzylanilines

Inducer	P-450[a]	*N*-benzylaniline α,*N*-diphenylnitrone	*N*-benzyl-4-chloroaniline α-phenyl-*N*-(4-chloro)phenylnitrone	*N*-benzyl-*N*-(4-chloro)phenylhydroxylamine
Control	1.18 ± 0.07	3.96 ± 0.41	0.99 ± 0.02	6.14 ± 0.29
Phenobarbitone	2.21 ± 0.32*	1.58 ± 0.15*	2.05 ± 0.02*	2.88 ± 0.33*
3-Methylcholanthrene	2.03 ± 0.27*	1.82 ± 0.22*	2.24 ± 0.19*	—
Arochlor 1254	2.52 ± 0.17*	0.59 ± 0.20*	3.23 ± 0.02*	—

Values are mean ± SE, $n = 3$, nmole product formed per mg microsomal protein per 15 min incubation.
[a] nmole P-450 per mg microsomal protein.
* Significantly different from control ($p < 0.05$).
Male Syrian hamsters were injected (i.p.) with phenobarbitone (80 mg/Kg) for 3 days and killed on day 5, or 3-methylcholanthrene (40 mg/Kg) and killed 2 days later, or arochlor 1254 (500 mg/Kg) and killed 5 days later. Incubation conditions were as described in Table 2.

Table 5

Effect of some enzyme inhibitors on the *N*-oxidation of *N*-benzyl-
aniline by hamster hepatic microsomal preparations

inhibitor	concentration	α,*N*-diphenylnitrone formed (% of control)
Control		100.0[a]
SKF 525A	10^{-5}M	115.9 ± 11.9
	10^{-4}M	93.3 ± 9.3
	10^{-3}M	107.2 ± 6.1
DPEA	10^{-5}M	87.6 ± 11.8
	10^{-4}M	84.6 ± 12.2
	10^{-3}M	71.7 ± 15.2*
N-Octylamine	10^{-5}M	96.9 ± 12.4
	10^{-4}M	147.2 ± 10.2*
	10^{-3}M	164.8 ± 18.1*
Methimazole	10^{-5}M	58.5 ± 12.0*
	10^{-4}M	22.0 ± 8.7*
	10^{-3}M	15.7 ± 6.2*
1-Naphthyl thiourea	10^{-5}M	73.3 ± 11.6*
Carbon monoxide	$N_2:O_2:CO(9:1:0)$	100.0[b]
	$N_2:O_2:CO(4:1:5)$	144.7 ± 14.4*

[a]Control value, 10.09 ± 0.81 nmole α,*N*-diphenylnitrone formed/mg microsomal protein/15 min incubation.
[b] Control value, 9.01 ± 0.96 nmole α,*N*-diphenylnitrone formed/mg microsomal protein/10 min incubation.
*Significantly different from control ($p < 0.05$).
Incubation conditions as described in Table 2.

The flavin-containing monooxygenase has previously been implicated in microsomal secondary aromatic amine *N*-oxidation (Uehleke, 1971, 1973; Das & Ziegler, 1970; Kadlubar *et al.*, 1976). Furthermore, studies using purified flavin-containing monooxygenase in reconstituted systems have shown it to be capable of *N*-oxidizing secondary aromatic amines (Ziegler *et al.*, 1969, 1973; Kadlubar *et al.*, 1976). Reviewing the enzymes involved in *N*-oxidation Gorrod (1973, 1978) has suggested that secondary aromatic amines may be substrates for *N*-oxidation by both flavin-containing monooxygenase and cytochrome P-450 enzyme systems. We reiterate this concept here and propose that their relative contribution to secondary aromatic amine *N*-oxidation is determined by physico-chemical characteristics of the substrate and by the relative amounts of each enzyme in the tissue being examined.

Table 6
Effect of some enzyme inhibitors on the _N_-oxidation of
_N_-benzyl-4-chloroaniline by hamster hepatic microsomal preparations

Inhibitor	Concentration	α-phenyl-_N_-(4-chloro) phenylnitrone (% of control)	_N_-benzyl-_N_-(4-chloro) phenylhydroxylamine (% of control)
Control		100.0[a]	100.0[b]
SKF 525A	10^{-5}M	91.7 ± 5.0	100.4 ± 6.1
	10^{-4}M	76.7 ± 8.4*	96.1 ± 8.2
	10^{-3}M	63.1 ± 7.2*	100.4 ± 9.3
DPEA	10^{-5}M	36.0 ± 13.6*	76.2 ± 10.9*
	10^{-4}M	33.6 ± 13.1*	70.1 ± 10.2*
	10^{-3}M	35.4 ± 11.2*	39.2 ± 7.5*
_N_-Octylamine	10^{-5}M	88.6 ± 6.7	100.9 ± 6.6
	10^{-4}M	82.0 ± 11.2	99.4 ± 7.2
	10^{-3}M	48.1 ± 8.3*	67.0 ± 12.3*
Methimazole	10^{-5}M	77.5 ± 4.1*	121.0 ± 15.2
	10^{-4}M	57.4 ± 7.9*	115.4 ± 7.5
	10^{-3}M	46.8 ± 7.6*	106.1 ± 8.9

[a]Control value, 1.24 ± 0.11 nmole α-phenyl-_N_-(4-chloro)phenylnitrone formed/mg microsomal protein/15 min incubation.
[b]Control value, 7.20 ± 0.46 nmole _N_-benzyl-_N_-(4-chloro)phenylhydroxyamine formed/mg microsomal protein/15 min incubation.
*Significantly different from control ($p < 0.05$).
Incubation conditions as described in Table 2, 15 min incubation.

REFERENCES

Appel, W., Graffe, W., Kampffmeyer, H. & Kiese, M. (1965), _Arch exp. Path. Pharmak._, **251**, 88.

Das, M. L. & Ziegler, D. M. (1970), _Arch. Biochem. Biophys._, **140**, 300.

Devereux, T. R. & Fouts, J. R. (1974), _Chem. Biol. Interact._, **8**, 91.

Gooderham, N. J. & Gorrod, J. W. (1984), _J. Chromatog._, **309**, 339.

Gorrod, J. W. (1973), _Chem. Biol. Interact._, **7**, 289.

Gorrod, J. W. (1978), in _Biological Oxidation of Nitrogen_, ed. Gorrod, J. W. Elsevier/North Holland, Amsterdam, p. 201.

Gorrod, J. W. & Gooderham, N. J. (1985a), _Xenobiotica_, in press.

Gorrod, J. W. & Gooderham, N. J. (1985b), _Arch. Pharmazie_ (Weinheim) in press.

Gorrod, J. W. & Patterson, L. H. (1983), _Xenobiotica_, **13**, 521.

Heinze, E., Hlavica, P., Kiese, M. & Lipowsky, G. (1970), _Biochem Pharmacol._, **19**, 641.

Herr, F. & Kiese, M. (1954), *Arch. exp. Path. Pharmak.*, **235**, 351.

Hlavica, P. (1970), *Biochem. Biophys. Res. Comm.*, **40**, 212.

Hlavica, P. & Kiese, M. (1969), *Biochem. Pharmac.*, **18**, 1501.

Hoffman, A. K., Feldman, A. M., Gelblum, E. & Hodgson, W. E. (1964), *J. Am. Chem. Soc.*, **86**, 639.

Kadlubar, F. F., Miller, E. C. & Miller, J. A. (1976), *Can. Res.*, **36**, 1196.

Kampffmeyer, H. & Kiese, M. (1964), *Arch. exp. Path. Pharmak.*, **246**, 397.

Kampffmeyer, H. & Kiese, M. (1965), *Arch. exp. Path. Pharmak.*, **250**, 1.

Kiese, M. (1966), *Pharmac. Rev.*, **18**, 1091.

Kiese, M. & Uehleke, H. (1961), *Arch. exp. Path. Pharmak.*, **242**, 117.

Klages, F., Heinle, R., Sitz, H. & Spect, E. (1963), *Ber. dtsch. chem. Ges.*, **96**, 154.

Lange, G. (1967), *Arch. exp. Path. Pharmak.*, **257**, 230.

Lange, G. (1968), *Arch. exp. Path. Pharmak.*, **259**, 221.

Pan, H. P., Fouts, J. R. & Devereux, T. R. (1979), *Xenobiotica*, **9**, 441.

Patterson, L. H., Damani, L. A., Smith, M. R. & Gorrod, J. W. (1978), in *The Biological Oxidation of Nitrogen*, ed. Gorrod, J. W. Elsevier/North Holland, Amsterdam, p. 213.

Patterson, L. H. & Gorrod, J. W. (1978), in *The Biological Oxidation of Nitrogen*, ed. Gorrod, J. W. Elsevier/North Holland, Amsterdam, p. 471.

Poulsen, L. L., Kadlubar, F. F. & Ziegler, D. M. (1974), *Arch. Biochem. Biophys.*, **164**, 774.

Renner, G. (1963), *Z. Anal. Chem.*, **193**, 92.

Smith, M. R. & Gorrod, J. W. (1978), in *The Biological Oxidation of Nitrogen*, ed. Gorrod, J. W. Elsevier/North Holland, Amsterdam, p. 65.

Tsen, C. C. (1961), *Anal. Chem.*, **33**, 849.

Uehleke, H. (1966), *Life Sci.*, **5**, 1489.

Uehleke, H. (1969), *Proc. Eur. Soc. Study Drug Tox.*, **10**, 94.

Uehleke, H. (1971), *Xenobiotica*, **1**, 327.

Uehleke, H. (1973), *Drug Metab. Disp.*, **1**, 299.

Utzinger, G. E. & Regenass, F. A. (1954), *Helv. Chim. Acta.*, **37**, 1885.

Ziegler, D. M., McKee, E. M. & Poulsen, L. L. (1973), *Drug Metab. Disp.*, **1**, 314.

Ziegler, D. M., Mitchell, C. H. & Jollow, D. (1969), in *Microsomes and Drug Oxidations*, eds. Gillette, J. R., Conney, A. H., Cosmides, G. J., Estabrook, R. W., Fouts, J. R. & Mannering, G. J. Academic Press, New York, p. 173.

Chapter 8

N-DEMETHYLATION AND N-OXIDATION OF N,N-DIMETHYLANILINE BY RAT LIVER MICROSOMES

A. A. Houdi and **L. A. Damani**, Department of Pharmacy, University of Manchester, Manchester, M13 9PL, UK

1. N-Oxidation of N,N-dimethylaniline is not enhanced in microsomes from rats pretreated with phenobarbitone and β-naphthoflavone.
2. Cytochrome P-450 inhibitors have no effect on N-oxidation, but inhibit N-demethylation of N,N-dimethylaniline.
3. N-Oxidation is stimulated by n-octylamine and strongly inhibited by naphthylthiourea.
4. Purified rat liver cytochrome P-450 and P-448 mediate N-demethylation, but not N-oxidation of N,N-dimethylaniline. On the other hand, purified hog liver flavin-containing monooxygenase mediates N-oxidation, but not N-demethylation.

INTRODUCTION

Most drug oxidations are mediated via membrane-bound microsomal mixed function oxidases. These either utilize cytochrome P-450 isozymes or the flavin-containing monooxygenases (EC 1.14.13.8) as their terminal oxidase. Numerous substrates are available for monitoring cytochrome P-450 activity

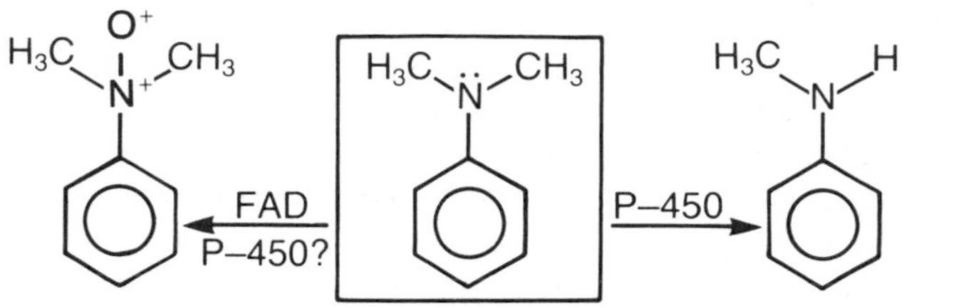

Fig. 1 – *N*-Oxidation and *N*-demethylation of *N,N*-dimethylaniline.

in cellular and subcellular preparations. For example 7-ethoxycoumarin *O*-de-ethylation (Greenlee & Poland, 1978), 7-ethoxyresorufin *O*-de-ethylation (Burke & Mayer, 1974) and aniline hydroxylation (Imai *et al.*, 1966). The number of model compounds available for measuring activity of the flavin-containing monooxygenase are however limited, the reactions most widely used being *N,N*-dimethylaniline *N*-oxidation (Ziegler & Pettit, 1964) and methimazole *S*-oxygenation (Poulsen *et al.*, 1974). The metabolism of *N,N*-dimethylaniline and other *N,N*-dialkylanilines has been extensively studied with respect to *N*-dealkylation and *N*-oxidation, both *in vitro* and *in vivo* (Bickel, 1971; Cho & Miwa, 1974; Das & Ziegler, 1970; Gorrod *et al.*, 1975a,b; Gorrod *et al.*, 1979). The accumulated evidence appeared to suggest that *N,N*-dimethylaniline is *N*-demethylated by a phenobarbitone-inducible cytochrome P-450, and *N*-oxidized by the flavin-containing monooxygenase (Gold & Ziegler, 1973) (see Fig. 1).

However, reports by other workers (Hlavica & Kehl, 1977; Hlavica & Hulsman, 1979) indicate the presence of two distinct *N,N*-dimethylaniline *N*-oxidases, i.e. cytochrome P-450 and flavin-containing monooxygenase, in the microsomal fractions prepared from several animal species.

We had proposed using *N,N*-dimethylaniline routinely in our laboratory for monitoring the activity of the flavin-containing monooxygenase in rat hepatocytes monolayer cultures (Sherratt & Damani, this volume). In view of the above reports claiming cytochrome P-450 involvement in the *N*-oxidation of this substrate, it was essential that we ascertained the relative contributions of cytochrome P-450 and flavin-containing monooxygenases in the rat hepatic microsomal *N*-oxidation and *N*-demethylation of *N,N*-dimethylaniline.

EXPERIMENTAL

Microsomes from control or phenobarbitone-pretreated, or β-naphthoflavone-pretreated male Sprague–Dawley rats were prepared as described in the literature (Gorrod *et al.*, 1975a). Typical reaction mixtures contained 2 ml Tris/KCl buffer of pH 7.4, 5 μmole substrate and about 6–8 mg microsomal protein in the presence of an NADPH-regenerating system. Cofactor concentrations were optimized and product formation was linear with respect to time and protein concentration under these conditions. Reactions were terminated by addition of alkali (1.0 M-NaOH, 0.5 ml), and a solution of internal standard (*N*-ethylaniline) added. The unreacted substrate and its *N*-dealkylated metabolites were recovered by exhaustive ether extraction and analysed by an h.p.l.c assay developed in our laboratory. The *N*-oxide remaining in the aqueous phase was reduced back to *N*,*N*-dimethylaniline prior to quantification by h.p.l.c. Full details of the h.p.l.c assay are given elsewhere (Damani *et al.*, 1984).

RESULTS AND DISCUSSION

Initial studies on the effect of pretreatment of animals with the inducers phenobarbitone (PB) and β-naphthoflavone (BNF) showed that PB-pretreatment has a marked stimulatory effect on *N*-demethylation but no significant effect on *N*-oxidation of *N*,*N*-dimethylaniline (see Table 1). BNF-pretreatment had no significant effect on *N*-demethylation, and in fact *N*-oxidation was significantly lower in microsomes from BNF-pretreated rats. Table 1 also shows the effects of incorporating in the incubations certain inhibitors and activators. Naphthylthiourea had a marked inhibitory effect on the *N*-oxidation pathway without affecting the *N*-demethylation. *n*-Octylamine had a stimulating effect on *N*-oxidation but a marked inhibitory effect on the *N*-demethylation. The cytochrome P-450 inhibitors metyrapone and SKF 525A inhibited *N*-demethylation without effect on the *N*-oxidation

Table 1
Effect of inducers, inhibitors and activators on rat hepatic microsomal catalysed *N*-oxidation and *N*-demethylation of *N*,*N*-dimethylaniline

Inducer/inhibitor	Enzyme activity (% of control)[*]	
	N-Oxidation	*N*-Demethylation
None	100	100
Phenobarbitone	95	250
β-Naphthoflavone	50	110
Naphthylthiourea (4×10^{-3}M)	10	100
n-Octylamine (5×10^{-3}M)	160	30
Metyrapone (2×10^{-3}M)	100	50
SKF 525A (4×10^{-3}M)	95	50

[*]Control *N*-oxidation = 1.04 ± 0.04 nmol/min/mg protein.
Control *N*-demethylation = 2.17 ± 0.1 nmol/min/mg protein.

Table 2

Role of purified cytochrome P-450 isozymes and flavin-containing monooxygenases in the N-oxidation and N-demethylation of N,N-dimethylaniline

Enzyme	N-Oxidation	N-Demethylation
None	0	0
Cytochrome P-450	0	51.3 ± 0.5^a
Cytochrome P-448	0	14.5 ± 0.7^a
Flavine-monooxygenase	85 ± 0.7^b	0

^{a} nmol product/min/nmol P-450
^{b} nmol product/min/mg protein.

pathway. These results suggest that in the rat, N-oxidation of N,N-dimethylaniline is not mediated by cytochrome P-450.

In order further to confirm the above observations the metabolism of N,N-dimethylaniline was studied using purified enzyme systems. Incubation of N,N-dimethylaniline with a reconstituted system which consisted of cytochrome P-450 and NADPH cytochrome P-450 reductase purified from phenobarbitone-induced rats was found to mediate N-demethylation but not N-oxidation of N,N-dimethylaniline (see Table 2). Essentially the same results were obtained with cytochrome P-448 isolated from β-naphthoflavone-pretreated rats. On the other hand, hog liver flavin-containing monooxygenase was found to N-oxidize but not N-demethylate N,N-dimethylaniline. On the basis of this data we conclude that in the rat at least, N-oxidation of N,N-dimethylaniline is mediated exclusively via the flavin-containing monooxygenase, whereas N-demethylation is mediated via a cytochrome P-450 system.

ACKNOWLEDGEMENTS

The authors thank Dr Gordon G. Gibson of the University of Surrey for supplying the purified P-450 isozymes, and Professor D. M. Ziegler for the generous gift of a sample of purified hog liver flavin-containing monooxygenase.

REFERENCES

Bickel, M. H. (1971), *Arch. Biochem. Biophys.*, **148**, 54.
Burke, M. D. & Mayer, R. T. (1974), *Drug Metab. Dispos.*, **2**, 583.
Cho, A. K. & Miwa, G. T. (1974), *Drug Metab. Dispos.*, **2**, 477.

Damani, L. A., Houdi, A. A., Sherratt, A. J. & Waghela, D. (1984), *J. Pharm. Pharmacol.*, **36** (supplement), 30P.

Das, M. L. & Ziegler, D. M. (1970), *Arch. Biochem. Biophys.*, **140**, 300.

Greenlee, W. F. & Poland, A. (1978), *J. Pharmacol. Exp. Ther.*, **205**, 596.

Gorrod, J. W., Temple, D. J. & Beckett, A. H. (1975a), *Xenobiotica*, **5**, 453.

Gorrod, J. W., Temple, D. J. & Beckett, A. H. (1975b), *Xenobiotica*, **5**, 465.

Gorrod, J. W., Temple, D. J. & Beckett, A. H. (1979), *Xenobiotica*, **9**, 17.

Gold, M. S. & Ziegler, D. M. (1973), *Xenobiotica*, **3**, 179.

Hlavica, P. & Hulsman, S. (1979), *Biochem. J.*, **182**, 109.

Hlavica, P. & Kehl, M. (1977), *Biochem. J.*, **164**, 487.

Imai, Y., Ito, A. & Sato, R. (1966), *J. Biochem.*, **60**, 417.

Poulsen, L. L., Hyslop, R. M. & Ziegler, D. M. (1974), *Biochem. Pharmacol.*, **23**, 3431.

Ziegler, D. M. & Pettit, F. H. (1964), *Biochem. Biophys. Res. Commun.*, **15**, 188.

Chapter 9

THE METABOLISM OF N,N-DIMETHYLANILINE BY RAT HEPATOCYTES

Amanda J. Sherratt and **L. A. Damani**, Department of Pharmacy, University of Manchester, Manchester, M13 9PL, UK

1. Incubation of N,N-dimethylaniline with rat hepatocytes results in N-demethylation, N,N-didemethylation and N-oxidation.
2. In addition, a major unknown metabolite was observed with hepatocytes, which was not seen in experiments with rat hepatic microsomes. This metabolite has tentatively been identified as the N-glucuronide of N-methylaniline.
3. No attempts were made in this study to isolate or quantify any ring hydroxylated products; such metabolites have been reported to occur with microsomes.
4. Cytochrome P-450 activity declines rapidly in rat hepatocyte monolayer cultures; this is as measured by N,N-dimethylaniline N-demethylation or 7-ethoxycoumarin O-de-ethylation.
5. Flavin-containing monooxygenase activity is however more stable in monolayer cultures; this is as measured by N,N-dimethylaniline N-oxidation.

INTRODUCTION

Isolated rat hepatocytes in suspension culture, and more recently in primary monolayer culture, have been used to study mechanisms of drug metabolism and cytotoxicity (Sirica & Pitot, 1980; Holme *et al.*, 1982). However, a major limitation of using rat hepatocyte monolayer cultures for drug metabolism studies has been the uncertainty about the 'metabolic viability' of cells in culture; for example there is a rapid decline in cytochrome P-450 within the first 24 hours of culture (Bissell *et al.*, 1973). Many investigators have attempted to devise culture conditions to prevent this loss of cytochrome P-450 and maintain its activity (Michalopoulos *et al.*, 1976; Decad *et al.*, 1977; Paine & Hockin, 1980). In contrast to such studies on cytochrome P-450, the microsomal flavin-containing monooxygenase (EC 1.14.13.8) has been less well studied in hepatocytes maintained *in vitro*. This is surprising in view of the fact that this second major microsomal oxidase is responsible for the oxygenation of nucleophilic nitrogen and sulphur centres in a variety of xenobiotics (Ziegler, 1980). However, attempts to study the flavin-containing monooxygenase at the cellular level have been restricted by the instability of its reaction products or competing enzyme pathways for its substrates (McManus *et al.*, 1983).

The N-oxidation of N,N-dimethylaniline (DMA) has been used extensively as a marker for estimating the activity of the flavin-containing monooxygenase in subcellular preparations and whole tissue homogenates (Ziegler, 1980). DMA undergoes not only N-oxidation and N-demethylation, but also N,N-didemethylation and hydroxylation to afford aniline and various ring hydroxylated products (Gorrod & Gooderham, 1981). Recent investigations in our laboratory using purified enzymes and a highly sensitive and specific h.p.l.c. system for the quantitation of DMA metabolites have clearly demonstrated that N-oxidation of DMA in the rat is mediated exclusively by the flavin monooxygenase, whereas N-demethylation is a phenobarbitone-inducible cytochrome P-450 mediated reaction (Houdi & Damani, this volume). DMA was therefore selected as a model substrate for monitoring the activities of the two microsomal monooxygenases in rat hepatocyte monolayer culture. Since there is no information about DMA biotransformation in cellular systems, and in view of our decision to use this compound as a marker for both cytochrome P-450 and flavin-containing monooxygenase in rat hepatocytes, we have carried out studies on its metabolism by freshly isolated rat hepatocytes.

EXPERIMENTAL

Rat hepatocytes were isolated by collagenase perfusion (Seglen, 1976). Cell viability was always greater than 90% as assessed by trypan blue exclusion, and cytochrome P-450 content was in the order of 150–220 pmoles/mg protein. Cells designated for culture were seeded at a density of 1×10^6 cells in 2 ml complete tissue culture media (c.TCM = Williams E supplemented with 10% FCS, glutamine 2.0 mM, penicillin 100 iu/ml, streptomycin 100 μg/ml) in 35-mm petri dishes. After 2 hours' incubation at 37 °C the overlying media was removed, the attached cells washed and replaced with fresh c.TCM.

DMA metabolite identification was by cochromatography of metabolites with authentic standards on several chromatographic systems (e.g. t.l.c., g.l.c. and h.p.l.c.) and in some instances by mass spectrometry. Quantification of *N*-methylaniline (NMA) and *N*,*N*-dimethylaniline *N*-oxide (DMA *N*-oxide), the two major metabolites was by an h.p.l.c. assay developed in our laboratory (Damani *et al.*, 1984). In addition, 7-ethoxycoumarin *O*-de-ethylation was used as a marker of cytochrome P-450 activity in rat hepatocyte monolayer cultures; 7-hydroxycoumarin was measured fluorimetrically, essentially as described by Fry & Bridges (1980).

RESULTS AND DISCUSSION

Incubation of DMA with freshly isolated cells at 37 °C in conical flasks resulted in *N*-demethylation and *N*-oxidation, formation of NMA and DMA *N*-oxide being linear over a range of cell densities (0.5–4.0 million cells/flask) and over a 30-minute incubation period. Using 2 million cells per flask,

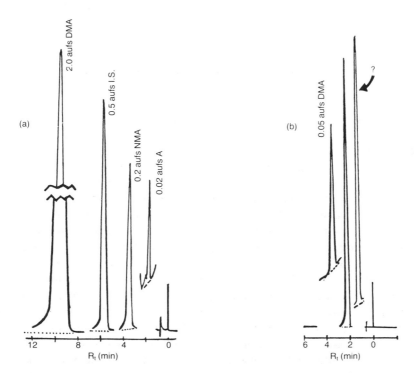

Fig. 1 – Typical h.p.l.c. Chromatograms: (a) first exhaustive ether extracts containing DMA, internal standard, NMA and aniline; (b) second ether extracts, after TiCl$_3$ treatment, containing DMA, internal standard and NMA.

h.p.l.c. conditions : column : Spherisorb 5 ODS (10 cm); Mobile phase : Methanol : phosphate buffer pH 7.4 (0.01 M); (a) 35 : 65%, v/v. 1 ml/min; (b) 50 : 50%, v/v, 1.5 ml/min.

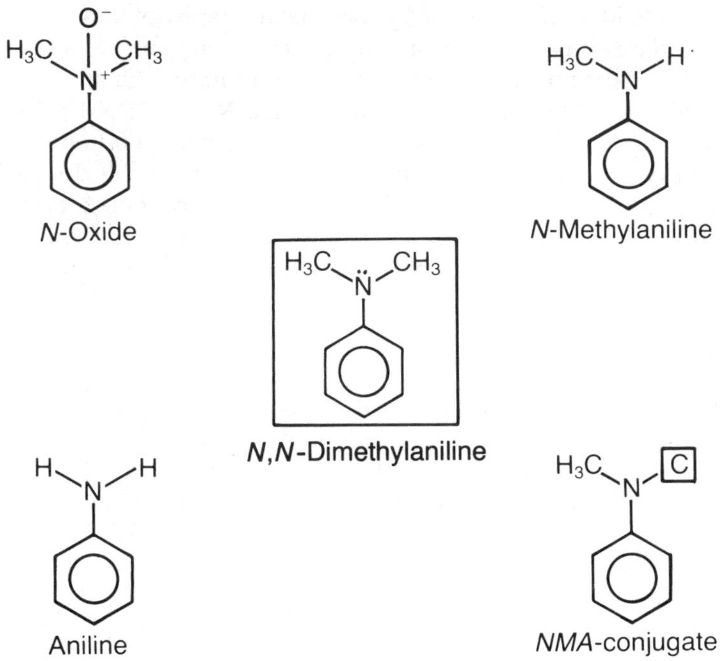

Fig. 2 – Metabolism of *N,N*-Dimethylaniline by Rat Hepatocytes.

concentrations of DMA in excess of 0.5 mM were found to be saturating for *N*-oxide formation, but concentrations greater than 2.5 mM were required to saturate NMA formation. The *N,N*-didemethylated product aniline was also observed as a metabolite, but it was formed in very small amounts, and was not quantified. In addition to NMA and DMA *N*-oxide, a third major unknown metabolite was observed with hepatocytes which was not seen in experiments with rat hepatic microsomes. This metabolite is very water soluble and like DMA *N*-oxide, was not extracted into ether, even on exhaustive extraction (4 × 4 ml). Examination of the 'first ether extracts' by h.p.l.c. revealed the presence of the unchanged substrate (DMA), the internal standard (*N*-ethylaniline), and the *N*-dealkylated products *N*-methylaniline and aniline (see Fig. 1a). On subsequent treatment of the remaining aqueous phase with TiCl₃ under acid conditions, whereas DMA *N*-oxide was reduced to DMA, the other metabolite was converted to NMA (Fig. 1b). Formation of this metabolite is unique to hepatocytes since it is never seen with rat hepatic microsomes. Conversion of the metabolite to NMA also occurred under acid treatment alone, or on treatment with β-glucuronidase. Formation of this metabolite was proportional to cell density, incubation time and substrate concentration. Incubation of NMA with hepatocytes resulted in the production of greater quantities of this metabolite than with equivalent concentrations of DMA. This metabolite has been tentatively identified as the *N*-glucuronide of NMA. In conclusion, DMA is metabolized by rat hepato-

Table 1

Changes in P-450 and flavin-containing monooxygenase activity of
rat hepatocytes maintained in monolayer culture

Time in culture (hr)	Cytochrome P-450 activity	FAD-monooxygenase activity
	(% of initial activity)[a]	
2	100	100
8	56 ± 21	75 ± 10
24	25 ± 8	74 ± 18
48	16 ± 4	56 ± 11

[a]7-EC 174 ± 27 pmole/min/mg protein.
DMA 400 ± 70 pmole/min/mg protein.
$n = 3$ or more animals, duplicated with each animal.

cytes to the N-oxide, NMA, aniline and to an N-methylaniline conjugate
which may be an N-glucuronide (see Fig. 2).

Optimal conditions for measuring DMA N-demethylase activity (reflecting
cytochrome P-450 activity) and DMA N-oxidase activity (reflecting activity of
the flavin-containing monooxygenase) were established as described above,
for monitoring changes in the activities of these two monooxygenases in rat
hepatocytes maintained in monolayer culture for various time periods.

In agreement with previous reports (e.g. Lake & Paine, 1982) we can
demonstrate a rapid decline in cytochrome P-450 activity as measured by
7-ethoxycoumarin O-de-ethylation. The activity at 24 hours is only around
25% of initial activity (Table 1). A similar decline profile to that of
7-ethoxycoumarin O-de-ethylase was also observed using DMA
N-demethylation as a marker for cytochrome P-450 activity (data not shown).
In contrast to cytochrome P-450 instability, the flavin-containing monooxygenase appears to be more stable, and at 24 hours only drops to around 75%
of initial. It would therefore seem that both microsomal monooxygenases
decrease on hepatocyte monolayer culture, but to different extents, the
flavoprotein apparently being more stable than cytochrome P-450. This
demonstrates the current limitations of rat hepatocyte monolayers as models
for metabolic and toxicological studies, until such time as culture conditions
are devised for the maintenance of these and other drug metabolizing enzyme
systems.

REFERENCES

Bissell, D. M., Hammaker, L. E. & Meyer, V. A. (1973), *J. Cell Biol.*, **59**,
722.
Damani, L. A., Houdi, A. A., Sherratt, A. J. & Waghela, D. (1984), *J.
Pharm. Pharmacol.*, **36** (supplement), 30P.

Decad, G. M., Hsich, D. P. H. & Byard, J. L. (1977), *Biochem. biophys. Res. Commun.*, **78**, 279.

Fry, J. R. & Bridges, J. W. (1980), *Arch. Pharmacol.*, **311**, 85.

Gorrod, J. W. & Gooderham, N. J. (1981), *Eur. J. Drug Metab. Pharmacokinet.*, **6**, 195.

Holme, J. A., Eek-Hansen, A. & Jervell, K. F. (1982), *Acta pharmacol. et toxicol.*, **50**, 272.

Hoodi, A. A. & Damani, L. A. (1985), this volume.

Lake, B. G. & Paine, A. J. (1982), *Biochem. Pharmacol.*, **31**, 2141.

McManus, M. E., Grantham, P. H., Cone, J. L., Roller, P. P., Wirth, P. J. & Thorgiersseon, S. S. (1983), *Biochem. biophys. Res. Commun.*, **112**, 437.

Michalopoulos, G., Sattler, G. L. & Pitot, H. C. (1976), *Life Sci.*, **18**, 1139.

Paine, A. J. & Hockin, L. J. (1980), *Biochem. Pharmacol.*, **29**, 3215.

Seglen, P. O. (1976), in *Methods in Cell Biology*, Vol. XIII (ed. Prescott, D. M.), Academic Press, New York, p. 29.

Sirica, A. E. & Pitot, H. C. (1980), *Pharmacol. Rev.*, **31**, 205.

Ziegler, D. M. (1980), in *Enzymatic Basis of Detoxication*, Vol. 1 (ed. Jakoby, W. B.), Academic Press, New York, p. 201.

Chapter 10

OXYGENATION OF PRIMARY ARYLAMINES BY A HYDROPEROXYFLAVIN: MODEL STUDIES FOR THE FLAVOPROTEIN MONOOXYGENASE

D. R. Doerge, Department of Agricultural Biochemistry, University of Hawaii, Honolulu, HI 96822, USA and **M. D. Corbett**, Pesticide Research Laboratory, University of Florida, Gainesville, FL 32611, USA

1. Amine nucleophilicity of p-substituted anilines determines the reactivity in oxygenation by a 4a-hydroperoxyflavin model compound.
2. Amine nucleophilicity is an important but not exclusive determinant of flavin monooxygenase (FMO) activity.
3. Arylnitroso compounds are the only products observed in the model reactions and may be primary products of FMO reactions.
4. These results do not support the participation of imino tautomers as reactive intermediates in primary arylamine oxygenations by FMO.

INTRODUCTION

Enzymatic N-oxygenation of primary arylamines has been shown to be an obligatory step in the mechanism of metabolic activation to toxic intermediates (Weisburger & Weisburger, 1973). Cytochromes P-450, prostaglandin H synthetase and the flavoprotein monooxygenase (FMO) have been shown to

effect the *N*-oxidation of primary arylamines (Hlavica, 1982). The oxygenated products were arylhydroxylamines and arylnitroso compounds and
these have been identified as proximate genotoxic metabolites (Weisburger &
Weisburger, 1973; Wirth *et al.*, 1980).

Ziegler and coworkers elucidated the mechanism of catalysis for the pig
liver FMO and demonstrated the formation of a hydroperoxy-FAD derivative
in an NADPH and O_2-dependent reaction (Poulsen & Ziegler, 1979). Subsequently, Bruice and coworkers demonstrated the oxygenation of secondary
and tertiary aliphatic amines by a synthetic 4a-hydroperoxyflavin model compound (Ball & Bruice, 1980). The results were consistent with a mechanism
involving attack of nitrogen nucleophiles on the terminal oxygen of the
hydroperoxide. The reaction products were identical to those from FMO
incubations.

Primary polynuclear arylamines such as 1- and 2-aminonaphthalene and
2-aminofluorene have been described as FMO substrates (Ziegler *et al.*,
1973; Frederick *et al.*, 1982). These results were interpreted in terms of
steady-state enzyme kinetics and consequently suffer from the limitation that
the substrate oxygenation step precedes the overall rate determining step of
the catalytic mechanism (Poulsen & Ziegler, 1979; Beaty & Ballou, 1980).
This results in identical values of V_{max} for all substrates albeit widely varying
K_m values. The present study describes the reaction of a synthetic hyd-
roperoxyflavin with primary arylamines and provides direct information
about substrate oxygenation in the effort to define further the substrate
specificity of FMO.

EXPERIMENTAL

All solvents were dried and purified by distillation according to standard
methods (Gordon & Ford, 1972). Anhydrous methanol was prepared by
refluxing analytical reagent methanol over magnesium and a few iodine cryst-
als, followed by distillation. Arylamines were purified by distillation at
reduced pressure, recrystallization or silica gel chromatography. Arylnitroso
compounds were synthesized as chromatographic standards by methods previously described (Corbett *et al.*, 1979). 3,8,10-Trimethylisoalloxazine was
synthesized by the method of Yoneda *et al.* (1976). The method of Ghisla
et al. (1973) was used to prepare 5-ethyl-3,8,10-trimethylisoalloxazinium
percholate ($FlEt_{ox}$) which was reacted with hydrogen peroxide in a procedure similar to that of Kemal and Bruice (1976) to yield 4a-hydroperoxy-5-
ethyl-3,8,10-trimethylisoalloxazine (FlEtOOH). FlEtOOH was characterized by its u.v.–vis spectrum, chemical ionization mass spectrum, melting
point (118–19 °C) and elemental analysis performed by Galbraith
Laboratories (found: C, 56.47; H, 5.81; N, 17.63; $C_{15}H_{18}N_4O_4$ requires: C,
56.58; H, 5.70; N, 5.70; N, 17.61).

Kinetic measurements were conducted under pseudo-first-order conditions
in anhydrous alcoholic solvents. Solutions of FlEtOOH were prepared in

anhydrous methanol and the peroxidative equivalent determined (Kemal & Bruice, 1976). FlEtOOH was added to methanolic amine solutions at 30 °C such that the final concentration was $1.5-2.5 \times 10^{-4}$ M. Kinetics were measured at 370 nm in a Beckman Model 35 recording spectrophotometer in stoppered cuvettes.

Arylamine oxidation products were analysed by h.p.l.c. on a μ-Bondapak C-18 column (Waters Associates). The solvent flow rate was 1.5 ml min^{-1} and component detections were made by u.v. absorbance at 313 or 340 nm. Quantitative calculations were made on the basis of component peak heights compared with peak heights generated by known amounts of authentic standards.

The flavin derived product (FlEtOH) was determined in anhydrous *tert*-butanol. The excess arylamine concentrations used precluded a direct h.p.l.c. determination of FlEtOH, this product was therefore quantitated following its conversion to $FlEt_{ox}$ by reaction with 1 M-HCl (Ball & Bruice, 1980). To 0.1 ml of the reaction mixture in *tert*-butanol was added 0.9 ml 1 M-HCl, and the product measured spectrophotometrically at 555 nm.

RESULTS AND DISCUSSION

The reaction kinetics of FlEtOOH and a series of *para*-substituted anilines were measured in anhydrous methanol by measuring the disappearance of FlEtOOH at 370 nm. Under the conditions where [arylamine]>>-[FlEtOOH], the reaction was first order in FlEtOOH concentration. The pseudo-first-order rate constant (k_{obs}) was determined from the slope of the plot of ln $[FlEtOOH]_t/[FlEtOOH]_0$ vs time and k_{obs} was found to be linearly dependent on arylamine concentrations. Plots of k_{obs} vs arylamine concentration showed a positive non-zero intercept whose value agreed with the independently determined rate constant (k_o) for the spontaneous decomposition of FlEtOOH in anhydrous alcoholic solvents. Product yields were consistent with a scheme of two consecutive, competing pathways for FlEtOOH decomposition as shown in equation (1).

$$-d/dt[FlEtOOH] = (k_o + k_{obs}) [FlEtOOH] \qquad (1)$$

The reaction products were the corresponding arylnitroso compound and FlEtOH, the 4a-hydroxyflavin pseudobase (cf. Fig. 1). Table 1 shows the concurrent increases in the product yields as the concentration of arylamine increases. The products were formed in a ratio of 1 mole arylnitroso compound to 2 mole FlEtOH, and both products were formed in quantitative yield at high concentrations of arylamine. The ratio of arylnitroso product to FlEtOH was constant under aerobic and anaerobic conditions. This lack of an oxygen effect on the stoichiometry excludes the possibility of air oxidation to the arylnitroso derivative of an arylhydroxylamine formed by an FlEtOOH-dependent oxygenation.

Table 1

Dependence of product yields on p-phenetidine concentration

[p-Phenetidine] (μM)	[Nitrosophenetole] (μM)	[FlEtOH] (μM)	[FlEtOH]/[Nitrosophenetole]
0.2	50	95	1.9
0.4	62	129	2.1
0.6	88	181	2.1
0.8	96	191	2.0
1.0	96	199	2.1

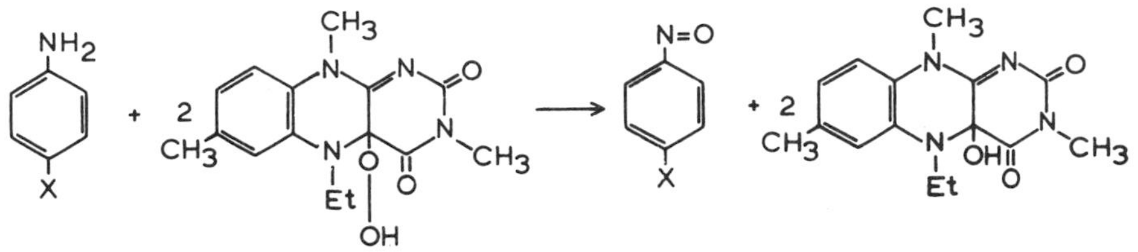

Fig. 1 – Reaction of FlEtOOH with *p*-substituted aniline.

The formation of arylhydroxylamines could not be demonstrated by h.p.l.c. in reactions containing FlEtOOH and arylamines utilizing conditions which readily separated arylhydroxylamines from arylamine precursors and arylnitroso products. Arylnitroso compounds were the sole products under the conditions where [arylamine]>>[FlEtOOH].

Arylhydroxylamines were oxygenated by FlEtOOH to produce arylnitroso compounds under the conditions described in the experimental section. The reaction proceeded very rapidly but extensive side reactions prevented detailed kinetic measurements. Arylnitro compounds were not observed from reactions containing FlEtOOH and arylamines or arylnitroso derivatives.

Equation (2) shows a scheme which is consistent with the kinetic and product studies described above.

$$ArNH_2 + FlEtOOH \xrightarrow{k_1} ArNHOH + FlEtOH$$

$$ArNHOH + FlEtOOH \xrightarrow{k_2} ArN{=}O + FlEtOH + H_2O$$

where $k_2 \gg k_1$. Under these conditions, [ArNHOH] is at steady state and $-d/dt\,[FlEtOOH] = 2k_1\,[ArNH_2]\,[FlEtOOH] = k_{obs}\,[FlEtOOH]$.

The value of k_1 was determined for a series of *para*-substituted anilines and the logarithm was plotted vs the Hammett substituent, σ, as shown in Fig. 2. We interpret the rate enhancement by electron donating substituents to reflect the influence of electronic charge localization on amino-group nucleophilicity.

Figure 3 shows a Bronsted-type plot of log k_1 vs the pK_a for the series of *para*-substituted anilines (Briggs & Robinson, 1961). It can be concluded from the slope of the line ($\beta_{nuc} = 0.96$) that within this series of compounds, reactivity towards FlEtOOH is highly dependent upon basicity.

Hückel molecular orbital calculations have shown a direct correlation between total amino-nitrogen charge and pK_a for the aminonaphthalenes and aminoazulenes (Schulze & Heilbronner, 1958; Coulsen & Streitweiser, 1965). These results imply an increased nucleophilicity for 2-aminonaphthalene relative to 1-aminonaphthalene. 2-Aminonaphthalene has been shown to undergo oxygenation by FMO with a greater specificity than 1-aminonaphthalene (Ziegler *et al.*, 1973). The participation of imino tautomers in the reaction, as previously advanced (Ziegler *et al.*, 1980), would predict 1-aminonaphthalene as the preferred substrate. The relative

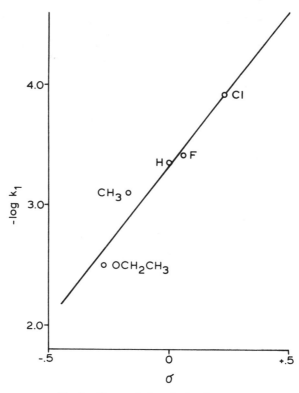

Fig. 2 – Hammett plot of $-\log k_1$ vs σ.

equilibrium contribution of imino tautomers to the isomeric aminonaph-thenes as estimated by infrared stretching frequencies of the N—H bond is consistent with a larger fraction of the 1-isomer in the imino form (Bryson, 1960). These results agree with those of the present study and suggest that amino-group nucleophilicity is a more relevant parameter for predicting primary arylamine oxygenation by FMO.

However, basicity considerations alone are insufficient in explaining FMO substrate specificity for primary arylamine N-oxygenations. Table 2 shows the pK_a values for selected arylamines (Briggs & Robinson, 1961; Schulze &

Table 2

pK_a values for arylamines (25 °C, H_2O)

Arylamine	pK_a
Aniline	4.6[a]
2-Aminonaphthalene	4.2[b]
1-Aminonaphthalene	3.9[b]

[a]Briggs & Robinson (1981).
[b]Schulze & Heilbronner (1958).

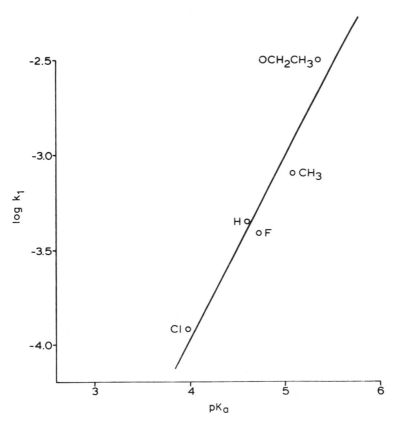

Fig. 3 – Bronsted-type plot of log k_1 vs arylamine pK_a (25 °C, H_2O).

Heilbronner, 1958). Although aniline is more basic than either aminonaphthalene isomer, it is not a substrate for FMO (Ziegler *et al.*, 1980). While a slight preference for polyaromatic substrates is seen in the oxygenation of substituted thiourea derivatives by FMO (Poulsen, 1981), this cannot totally explain the radical difference between aniline and the aminonaphthalenes in substrate specificity.

The products from incubations of FMO with primary arylamines have been described as mixtures of arylhydroxylamines and arylnitroso compounds (Ziegler *et al.*, 1973; Frederick *et al.*, 1982). The exclusive formation of aryl-nitroso products from the oxygenation of primary arylamines by FlEtOOH suggests that arylnitroso compounds may be the primary products of FMO catalysis. It is not clear, however, if arylhydroxylamines are recalcitrant to oxidation by FMO in the presence of NADPH and oxygen or whether the expected arylnitroso product is reduced back to the arylhydroxylamine by NADPH or a similar reducing agent (Becker & Sternson, 1980).

A comparison of the results from the present study with FMO substrate specificity data in the literature suggests that enzyme-substrate binding is a large determinant of enzymatic reactivity and that the binding properties favour the oxygenation of polynuclear arylamines. Metabolic conversion of primary arylamines to arylhydroxylamines and arylnitroso derivatives probably represents a significant pathway in the production of proximate genotoxic intermediates *in vivo*.

REFERENCES

Ball, S. & Bruice, T. C. (1980), *J. Am. Chem. Soc.*, **102**, 6498.
Beaty, N. B. & Ballou, D. P. (1980), *J. Biol. Chem.*, **255**, 3817.
Becker, A. R. & Sternson, L. A. (1980), *Bioorg. Chem.*, **9**, 305.
Briggs, A. I. & Robinson, R. A. (1961), *J. Chem. Soc.*, 388.
Bryson, A. (1960), *J. Am. Chem. Soc.*, **82**, 4862.
Corbett, M. D., Baden, D. G. & Chipko, B. R. (1979), *Bioorg. Chem.*, **8**, 91.
Coulsen, C. A. & Streitweiser, A. (1965), *Dictionary of π-Electron Calculations*, W. H. Freeman and Co., San Francisco, p. 429.
Frederick, C. B., Mays, J. B., Ziegler, D. M., Guengerich, F. P. & Kadlubar, F. F. (1982), *Cancer Res.*, **42**, 2671.
Ghisla, S., Hartmann, U., Hemmerich, P. & Muller, F. (1973), *Liebigs Ann. Chem.*, 1388.
Gordon, A. J. & Ford, R. A. (1972), *The Chemist's Companion*, Wiley–Interscience, New York, p. 429.
Hlavica, P. (1982), *CRC Crit. Rev. Biochem.*, **12**, 39.
Kemal, C. & Bruice, T. C. (1976), *Proc. Nat. Acad. Sci. USA*, **73**, 995.
Poulsen, L. L. (1981), in *Reviews in Biochemical Toxicology*, Vol. 3 (eds. Hodgson, E., Bend, J. R. & Philpot, R. M. Elsevier/North Holland, New York, p. 33.
Poulsen, L. L. & Ziegler, D. M. (1979), *J. Biol. Chem.*, **254**, 6449.
Schulze, J. & Heilbronner, E. (1958), *Helv. Chim. Acta*, **41**, 1492.
Weisburger, J. H. & Weisburger, E. K. (1973), *Pharmacol. Rev.*, **25**, 1.
Wirth, P. J., Dybing, E., von Bahr, C. & Thorgeirsson, S. S. (1980), *Molec. Pharmacol.*, **18**, 117.
Yoneda, S., Sukuma, Y., Ichiba, M. & Shinomura, K. (1976), *J. Am. Chem. Soc.*, **98**, 830.
Ziegler, D. M., McKee, E. M. & Poulsen, L. L. (1973), *Drug Metab. Disp.*, **1**, 314.
Ziegler, D. M., Poulsen, L. L. & Duffel, M. W. (1980), in *Microsomes, Drug Oxidations and Chemical Carcinogenesis* (eds. Coon, M. J., Cooney, A. H., Estabrook, R. W., Gelboin, H. V., Gillette, J. R. & O'Brien, P. J., Academic Press, New York, p. 637.

Chapter 11

OXIDATION OF CIMETIDINE AND RANITIDINE BY THE FLAVIN-CONTAINING MONOOXYGENASE IN PIG LIVER MICROSOMES

Harriet G. Oldham and **R. J. Chenery**, Smith Kline & French Research Limited, The Frythe, Welwyn, Herts, UK

1. The *in vitro* interactions of cimetidine and ranitidine with pig hepatic microsomal flavin-containing monooxygenase have been compared.
2. Flavin-containing monooxygenase activity in pig liver microsomes was measured by the stimulation of oxygen consumption caused by methimazole, a specific substrate for this enzyme.
3. Methimazole exhibited a K_m of 58.9 μM, whereas cimetidine exhibited a K_m of 1.93 mM and ranitidine 2.60 mM.
4. Mixing experiments suggested that cimetidine and ranitidine interact with the same enzyme system as methimazole.

INTRODUCTION

The purified flavin-containing monooxygenase (EC 1.14.13.8) from pig liver catalyses the oxygenation of nucleophilic nitrogen and sulphur compounds and in particular secondary and tertiary amines and divalent sulphur-containing compounds to form the corresponding hydroxylamines and nitrones, amine oxides and sulphoxides. Both cimetidine and ranitidine contain divalent sulphur, each also contains two secondary amine groups and in addition, ranitidine also contains a tertiary amine group. About 10% of an administered dose of cimetidine is recovered in urine as its sulphoxide (Griffiths *et al.*, 1977). Both the *N*-oxide and *S*-oxide of ranitidine have been identified in urine after oral or intravenous administration of ranitidine and account for approximately 5% and 2% respectively of the administered dose (Carey *et al.*, 1981).

Previous studies (Breen *et al.*, 1982) have indicated that in comparison with cimetidine, ranitidine is far less potent at interacting with the cytochrome P-450 system although both form ligand complexes with rat liver microsomes (Rendic *et al.*, 1982). The studies reported here were designed to examine the *in vitro* interaction of cimetidine and ranitidine with the pig hepatic flavin-containing monooxygenase, both to compare their abilities to act as substrates for this system and to determine whether they could competitively inhibit the oxygenation of methimazole, a substrate which, at concentrations below 2 mM, has been reported to be metabolized solely by this enzyme system (Ziegler, 1980).

EXPERIMENTAL

Pig livers were obtained from Playle and Sons, Butchers (Bassingbourn, Royston). Six-month-old female pigs were chosen and livers were removed within 5 minutes of death. Microsomes were prepared essentially as described by Ziegler & Poulsen (1978) and protein measured by the method of Lowry *et al.* (1951).

Flavin monooxygenase activity was assayed in pig liver microsomes by measuring oxygen consumption in the presence or absence of substrate using an oxygen electrode (Rank Bros., Bottisham, Cambridge).

A calibration curve was constructed by adding known concentrations of phenylhydrazine hydrochloride (using a stock solution of 10 mM in water) to 2 ml potassium ferricyanide (1.0 mM) in potassium phosphate buffer (50 mM) according to the method of Misra & Fridovich (1975). Kinetic data were analysed by various computer curve fitting routines. The Michaelis–Menten kinetic parameters, K_m and V_{max}, were determined by a rectangular hyperbolae iterative fit programme from Wilkinson (1961). The ALLFIT programme (NICHD, NIH, Bethesda, USA) was utilized to estimate IC_{50} values.

RESULTS AND DISCUSSION

The effect of substrate concentration on oxygen consumption by pig liver microsomes when methimazole, cimetidine and ranitidine are incubated singly as substrates in the flavin monooxygenase assay system was examined. Kinetic estimates were repeated on several occasions throughout the study

Table 1
K_M and V_{max} estimates for methimazole, cimetidine and ranitidine
(means ± SEM)

Substrate	n	$K_m(\mu M)$	$V_{max}(nmol.min^{-1}.mg\ protein^{-1})$
Methimazole	2	58.9 ± 7.0	8.69 ± 1.69
Cimetidine	4	1928.0 ± 522.0	3.34 ± 0.20
Ranitidine	4	2600.0 ± 790.0	3.15 ± 0.48

period and the values obtained are shown in Table 1. The K_m values for cimetidine and ranitidine are 33 and 44 times greater respectively than the value for methimazole, suggesting that neither of the two histamine H_2-antagonists have such a high affinity for the flavin monooxygenase as does methimazole. V_{max} values for cimetidine and ranitidine are approximately one third of the value obtained for methimazole. Although K_m values for cimetidine and ranitidine are high they are comparable with values obtained for antipyrine (900 μM) (Andreason et al., 1977) in pig liver microsomes in vitro.

The assay was carried out at pH 8.4 to minimize interference by other microsomal oxygenases and also because this pH has been reported to be the pH optimum of the flavin monooxygenase (Ziegler, 1980). In addition, all substrate-induced oxygen consumption was measured in the presence of n-octylamine, which is an inhibitor of cytochrome P-450 mediated oxidations. However, despite these precautions, the measurement of enzyme activity by stimulation of oxygen consumption is relatively unspecific. Oxygen consumption may be the result of numerous processes including metabolism of endogenous substrates by both flavin and cytochrome P-450 monooxygenases and by various peroxidative mechanisms. Two further experiments were therefore carried out to test the hypothesis that cimetidine and ranitidine are metabolized by the same enzyme system as methimazole.

Table 2 shows the results of an experiment in which cimetidine or ranitidine was added in combination with methimazole at their K_m concentrations. If the substrates are metabolized by independent enzyme systems, then mixing will result in a total rate of reaction (V_T) which is the sum of the two rates determined separately (equation 1). Conversely, if the drugs are metabolized by the same enzyme system then the total rate on mixing will be two thirds of the rate observed with separate enzymes (equation 2).

$$\text{Separate enzymes: } V_T = \left[\frac{V_{max}^A}{2} + \frac{V_{max}^B}{2}\right] \tag{1}$$

$$\text{Same enzyme: } V_T = 2/3\left[\frac{V_{max}^A}{2} + \frac{V_{max}^B}{2}\right] \tag{2}$$

Table 2

A comparison of observed and predicted rates of oxygen consumption by pig liver microsomes with methimazole, cimetidine and ranitidine at K_M concentrations both alone and on mixing.

| Compound | Oxygen consumption (nmol.min^{-1}.mg protein^{-1}) (Mean ± SEM) n = minimum of 2 | | |
| | Predicted | | Observed |
	Independent[a]	Dependent[b]	
Methimazole		4.34[c]	4.66 ± 0.00
Cimetidine		1.67[c]	1.38 ± 0.02
Ranitidine		1.57[c]	1.59 ± 0.21
Methimazole + cimetidine	6.01	4.01	4.72 ± 0.81 (range 3.1–5.8)
Methimazole + ranitidine	5.91	3.94	4.84 ± 0.52 (range 4.3–5.4)

[a]Independent = Separate enzyme systems metabolizing the various substrates.
[b]Dependent = the same enzyme system metabolizing the various substrates.
[c]Values calculated using $V_{max}/2$ from Table 1.

Table 2 shows that the observed value of oxygen consumption in the presence of methimazole was 78.5% of the predicted rate for separate enzymes with cimetidine and 81.9% of the predicted rate for separate enzymes with ranitidine. The spread of the observed values for cimetidine + methimazole and ranitidine + methimazole suggest that they are more likely to be metabolized by the same enzyme system than by independent systems. However, considerable variability was noted in our determinations of oxygen consumption and therefore a more rigorous approach was adopted.

If it is assumed that these drugs interact at the same site on the enzyme, that is at the catalytic site, then it is possible to construct equations which describe the overall rate of reaction when a fixed concentration of methimazole is present initially, and the concentration of either cimetidine or ranitidine is increased in a step-wise manner (equation 3).

$$V_{total} = \frac{a/b - V_{max}^B}{1 + [B]/b} + V_{max}^B \tag{3}$$

$$\text{where } a = \frac{V_{max}^A \cdot [A] \cdot K_M^B}{K_M^A}, \qquad b = \frac{K_M^B(K_M^A + [A])}{K_M^A}$$

A = methimazole   B = test compound.

In this experiment, one concentration of methimazole (equal to $5 \times K_m$ concentration = 294.5 μM) was chosen so that the rate of oxygen consumption would be 83.3% of V_{max}. After background oxygen consumption was meas-

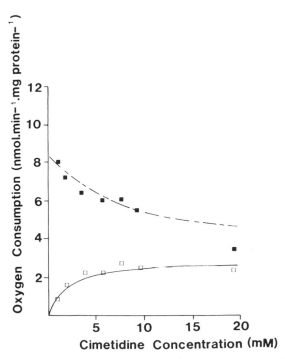

Fig. 1 – Total oxygen consumption as a consequence of varying cimetidine concentration in the presence and absence of a fixed methimazole concentration (294.5 μM)
Cimetidine plus methimazole: — – — predicted, ■ observed, Cimetidine alone: □.

ured in the presence of enzyme alone, the test compound was introduced into the incubation system and oxygen consumption measured. Finally methimazole was added to the incubation chamber and the total oxygen consumption due to the test compound and methimazole determined. It was observed that as increasing concentrations of cimetidine or ranitidine were introduced into the incubation, total oxygen consumption decreased from the level of oxygen consumption due to methimazole alone towards the V_{max} oxygen consumption for the test compound due to competition between cimetidine or ranitidine with methimazole for enzyme active sites. The increasing concentrations of test compound also provided another estimate of K_m and V_{max} values for cimetidine and ranitidine. By using these values it is possible to predict, using equation (3), the curves for total oxygen consumption assuming a K_m for methimazole of 58.9 μM. Figures 1 and 2 compare experimentally determined values and predicted values for cimetidine and ranitidine respectively. IC_{50} values for the inhibition of methimazole metabolism by cimetidine (Fig. 1), calculated by the ALLFIT programme, gave a predicted value of 8.9 mM and an experimentally determined value of 7.5 mM. This close agreement suggests that cimetidine does indeed interact

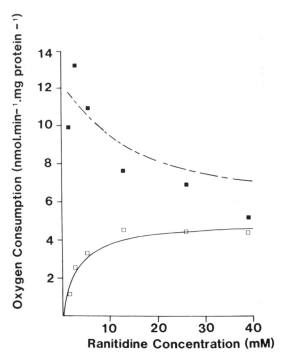

Fig. 2 – Total oxygen consumption as a consequence of varying ranitidine concent-ration in the presence and absence of a fixed methimazole concentration (294.5 μM)
Ranitidine plus methimazole: — – — predicted, ■ observed, Ranitidine alone: □.

with the catalytic site of the flavin monooxygenase. The predicted IC_{50} value for the inhibition of methimazole by ranitidine was 16.7 mM compared with the experimentally determined value of 11.1 mM. It is apparent from Fig. 2 that the experimentally determined points for total oxygen consumption due to ranitidine and methimazole do not closely fit the predicted values as ranitidine appears to decrease total oxygen consumption more than pre-dicted.

These results strongly suggest that both cimetidine and ranitidine are sub-strates for the flavin monooxygenase. Consequently this enzyme system may be important in the metabolism of both drugs and as a site of potentially important drug interactions.

REFERENCES

Andreasen, P. B., Tonnesen, K., Rabol, A. & Keiding, S. (1977), *Acta Pharmacol. Toxicol.*, **40**, 1.
Breen, K. J., Bury, R., Desmond, P. V., Mashford, M. L., Morphett, B., Westwood, B. & Shaw, R. G. (1982), *Clin. Pharmacol. Ther.*, **31**, 297.

Carey, P. F., Martin, L. E. & Owen, P. E. (1981), *J. Chromatogr.*, **225**, 161.

Griffiths, R., Lee, R. M. & Taylor, D. C. (1977) in *Cimetidine*, proceedings of the second International Symposium on histamine H_2-receptor antagonists, eds. Burland, W. L. & Alison Simkins, M. Excerpta Medica, p. 38.

Lowry, O. H., Rosebrough, N. J., Farr, A. L. & Randall, R. J. (1953), *J. Biol. Chem.*, **193**, 265.

Misra, H. P. & Fridovich, T. (1975), *Analyt. Biochem.*, **70**, 632.

Rendic, S., Alebic-Kolbah, T., Kajfez, F. & Ruf, H.-H. (1982), *Xenobiotica*, **12**, 9.

Wilkinson, G. N. (1961), *Biochem. J.*, **60**, 324.

Ziegler, D. M. (1980), in *Enzymatic Basis of Detoxication*, Vol. 1, ed. Jakoby, W. B., Academic Press, New York, p. 201.

Ziegler, D. M. & Poulsen, L. L. (1978), in *Methods in Enzymology*, LII, eds. Fleischer, S. & Packer, L. Academic Press, New York, p. 142.

Chapter 12

THE *IN VITRO* METABOLISM OF S-(−)-NICOTINE BY LIVER PREPARATIONS FROM PREGNANT AND LACTATING RATS

G. R. Keysell and **M. W. Davies**[†], School of Pharmacy, Portsmouth Polytechnic, White Swan Road, Portsmouth, Hants PO1 2DT, UK

1. The *in vitro* formation of cotinine and nicotine 1-*N*-oxide was depressed, relative to controls, during pregnancy and lactation.
2. The two metabolites of nicotine were not affected to the same extent so that the cotinine/*N*-oxide ratio increased during pregnancy and lactation.
3. The use of reconstituted liver preparations indicated the presence of an endogenous inhibitor(s) in the cell cytosol of pregnant rats.
4. There may be more than one inhibitor and/or enzyme inducer because cotinine production returns to control levels after the third day of lactation, whereas *N*-oxide formation remains depressed.
5. The endogenous inhibitor(s) do not appear to be either oestradiol, progesterone or their metabolites.

INTRODUCTION

Man is exposed to a wide variety of exogenous chemicals and the ability to remove these compounds from the body is closely related to the rate and

---

[†]Present address: Winthrop Laboratories, Sterling Winthrop House, Guildford, Surrey.

extent that they are metabolized. A variety of physiological factors have been reported as affecting the qualitative and quantitative metabolism of drugs.

Pregnancy and lactation are generally associated with a reduction in the rate of metabolism of compounds such as the corticosteroid, cortisone (Hench et al., 1950), the hypnotic pentobarbitone (Feuer & Liscio, 1969) and the analgesic, pethidine (Crawford & Rudofsky, 1966). However, not all routes of metabolism are equally affected since Ramsey et al. (1978) have reported that epileptic women, previously stabilized on anticonvulsants, tend to show an increase in the incidence of seizures during pregnancy. This increase has been linked to the increased metabolism and plasma clearance of the anticonvulsants phenytoin and carbamazepine (Damm et al., 1979).

EXPERIMENTAL

S-(−)-Nicotine was incubated with rat liver preparations (10,000 g supernatant) obtained from either pregnant (10–22 days gestation), lactating (1–40 days) or virgin sister controls. Incubations were also carried out using hybrid homogenates consisting of microsomal and cell cytosol fractions (120,000 g) obtained from either pregnant or control rats. Analysis of the incubates for nicotine and its two major metabolites, nicotine 1′-N-oxide (N-oxide) and cotinine was a modification of the method described by Beckett et al. (1971). Nicotine and cotinine were extracted together with the internal markers, phendimetrazine and lignocaine, using a dichloromethane–diethylether mixture (45:55, v/v). The extracts were pooled and concentrated by evaporation and the metabolites estimated by g.l.c. using a column (2 m length, 3 mm i.d.) containing 5% Carbowax 20M and 2% KOH on Chromosorb W (HMDS) and a nitrogen/phosphorus detector system.

RESULTS AND DISCUSSION

At the earliest gestation period investigated (10 days), no alterations were observed in the formation of the N-oxide or cotinine when compared to controls. However significant reductions in the amounts of both metabolites occurred after day 14. The amount of N-oxide and cotinine formed by pregnant liver samples gradually decreased to 35% and 44% respectively of those amounts formed by controls at day 20 (Fig. 1). However at day 16, cotinine formation briefly returned to control levels, a reversal that is not observed in the formation of the N-oxide. This may reflect a reduction in the amount of an endogenous inhibitor that either affects cytochrome P-450 and/or the aldehyde oxidase or a compensatory change in the amount of these enzyme(s) concerned with cotinine production.

Lactation was also associated with decreased nicotine metabolism. The depression in the formation of the N-oxide continued throughout lactation except at day 3. Cotinine production by day 3 of lactation was significantly

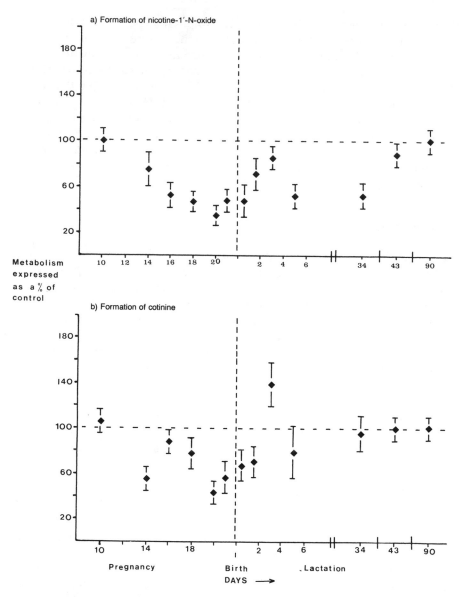

Fig. 1 – The *in vitro* metabolism of nicotine by liver preparations obtained from pregnant and lactating rats.

higher than controls and may be a consequence of the reduction in the concentration of endogenous steroids that occurs at this time. After day 5 cotinine production has returned to control values resulting in a highly significant increase in the cotinine/N-oxide ratio. Changes in this ratio and a reduction in the amount of these two metabolites has also been observed during pregnancy in women smokers (Klein & Gorrod, 1978). These authors noted

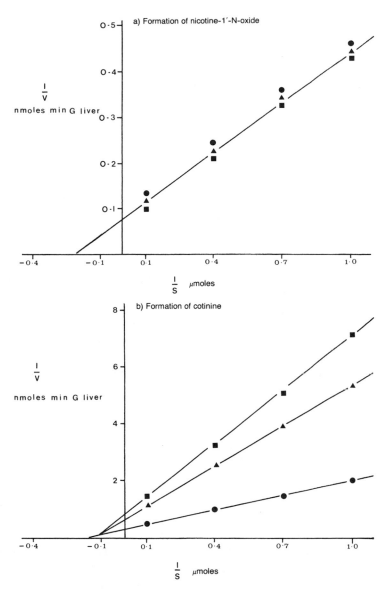

Fig. 2 – Lineweaver–Burk plot of the formation of metabolites from nicotine, alone (●) and in the presence of the steroids, oestradradiol (▲) and progesterone (■), at 10^{-4} M concentration.

that the increased ratio was due to a reduction in the formation of the N-oxide.

The alterations in the relative amounts of the N-oxide and cotinine during pregnancy and lactation may be for one of two reasons. Thee may be endogenous compounds that inhibits the activity of the amine oxidase and cytochrome P-450 mixed function, or alternatively there may be an alteration in metabolism due to a change in the amounts of these enzymes. A convenient method for investigating these hypotheses was the formation of hybrid rat liver preparations consisting of washed microsomal fraction and cell cytosol from both control and pregnant animals.

Cotinine is formed by a two-step process. The intermediary iminium ion is formed from nicotine by the NADPH-dependent cytochrome P-450 pathway and is converted to cotinine by an aldehyde oxidase present in cell cytosol. Preliminary investigations with hybrid mixtures have established that up to 60% of cell cytosol can be replaced by buffer with no adverse effect on the formation of cotinine. This suggests that the aldehyde oxidase enzyme is present in excess and that the rate limiting step in cotinine production is the cytochrome P-450. Therefore small variations in the content of cell cytosol would not be responsible for changes in the production of cotinine by hybrid mixtures.

This view was confirmed by comparing the amounts of both metabolites formed from either liver homogenate (10,000 g supernatant) or a hybrid mixture of microsomes and cell cytosol obtained from virgin rats. In both cases, the amount of cotinine and N-oxide were the same. However, when metabolism by the homogenate and hybrid mixture obtained from pregnant rats were compared there was a significant rise (94%) in the amount of the N-oxide formed by the hybrid mixture. This increase in metabolism may have been caused by the removal of an endogenous inhibitor during the washing of the microsomes. The amount of the cotinine and the N-oxide formed when nicotine was incubated with 'pregnant microsomes' and 'virgin cell cytosol' was significantly higher (44% and 34% respectively) than the amounts formed by 'pregnant microsomes' and 'pregnant cell cytosol'. Furthermore, both metabolites were significantly reduced (25%) when 'pregnant cell cytosol' was combined with 'virgin microsomes'. These results indicate that there is an endogenous substance present in 'pregnant cell cytosol' that can inhibit the formation of both cotinine and the N-oxide.

Kinetic studies (Fig. 2) show that two steroids associated with pregnancy, oestradiol and progesterone, inhibit the formation of cotinine but not that of the N-oxide. Preincubation of the two steroids produced the same effect indicating that the steroid metabolites act in the same way. These steroids are probably not responsible for the inhibition observed during pregnancy and lactation since they do not depress N-oxide formation. However, there may be more than one substance involved because N-oxide production remains depressed during lactation whereas cotinine formation returns to control values.

REFERENCES

Beckett, A. H., Gorrod, J. W. & Jenner P. (1971), *J. Pharm. Pharmac.*, **22**, 722.
Crawford, J. S. & Rudofsky, S. (1966), *Brit. J. Anaesthet.*, **38**, 446.

Damm, M., Christansen, J., Munck, O. & Mygind, K. I. (1979), *Clin. Pharmacokinet.*, **4**, 53.

Feuer, G. & Liscio, A. (1969), *Nature*, **223**, 68.

Hench, P. S., Kendall, E. C., Slocumb, C. H. & Polley, H. S. *Arch. Intern. Med.*, **85**, 545.

Klein, A. E. & Gorrod, J. W. (1979), *Eur. J. Drug Metab. Pharmacokinet.*, **2**, 87.

Ramsey, R. E., Strouth, R. G., Wilder, B. J. & Wilmore, L. J. (1978), *Neurology,* **28**, 85.

Part 3

AROMATIC AMINE OXIDATION AND FORMATION

Chapter 13

N-OXIDATION OF PRIMARY AROMATIC AMINES IN RELATION TO CHEMICAL CARCINOGENESIS

C. B. Frederick,[†] G. J. Hammons,[‡] F. A. Beland,[‡] Y. Yamazoe,[‡] F. P. Guengerich,[§]
T. V. Zenser,[‖] D. M. Ziegler[¶] and F. F. Kadlubar[‡]
[†]Metabolism and Environmental Chemistry Section, Uniroyal Chemical, Elm St.,
Naugatuck, Connecticut 06770, USA
[‡]Carcinogenesis Division, National Centre for Toxicological Research, Jefferson,
Arkansas 72079, USA
[§]Dept. of Biochemistry and Centre in Environmental Toxicology, Vanderbilt
University School of Medicine, Nashville, Tennessee 37232, USA
[‖]Geriatric Research, Education and Clinical Centre, St. Louis VA Medical Centre and
Depts. of Medicine and Biochemistry, St. Louis University, St. Louis, Missouri 63104,
USA
[¶]Clayton Foundation Biochemical Institute, Dept. of Chemistry, The University of
Texas at Austin, Austin, Texas 78712, USA

1. Metabolism studies have been conducted with the purified flavin-containing monooxygenase (FMO), purified cytochrome P-450 isozymes, and microsomal prostaglandin H synthetase (PHS). The oxidation products isolated from the incubation of these enzymes with 1-naphthylamine (1-NA), 2-naphthylamine (2-NA), and 2-aminofluorene (2-AF) indicate a preferential ring hydroxylation of 1-NA and 2-NA and N-hydroxylation of 2-AF.

2. These results have been used to discriminate between possible chemical mechanisms for the oxidation catalysed by each enzyme. Analysis of π-electron densities for various intermediates suggest that a nitrenium/

carbenium ion intermediate is common to both cytochrome P-450- and FMO-catalysed arylamine oxidations.

3. For the PHS-catalysed reactions, a comparison of the extent of metabolic activation to the calculated ionization potential suggests that the removal of an electron from the lone pair of the nitrogen is a rate-determining step in the formation of reactive intermediates from arylamines.

INTRODUCTION

The *N*-oxidation of arylamines is a subject of interest to many investigators concerned with the metabolism of biologically active organic compounds. A wide variety of pharmaceutical products, pesticides, environmental contaminants, and natural products to which humans are exposed contain an arylamine moiety and are transformed within the body to a variety of toxic and non-toxic metabolites.

Human exposure to arylamines may occur quite frequently. For example, Patrianakos & Hoffman (1979) have shown that 2-naphthylamine and 4-aminobiphenyl are present in both mainstream and sidestream cigarette smoke. Haugen *et al.* (1982) have detected a variety of arylamines in coal-derived synthetic fuels, and many studies have now documented the formation of heterocyclic arylamines in broiled or fried meat (Sugimura & Sato, 1983).

Based on epidemiological studies of workers in the dye and rubber industries (e.g. Case *et al.*, 1954), arylamines are typically associated with an increased incidence of tumours of the urinary bladder. More recent studies of the mutagenic and carcinogenic heterocyclic arylamines found in broiled meat have led to the suggestion that these compounds may contribute to an increased incidence of colon cancer (Weisburger *et al.*, 1982). In animal models, rodents have been found to form primarily hepatic tumours following dosing with most arylamines, although some bladder tumours are observed in rats following ingestion of 2-naphthylamine (Hicks & Chowaniec, 1977). Dogs have been shown to produce primarily bladder tumours following arylamine administration in a response very similar to that of humans.

In this review, we emphasize metabolic studies conducted with three arylamines that exhibit a wide range of metabolic and toxicological responses: 2-aminofluorene (2-AF), 2-naphthylamine (2-NA), and 1-naphthylamine (1-NA). 2-AF is found in synthetic fuels and is the deacetylated derivative of the extensively studied model liver carcinogen, 2-acetylaminofluorene. 2-AF is a metabolite of 2-acetylaminofluorene, *in vivo*, and *N*-hydroxy-AF has been suggested to be the ultimate mutagen derived from 2-acetylaminofluorene in a variety of mutagenicity assays (e.g. Stout *et al.*, 1976; Schut *et al.*, 1978). 2-NA is a classic human bladder carcinogen and was originally used to develop the dog as a model for the induction of bladder cancer by arylamines. 1-NA is a geometrical isomer of 2-NA with a quite different toxicological profile because it has not been shown to be a carcinogen in a variety of animal models. However, some epidemiological evidence has been presented that suggests that 1-NA may be a human carcinogen (Case *et al.*, 1954).

OXIDATION OF ARYLAMINES—CRITICAL ENZYMES AND PRODUCTS

Microsomal metabolism

Analysis of the metabolites produced from 2-AF, 1-NA, and 2-NA by the oxidative enzymes present in rat and dog hepatic microsomes indicates that a similar profile of ring and *N*-hydroxylated metabolites are derived from each substrate by both species (Fig. 1; Hammons *et al.*, 1985). The structures of the metabolites derived from each substrate are shown in Fig. 2a–c. Note that the *N*-oxidation of 1-NA was not detected in either species (the bar shown for N-OH-1-NA in Fig. 1 represents the limits of detection of the analytical system). However, *N*-oxidation of 2-NA was observed, although ring hydroxylation was more extensive (N_{ox}/C_{ox} = 0.4–0.5). In contrast, *N*-hydroxylated metabolites predominated during the metabolism of 2-AF (N_{ox}/C_{ox} = 2.7–6.6). These substrates, therefore, present a range of carcinogenicity and oxidative metabolism that may provide insight on the primary activation step in susceptible species.

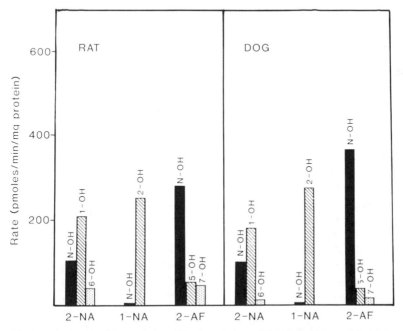

Fig. 1 – The rates of *N*- and ring *C*-hydroxylation of 2-NA, 1-NA, and 2-AF by rat and dog hepatic microsomes. The experimental methods are described in Hammons *et al.* (1985).

(a)

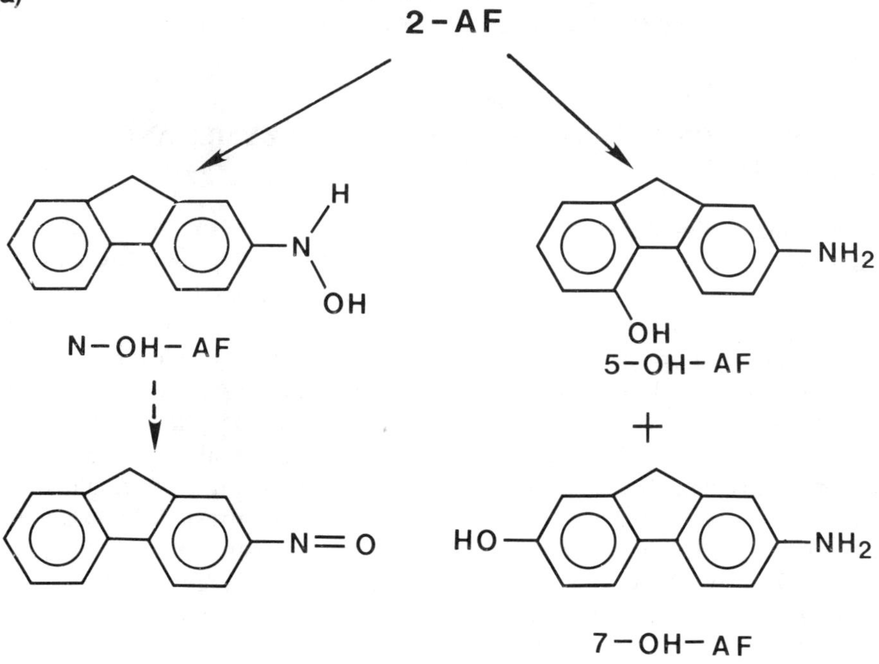
2-AF
H
N
OH
N-OH-AF
NH2
OH
5-OH-AF
+
N=O
HO
NH2
7-OH-AF

(b)

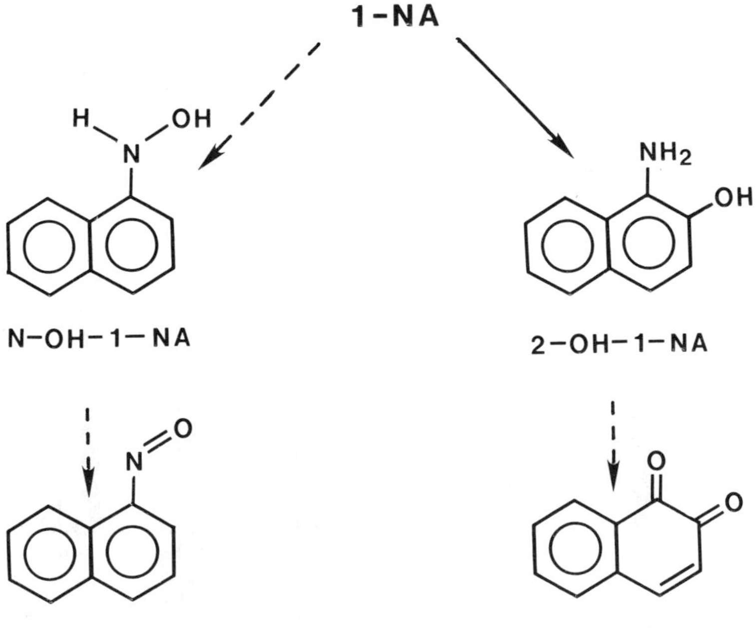
1-NA
H
OH
N
NH2
OH
N-OH-1-NA
2-OH-1-NA
N
O
O
O

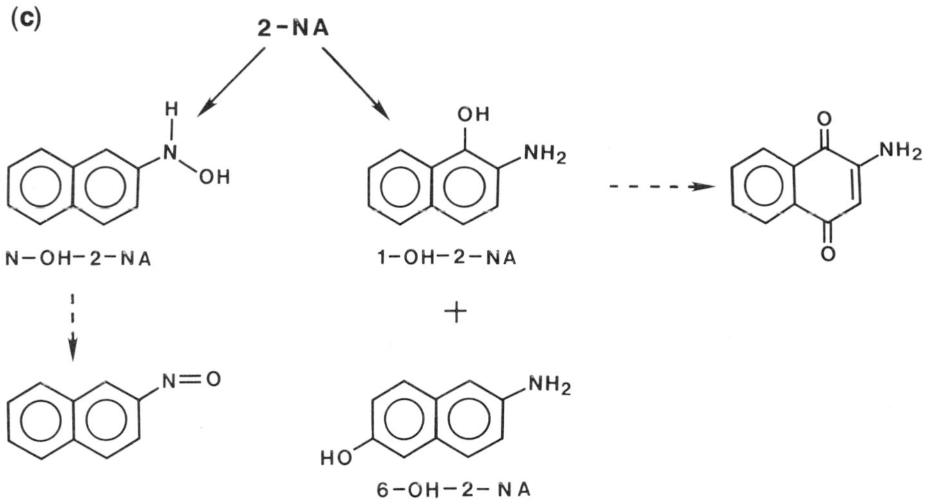

(c)

Fig. 2 – The structures of the primary oxidative metabolites derived from 2-AF (a), 1-NA (b), and 2-NA (c) in the presence of rat and dog hepatic microsomes. Secondary oxidation products that were detected are indicated by a dashed arrow from their presumed precursor.

A variety of *in vitro* studies have indicated that there are at least three possible enzymatic mechanisms by which arylamines may be oxidized. These include: (a) the microsomal flavin-containing monooxygenase (FMO); (b) specific inducible isozymes among the cytochrome P-450 monooxygenases; (c) and peroxidative enzymes, typified by prostaglandin H synthetase (PHS).

Microsomal flavin-containing monooxygenase

A.   2-Aminofluorene
The *N*-oxidation of 2-AF by the flavin-containing monooxygenase (FMO) was first suggested by Pelroy & Gandolfi (1980) in a survey of the mutagenic activation of various aromatic amines in a *Salmonella typhimurium* assay. Purified pig hepatic FMO was used for metabolic activation and AF was found to give a strong dose-related response in the assay. The specific formation of *N*-hydroxy-2-aminofluorene (N-OH-AF) was subsequently demonstrated by incubation of the purified porcine FMO with 2-AF followed by organic extraction and h.p.l.c. analysis of the oxidized metabolites (Frederick *et al.*, 1982; Hammons *et al.*, 1985). Ring-hydroxylated products were not detected in the organic extract as opposed to the other arylamines studied (Table 1). To provide an estimate of the relative contribution of FMO oxidation in microsomal oxidations, 2-(2,4-dichloro-6-phenyl) phenoxyethylamine (DPEA) was used to inhibit the cytochrome P-450-mediated oxidation and methimazole (a very low K_m substrate for FMO) was used to inhibit the remaining *N*-oxidative activity. These data suggest that FMO makes a signif-

icant contribution to 2-AF *N*-oxidation with both pig and human hepatic microsomes.

B.   2-Naphthylamine

The participation of FMO in the oxidation of 2-NA was first reported in a study of the substrate specificity of porcine hepatic FMO (Ziegler *et al.*, 1973). The reaction was monitored by means of oxygen uptake with a Clarke electrode, and a high rate of reaction (210 nmol/min/mg protein) was observed with the purified enzyme. Subsequently, Pelroy & Gandolfi (1980) reported that a variety of arylamines could be metabolized to mutagens in an assay with *S. typhimurium* and purified porcine hepatic FMO. In this assay 2-NA did not give a significant mutagenic response, thus implying that the proposed mutagenic metabolite of 2-NA, *N*-hydroxy-2-naphthylamine (N-OH-2-NA), is not the major enzymatic product.

We have recently examined the metabolites produced from the purified porcine enzyme by h.p.l.c. and found that 2-amino-1-naphthol was the principle metabolite derived from 2-NA with no *N*-oxidation being detected (Table 1). 2-Amino-1-naphthol was subject to further oxidation in the reaction medium to 2-amino-1,4-naphthoquinone (Fig. 2c). When this reaction was conducted in the presence of $[^{18}O]$oxygen (99% enrichment) and $[^{16}O]$water, mass spectral analysis of the 2-amino-1,4-naphthoquinone product, indicated quantitative incorporation of a single ^{18}O atom. This suggests that the product may be derived from addition of water to an intermediate iminoquinone followed by a second oxidation step. It should be noted that this reaction represents unusual example of ring-carbon oxidation induced by FMO, which is generally associated with the oxidation of nitrogen and sulphur (Ziegler, 1980).

C.   1-Naphthylamine

In the same study in which Ziegler *et al.* (1973) observed the oxidation of 2-NA by the purified porcine FMO, a slow rate of oxidation of 1-naphthylamine was also reported. When the metabolites of this reaction

Table 1
Comparative metabolism of 2-AF, 1-NA, and
2-NA by purified porcine liver FMO.

Substrate	*N*-Oxidation	Ring-oxidation
1-NA	ND[a]	1070 (C-2)
2-NA	ND	1505 (C-1)
2-AF	757	ND

[a]Rates in pmol/min/nmol enzyme; ND = not detected with a minimal detectable rate of 100 pmol/min/nmol enzyme (Hammons *et al.*, 1985).

were specifically analysed (Table 1), 1-amino-2-naphthol and its secondary oxidation product, 1,2-naphthoquinone, were found to be the dominant metabolic products; N-oxidation was not detected (Hammons *et al.*, 1985).

Cytochrome P-450 monooxygenases

A.   2-Aminofluorene

Although the hepatic microsomal N-oxidation of 2-AF was noted as early as 1961 (Uehleke, 1961), the specific association of the reaction with cytochrome P-450 catalysis has only been explored in the last few years. Aune *et al.* (1980) demonstrated the mutagenic activation of 2-AF for *S. typhimurium* with mouse and rat hepatic microsomes and nuclei. The response was decreased with the cytochrome P-450 inhibitor, cobaltous chloride. These observations were consistent with other studies that indicated that N-OH-AF is a potent mutagen derived from the arylhydroxamic acid, N-hydroxy-2-acetylaminofluorene (N-OH-AAF; e.g., Stout *et al.*, 1976; Schut *et al.*, 1978), and strongly suggested that N-OH-AF formed from 2-AF by cytochrome P-450 enzymes was responsible for the mutagenic response. Robertson *et al.* (1981) used a similar assay with purified forms of cytochrome P-450 isolated from rabbit liver and lung. In this reconstituted system, a specific isozyme (P-450$_{II}$) of the lung enzyme catalysed the formation of a mutagenic metabolite—presumed to be N-OH-AF. These results must be interpreted with due consideration of the recent results of McCoy *et al.* (1983) which implicate a transacetylase in the further activation of N-hydroxyarylamines within the bacterial cell.

Using a unique derivatization method and a gas chromatography analytical procedure, Razzouk & Roberfroid (1982 and references cited therein) have analysed the metabolites formed from 2-AF in the presence of hepatic microsomes from rats, mice, hamsters, guinea pigs, and monkeys. Their data indicate that N-oxidation of 2-AF occurs readily in all species examined, and that the enzymes catalysing the reaction are inducible by 3-methylcholanthrene (3-MC) in rodents. The formation of N-OH-AF was inhibited by the cytochrome P-450 inhibitors, 3-MC, 7,8-benzoflavone, miconazole, and metyrapone.

We have conducted similar metabolic studies with rat, dog, pig, and human hepatic microsomes using an h.p.l.c. analytical method and direct injection of an organic extract of the reaction mixture (Frederick *et al.*, 1982; Hammons *et al.*, 1985). The results for rat and dog hepatic microsomes (Fig. 1) indicate that N-hydroxylation of 2-AF predominates in both species. Hepatic microsomes from all animal species analysed to date catalyse the N-oxidation of AF relatively efficiently; although assays with seven samples of human hepatic microsomes have indicated a highly variable N-oxidative capacity (Table 2). The total cytochrome P-450 content of the microsomes varied over a threefold range, but the capacity for N-oxidation varied over a tenfold range; probably reflecting the specific isozyme content of the individual's liver (i.e.

Table 2

The rate of formation of hydroxylated products from 2-AF, 1-NA, and 2-NA by human liver microsomes.

Tissue	Cytochrome P-450[a]	2-AF[b]			1-NA[b]		2-NA[b]		
		N-OH	5-OH	7-OH	N-OH	2-OH	N-OH	1-OH	6-OH
No. 21	0.36	ND	ND	ND	ND	233	ND	ND	49
No. 23	0.40	214	17	16	ND	182	151	130	69
No. 25	0.92	256	ND	ND	ND	1194	40	246	93
No. 26	0.38	213	ND	ND	ND	358	44	129	33
No. 27	0.44	49	ND	ND	ND	144	63	111	16
No. 28	0.41	519	19	14	ND	281	352	308	ND
No. 29	0.41	234	ND	ND	ND	113	161	131	ND

[a]Expressed as nmol/mg microsomal protein.
[b]Rate in pmol/mg protein with a limit of detection of 10 pmol/min/mg protein (Hammons et al., 1985).
ND = not detected.

state of induction based on diet, age, heredity, smoking habits, etc.). To determine the contribution of specific forms of cytochrome P-450 in the N-oxidation of 2-AF, we have used reconstituted enzyme systems composed of purified rat hepatic cytochrome P-450 isozymes (Guengerich *et al.*, 1982). The results, summarized in Table 3, indicate that 2-AF is an excellent substrate of six of the eight isozymes analysed to date. Interestingly, the relative rate of N-oxidation to ring-carbon oxidation varies widely (ranging from a ratio of 1.3 to 21.2) suggesting that particular isozymes may provide an active site or reaction pathway that favours oxidation at a specific substrate site.

B. 2-Naphthylamine

Studies on the N-oxidation of 2-NA *in vitro* have been hampered by the formidable analytical problems encountered when trying to quantitate the oxidatively unstable N-OH-2-NA in the presence of a variety of oxidatively unstable ring-hydroxylated metabolites. Early studies estimated N-OH-2-NA formation by oxidizing the metabolite to 2-nitrosonaphthalene (2-NON) and using a spectrophotometric determination of the 2-NON in an organic extract. This approach was used in the first report of the N-oxidation of 2-NA by rat liver microsomes by Uehleke (1963). Subsequently, the same group showed the reaction to be inhibited by the cytochrome P-450 inhibitor, metyrapone.

Data collected by improved analytical techniques for rat and dog hepatic microsomes (Fig. 1) indicate that N-hydroxylation accounts for approximately one-half of the total oxidative metabolism detected for 2-NA (Hammons *et al.*, 1985). Human microsomes also N-oxidize 2-NA, but at a rate somewhat less than that observed with 2-AF (Table 2). Interestingly, when 2-NA is used as a substrate for a variety of purified cytochrome P-450 isozymes (Table 3), only one form (ISF-G) is found to catalyse N-oxidation; and it is the same form as that which N-oxidizes 2-AF most effectively. The ratio of N-hydroxylation to ring hydroxylation for these isozymes varies from 0.1 to 0.7 and provides a distinct contrast to the high rate of N-hydroxylation of 2-AF. The efficient N-oxidation of 2-AF relative to 2-NA may be the explanation for the higher mutagenic potency of 2-AF in *S. typhimurium* assays when microsomes from a variety of species are used for metabolic activation (Phillipson & Ioannides, 1983), although a difference in the further activation of the N-hydroxyarylamines within the bacterial cell cannot be excluded (McCoy *et al.*, 1983).

C. 1-Naphthylamine

Relatively few *in vitro* metabolic studies have been reported for this substrate—possibly reflecting the analytical difficulties in detecting N-hydroxy-1-naphthylamine (N-OH-1-NA). Nakayama *et al.* (1982) do not detect an e.s.r. signal characteristic of N-OH-1-NA in rat hepatic microsomal metabolism mixtures. Studies in our laboratory with rat, dog, and human microsomes, have indicated that 1-NA is not significantly N-hydroxylated

Table 3

The rate of *N*-oxidation and ring hydroxylation of 2-AF, 1-NA, and 2-NA by purified forms of rat hepatic cytochrome P-450 in reconstituted metabolic assays[a]

Reconstituted cytochrome P-450	2-NA[b]			1-NA[b]		2-AF[b]		
	N-OH	1-OH	6-OH	N-OH	2-OH	N-OH	5-OH	7-OH
UT-A	ND	152	ND	ND	274	543	ND	ND
UT-H	ND	ND	ND	ND	201	442	107	248
PCN/PB-E	ND	ND	ND	ND	ND	ND	ND	ND
PB-B	ND	773	ND	ND	2150	1062	ND	ND
PB-C	ND	ND	ND	ND	ND	ND	ND	ND
PB-D	ND	582	ND	ND	1661	456	ND	ND
BNF-B	ND	2544	360	ND	3910	5244	200	560
ISF-G	2517	2589	911	ND	3851	7421	2318	520

[a]Enzyme isolation and nomenclature as described by Guengerich *et al.*, (1982).
[b]Rates are expressed as pmol/min/nmol enzyme with a minimal detectable limit of 50 pmol/min/nmol.
ND = not detected.

when compared to 2-NA or 2-AF (Fig. 1 and Table 2). Furthermore, a variety of purified cytochrome P-450 isozymes did not N-hydroxylate 1-NA in reconstituted enzyme assays (Table 3). The observation that 1-NA is not susceptible to N-oxidation may serve to explain its lack of carcinogenicity because synthetic N-OH-1-NA is a potent mutagen and carcinogen.

Peroxidases

A variety of peroxidases have been identified in mammalian tissues, but the enzyme that has received the most attention in arylamine metabolism is PHS. This multifunctional enzyme oxidizes and cyclizes arachidonic acid to a hydroperoxide, prostaglandin G_2, and then reduces the hydroperoxide moiety to an alcohol, prostaglandin H_2. The hydroperoxidase activity provides a potent oxidizing agent which in studies *in vitro* has served to oxidize a variety of xenobiotics (reviewed by Eling *et al.*, 1983). We have analyzed a variety of arylamine substrates (Kadlubar *et al.*, 1982), and found that primary arylamines are generally good substrates for PHS (Table 4).

Table 4

PHS-mediated metabolism and covalent binding to protein of aromatic amines.

Aromatic amine	metabolism (%)	Protein binding	DNA binding
Benzidine	81 ± 1^a	84 ± 16^b	9800 ± 1500^c
p-Phenetidine	26 ± 5	28 ± 5	5
2-Aminofluorene	46 ± 4	20 ± 6	28 ± 7
N-Methyl-4-aminoazobenzene	33 ± 7	18 ± 2	1
2-Naphthylamine	14 ± 4	16 ± 4	22 ± 7
4-Aminobiphenyl	17 ± 1	8 ± 1	7 ± 2

[a]Extent of metabolism after a 5-min incubation measured as loss of substrate.
[b]Protein binding was measured as nmol substrate bound/mg microsomal protein.
[c]DNA binding was measured as pmol of substrate bound/mg of added calf thymus DNA (Kadlubar *et al.*, 1982).

A.   *2-Aminofluorene*

Following the report that 2-AF is oxidized by ram seminal vesicle microsomes (a tissue with high PHS activity and no detectable cytochrome P-450 activity) to metabolites that bind to cellular macromolecules (Kadlubar *et al.*, 1982), the principal organic-extractable metabolites derived from similar incubations were found to be 2-nitrofluorene and 2,2'-bisazofluorene (Boyd *et al.*, 1983). The same metabolites were observed when hydrogen peroxide was substituted for arachidonic acid in the incubation mixture. Horseradish peroxidase also gave a similar metabolic profile, whereas chloroperoxidase catalysed the formation of 2-nitrosofluorene rather than 2-nitrofluorene. No evidence for the ring hydroxylation of 2-AF was found under any of these incubation conditions. Robertson *et al.* (1983) used ram seminal vesicle microsomes and arachidonic acid as an activation fraction in a *S. typhimurium* mutagenicity assay with a variety of arylamines as substrates. 2-AF produced an arachidonic acid-dependent increase in mutagenicity under the assay conditions and this was inhibited by the PHS inhibitor, indomethacin.

B.   2-Naphthylamine

2-NA is also a substrate for PHS and is metabolized to two products identified as 2-nitrosonaphthalene and N^4-(2-naphthyl)-2-amino-1,4-naphthoquinoneimine. Robertson *et al.* (1983) found 2-NA to be activated to a mutagen by ram seminal vesicle microsomes and arachidonic acid in a *S. typhimurium* assay, and Wise *et al.* (1984) have recently shown that 2-NA is is oxidized by canine bladder mucosa microsomes in the presence of arachidonic acid to metabolites that bind to protein and tRNA. No NADPH-dependent metabolism of 2-NA could be detected in this tissue, thus suggesting that PHS may play a role in the activation of carcinogenic arylamines in a target tissue. Yamazoe *et al.* (1984) have presented data on the modified nucleoside adducts derived from the incubation of 2-NA with DNA in the presence of ram seminal vesicle microsomes and arachidonic acid. These modified nucleosides were principally derived from the addition of deoxyguanosine to the quinoneimine derived from 2-amino-1-naphthol. Evidence was also presented for the formation of these adducts in canine bladder mucosa following administration of 2-NA.

C.   1-Naphthylamine

Very little data are available on the metabolism of 1-NA by PHS. Robertson *et al.* (1983) were unable to detect metabolism of 1-NA to a mutagen in a *S. typhimurium* assay in the presence of ram seminal vesicle microsomes and arachidonic acid. Unpublished data from our laboratory indicate that 1-NA is a substrate for PHS (based on loss of substrate). Although product characterization has not been completed, *N*-oxidation products have not been detected.

PROPOSED ENZYMATIC MECHANISMS

The similarity of products resulting from the metabolic oxidation of the arylamines under consideration has led us to consider possible enzymatic mechanisms that may be common to each pathway. A brief summary of mechanistic possibilities will first be presented and then evidence in support of a specific mechanism will be discussed. The discussion to follow will emphasize the intermediates present on the substrate and will assume that, in the case of cytochrome P-450 mediated reactions, an electrophilic oxygen–porphyrin complex (perferryl oxygen) initiates an electrophilic attack on an electron-rich substrate. The basis of this assumption has been elaborated in detail by Guengerich & Macdonald (1984). With FMO-catalysed oxidations, a 4a-flavin hydroperoxide is assumed to be the electrophilic species present on the enzyme (as reviewed by Bruice, 1980). The peroxidase-catalysed reactions have not been as thoroughly studied, however evidence has been presented for an electrophilic haemoprotein radical as well as a reactive oxygen species released by the enzyme (such as superoxide, hydroxyl radical, etc.) as the reactive species mediating PHS oxidations (Eling *et al.*, 1983).

In cytochrome P-450 catalysed oxidations, three basic electronic states of the arylamine substrate may be proposed as participating in the transition state determining the site of substrate oxidation. In the first of these, a neutral substrate undergoes a direct addition of the perferryl oxygen to yield a sigma complex that collapses to the oxygenated product. This mechanism is equivalent to an electrophilic aromatic substitution reaction on an arylamine, and involves a transient Wheland intermediate in the formation of ring-hydroxylated products. Depending on whether the transition state was reactant-like (kinetic control) or product-like (thermodynamic control), the product distribution should reflect the π-electron density of the neutral substrate or the thermodynamic stability of the Wheland intermediate cation (Streitweiser, 1961; Lowe, 1978).

A second mechanism is that the perferryl oxygen abstracts an electron from a region of high electron density (presumably, the lone pair of the arylamine nitrogen) to form a radical cation. Transfer of a proton to the porphyrin–oxygen complex, followed by collapse of the neutral radical pair, would yield products dependent upon the electron density of the intermediate neutral substrate radical.

The third mechanism is similar to the second in that an initial rate determining electron-abstraction forms a radical cation, but then a second electron transfer occurs in a rapid reaction to yield the equivalent of hydroxyl anion on the porphyrin and the transition state of the substrate is a resonance-stabilized nitrenium/carbenium ion. Product distribution would be dependent on the distribution of positive charge (low electron density) on the substrate.

Each of these possibilities indicate that the π-electron distribution in the substrate should be an important factor in product distribution. By calculating the π-electron density of each transition state for each substrate and correlating these results with the products isolated from the reaction mixtures, an indication of the transition state of the substrate should be discernable.

Calculations of π-electron density at the atoms of interest (the arylamine nitrogen and the ring carbons that are hydroxylated) are summarized in Table 5. The original approach (Hammons *et al.*, 1985) was modified slightly in these calculations based on the Hückel heteroatom parameters suggested by Lowe (1978). The data were calculated by the program of Lowe (1978) as modified by Peake & Grauwmeijer (1981). These results do not provide a correlation between electron density and product distribution for either a neutral intermediate (calculated as either initial electron density or stability of a sigma-complex) or radical intermediate. However, the distribution of electron density in a resonance-stabilized nitrenium/carbenium ion does appar-

Table 5

Huckel calculations of π-electron density and localization energy for possible substrate intermediates in mixed function oxidations.

Arylamine	Intermediate	Electron density		Localization energy (β)	
		Nitrogen	Carbon	Nitrogen	Carbon
1-NA	Neutral	1.908	1.068	3.244	3.487
2-NA	Neutral	1.915	1.069	3.234	2.682
2-AF	Neutral	1.914	1.007; 1.006[a]	5.119	4.521; 4.578
1-NA	Radical	1.296	0.963		
2-NA	Radical	1.264	0.920		
2-AF	Radical	1.263	0.982; 0.978		
1-NA	Nitrenium ion	0.919	0.767		
2-NA	Nitrenium ion	0.813	0.668		
2-AF	Nitrenium ion	0.820	0.943; 0.937		

[a]The values given are for the five-and seven-ring carbons of 2-AF, respectively, which are the primary sites of metabolic ring hydroxylation.

ently correlate with the hydroxylated product distribution of the various substrates. Although this correlation does not prove the mechanism, it does provide a reasonable working hypothesis which is consistent with previous conclusions with other substrates. A general mechanism for the cytochrome P-450 catalysed oxidations based on these results is provided in Scheme 1.

Scheme 1

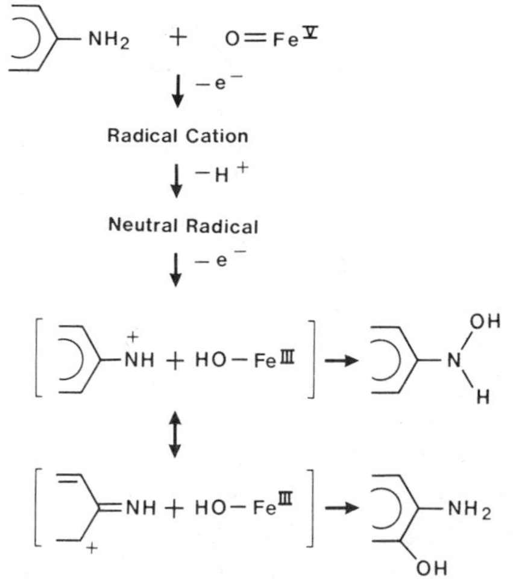

The mechanism for the FMO oxidation of arylamines follows from similar considerations. Previous mechanistic proposals have emphasized the attack of a lone pair of electrons, from a heteroatom, on the flavin hydroperoxide.

However, since this enzyme gave predominantly ring hydroxylation of 1-NA and 2-NA, a mechanism must be proposed that allows the addition of oxygen to a reactive site on a ring carbon. The [^{18}O]-oxygen incorporation results described above indicate that the oxygen atom is directly derived from molecular oxygen that has been activated on the flavin as opposed to oxygen from a water molecule. The pattern of metabolites for each substrate is consistent with a resonance-stabilized nitrenium/carbenium ion intermediate described by the charge density calculations of Table 5. A mechanism for the formation of this intermediate is presented in Scheme 2. A unique step in this

Scheme 2

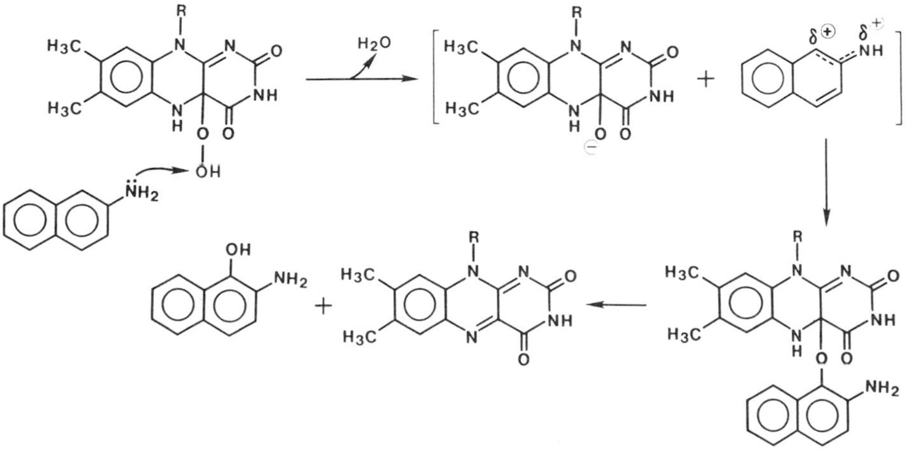

mechanism is the dehydration of the N-hydroxyarylamine that is initially formed in the reaction to yield the cation intermediate. The tertiary amine and sulphur-containing substrates that are generally associated with FMO oxidations would not have this step available to them, and would be released as nitrogen or sulphur oxides, respectively. The cation derived from an arylamine would be proposed to combine with the remaining oxygen atom on the flavin and a subsequent elimination reaction would release the hydroxylated product and regenerate the oxidized flavin cofactor. Hydroxylation would be favoured at sites on the substrate cation that have relatively low electron density (high positive charge).

The mechanism(s) of the PHS-catalysed oxidations is not obvious. As with the cytochrome P-450 and FMO-dependent reactions, oxidation occurs primarily on the nitrogen of 2-AF and ring carbon-2 of 2-NA (although significant N-oxidation also occurs). The structures of the products from PHS oxidation of 2-AF are different from those produced by the other enzymes, and this may be reflected in the product-determining intermediate for the substrate. We have attempted to correlate an index of metabolic rate for various substrates with a physical property to provide some insight into reaction mechanism. This correlation, described in Fig. 3, indicates that ionization potential (calculated with a MNDO program by Drs. John Scribner and

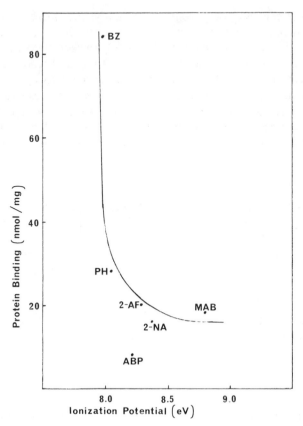

Fig. 3 – The relationship of arachidonic acid-dependent covalent protein binding of various aromatic amines (Kadlubar *et al.*, 1982) with their ionization potentials calculated by the method of Dewar & Thiel (1977). The abbreviations used for the amines are: benzidine (BZ), *p*-phenetidine (PH), 2-aminofluorene (2-AF), 4-aminobiphenyl (ABP), 2-naphthylamine (2-NA), and *N*-methyl-4-aminoazobenzene (MAB).

George Ford, Pacific Northwest Research Foundation, Seattle, Washington) and covalent binding of radioactive substrate to microsomal protein are related for the simple arylamines studied. The disubstituted arylamines, benzidine and phenetidine, have a lower ionization potential (presumably due to electron donation from the second functional group) and a much higher rate of binding. We interpret these results as indicating that the rate-determining step in arylamine oxidation by PHS is a one-electron abstraction from the nitrogen lone pair. The difference in product distribution for PHS versus the other enzymes studied may be indicative of a radical or radical cation intermediate that may react with atmospheric oxygen to yield a nitroso or nitroarene or react with a second substrate molecule at the relatively high substrate concentrations used in studies *in vitro* (note particularly the 2-AF coupling product). Studies with chloroperoxidase on the dealkylation of dimethylaniline suggest that other less reactive peroxidases may react in a manner similar to that proposed for cytochrome P-450 enzymes.

REFERENCES

Aune, T., Dybing, E. & Nelson, S. D. (1980), *Chem. Biol. Interact.*, **31**, 35.

Boyd, J. A., Harvan, D. J. & Eling, T. E. (1983), *J. Biol. Chem.*, **258**, 8246.

Bruice, T. C. (1980), *Acc. Chem. Res.*, **13**, 256.

Case, R. A. M., Hosker, M. E., McDonald, D. B. & Pearson, J. T. (1954), *Brit. J. Ind. Med.*, **11**, 75.

Dewar, M. J. S. & Thiel, W. (1977), *J. Am. Chem. Soc.*, **99**, 4899; 4907.

Eling, T., Boyd, J., Reed, G., Mason, R. & Sivarajah, K. (1983), *Drug Metab. Rev.*, **14**, 1023.

Frederick, C. B., Mays, J. B., Ziegler, D. M., Guengerich, F. P. & Kadlubar, F. F. (1982), *Cancer Res.*, **42**, 2671.

Guengerich, F. P., Dannan, G. A., Wright, S. T., Martin, M. V. & Kaminsky, L. S. (1982), *Biochemistry*, **21**, 6019.

Guengerich, F. P. & Macdonald, T. L. (1984), *Acc. Chem. Res.*, **17**, 9.

Hammons, G. J., Guengerich, F. P., Weis, C. C., Beland, F. A. & Kadlubar, F. F. (1985), *Cancer Res.*, in press.

Haugen, D. A., Peak, M. J., Suhrbier, K. M. & Stamoudis, V. C. (1982), *Anal. Chem.*, **54**, 32.

Hicks, R. M. & Chowaniec, J. (1977), *Cancer Res.*, **37**, 2943.

Kadlubar, F. F., Frederick, C. B., Weis, C. C. & Zenser, T. V. (1982), *Biochem. Biophys. Res. Comm.*, **108**, 253.

Lowe, J. P. (1978), *Quantum Chemistry*, Academic Press, Orlando, pp. 202, 538.

McCoy, E. C., Anders, M. & Rosenkranz, H. S. (1983), *Mutation Res.*, **121**, 17.

Nakayama, T., Kimura, T., Kodama, M. & Nagata, C. (1982), *Gann*, **73**, 382.

Patrianakos, C. & Hoffman, D. (1979), *J. Anal. Toxicol.*, **3**, 150.

Peake, B. M. & Grauwmeijer, R. (1981), *J. Chem. Ed.*, **58**, 692.

Pelroy, R. A. & Gandolfi, A. J. (1980), *Mutation Res.*, **72**, 329.

Phillipson, C. E. & Ioannides, C. (1983), *Mutation Res.*, **124**, 325.

Razzouk, C. & Roberfroid, M. B. (1982), *Chem. Biol. Interact.*, **41**, 251.

Robertson, I. G. C., Philpot, R. M., Zieger, E. & Wolf, C. R. (1981), *Mol. Pharmacol.*, **20**, 662.

Robertson, I. A. G., Sivarajah, K., Eling, T. E. & Zieger, E. (1983), *Cancer Res.*, **43**, 476.

Schut, H. A. J., Wirth, P. J. & Thorgeirsson, S. S. (1978), *Mol. Pharmacol.*, **14**, 682.

Stout, D. L., Hemminki, K. & Becker, F. F. (1976), *Cancer Lett.*, **1**, 269.

Streitweiser, A. (1961), *Molecular Orbital Theory for Organic Chemists*, John Wiley & Sons, New York, p. 307.

Sugimura, T. & Sato, S. (1983). *Cancer Res.*, **43**, 2415.

Uehleke, H. (1961), *Experientia*, **17**, 557.

Uehleke, H. (1963), *Biochem. Pharmacol.*, **12**, 219.

Weisburger, J. H., Wynder, E. L. & Horn, C. L. (1982), *Cancer*, **50**, 2541.

Wise, R. W., Zenser, T. V., Kadlubar, F. F. & Davis, B. B. (1984), *Cancer Res.*, **44**, 1893.

Yamazoe, Y., Miller, D. W., Gupta, R. C., Zenser, T. V., Weis, C. C. & Kadlubar, F. F. (1984), *Proc. Am. Assoc. Cancer Res.*, **25**, 91.

Ziegler, D. M. (1980) in *Enzymatic Basis of Detoxification*, Vol. I (ed. Jakoby, W. B.), Academic Press, New York, p. 201.

Ziegler, D. M., McKee, E. M. & Poulsen, L. L. (1973), *Drug Metab. Disp.*, **1**, 314.

Chapter 14

IDENTIFICATION OF HYDROXYLAMINOGLUTE-THIMIDE AS AN INDUCED URINARY METABOLITE OF AMINOGLUTETHIMIDE IN C57BL/6 MICE

Alison Seago, **M. Jarman**, **A. B. Foster** and **M. Baker**, Institute of Cancer Research, CRC Laboratory, Clifton Avenue, Belmont, Sutton, Surrey SM2 5PX, UK

1. A procedure is described for the derivatization of hydroxylamino-glutethimide using d_6-acetic anhydride, thus converting the molecule into a stable derivative. Mass spectrometry of the product confirms the metabolite to be excreted as hydroxylaminoglutethimide unaccompanied by the N-acetyl derivative.
2. Deuteroacetylation of metabolites of aminoglutethimide allows the one-step determination of the proportion of these urinary metabolites excreted in the free and acetylated form.
3. In the search for a suitable animal model for aminoglutethimide metabolism in humans it was found that the C57BL/6 mouse has a similar metabolic profile and also exhibits induced formation of hydroxylamino-glutethimide.

INTRODUCTION

Aminoglutethimide [3-(4-aminophenyl)-3-ethylpiperidine-2,6-dione; AG] is now widely used in the treatment of hormone-dependent metastatic breast cancer. It acts by inhibiting aromatase (conversion of androstenedione into oestrone, testosterone into oestradiol) and desmolase (conversion of cholesterol to pregnenolone) so blocking oestrogen synthesis and preventing stimulation of the cell receptor site. As has been previously reported, various urinary metabolites from patients undergoing AG therapy have been identified (Baker *et al.*, 1981; Coombes *et al.*, 1982; Foster *et al.*, 1984; Jarman *et al.*, 1983). One metabolite, hydroxylaminoglutethimide (N-OHAG) appeared to be absent on initial dosing but appeared between 2 and 8 days in the urine of patients on daily AG therapy (1 g). The appearance of the self-induced metabolite was accompanied by a decrease in the proportion of the dose excreted as N-acetylaminoglutethimide (N-AcAG) and with a decrease in the plasma half-life of AG. N-OHAG was not detected as a urinary metabolite in Wistar rats after 23 days of treatment (Jarman *et al.*, 1983). Moreover, in the rat, all metabolites were excreted as the N-acetylated products, which are inactive as inhibitors of the target enzymes. These findings cast doubt upon the validity of tests of AG and congeners against hormone-dependent mammary tumours in the rat as predictive models for their efficacy in patients.

Therefore the metabolism of AG in other animal species was explored in an attempt to find a model for its metabolism in humans. Using humans, the sample volume obtained is large, however this is not the case with small laboratory animals. Also, N-OHAG, the metabolite of interest, is relatively unstable, being oxidized to the nitroso derivative, as shown by thin-layer chromatography and h.p.l.c. Therefore it was desirable to develop a method of stabilizing N-OHAG to enable quantitative work-up and identification from small volumes of urine.

EXPERIMENTAL

A method of derivatization was developed with AG itself which involved N-trideuteroacetylation by reaction with d_6-acetic anhydride in dichloromethane. With N-OHAG d_6-N,O-diacetylhydroxylaminoglutethimide was formed quantitatively.

The animal species investigated was the mouse (C57BL/6 strain). Male C57BL/6 mice (8 weeks old) were dosed orally with ^{3}H-AG (20 mg/kg) (4.95 μCi/mg), and 24-hour urines were collected. The ^{3}H-AG was prepared by acid-catalysed exchange of protons *ortho* to the amine substituent using hydrochloric acid containing ^{3}H$_2$O. The label was stable to the isolation procedures reported here. For extended dose studies, animals were dosed with unlabelled AG in the drinking water (0.14 mg/ml) and ^{3}H-AG was given just prior to 24-hour urine collection. Urine was extracted with dichloromethane, the extract dried (anhydrous Na$_2$SO$_4$) and d_6-acetic anhydride (100 μl) added. After concentration on a rotary evaporator at 35 °C the residue was dissolved in dichloromethane (0.2 ml) and metabolites separated by reverse-phase thin-layer chromatography (r.p.-t.l.c., Whatman, KC$_{18}$F) in acetonitrile

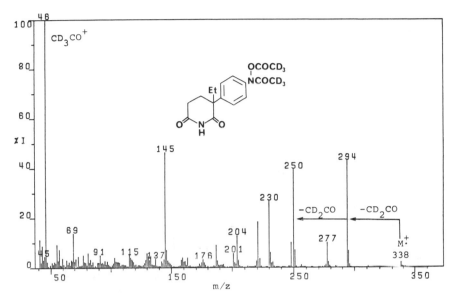

Fig. 1 – Mass spectrum of trideuteroacetylated N-OH AG obtained from C57BL/6 mouse urine extract, collected after repeated doses of AG, treated with d_6-acetic anhydride.

+ 5% w/v aq. NaCl (3:7). ³H-Activity was detected using a Berthold plate scanner, relevant areas on the chromatogram being removed and the compound eluted for mass spectrometry.

RESULTS AND DISCUSSION

d_6-N,O-Diacetylhydroxylaminoglutethimide was located as a discrete band on r.p.-t.l.c. (R_f 0.10) and the electron impact mass spectrum (Fig. 1) was compared with that obtained for synthetic N,O-diacetylhydroxylamino-glutethimide (Jarman et al., 1983). The spectrum contained ions at m/z 338 (M⁺·), 294 (M-CD₂CO) and 250 (M-2CD₂CO). The absence of an ion at m/z 291 confirmed that the metabolite was excreted as the free hydroxyl-amine since the molecule was completely deuteroacetylated.

As was earlier found in humans, N-OHAG could not be detected routinely in urine of C57BL/6 mice on initial dosing with AG. If present initially, there was always an increase in the output of this metabolite relative to AG after subsequent dosing. Therefore, it would appear that N-OHAG is an induced metabolite in this strain of mouse, as it is in humans.

More generally, the technique of reacting hydroxylamines, amines and their N-acetylated counterparts in metabolic extracts with d_6-acetic anhydride allows one to calculate the proportion of metabolite excreted in the acetylated form in one procedure. The area on the chromatogram corresponding to N-AcAG from urine extracts from C57BL/6 mice dosed with AG for various times was eluted and subjected to mass spectrometry. At all time points

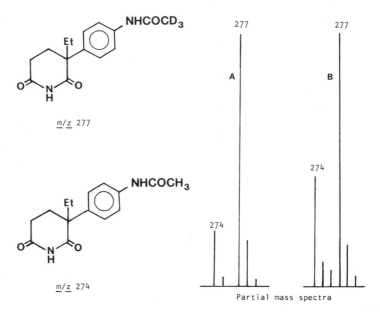

Fig. 2 – Molecular ions in the mass spectra of the mixture of d_3-N-AcAG (m/z 277) and N-AcAG(m/z 274) obtained by treating extracts from the urine of C57BL/6 mice collected after the second (A) and after the first (B) daily dose of AG, with d_6-acetic anhydride.

N-AcAG (M $\overset{+}{\cdot}$ 274) was present together with the corresponding deuterated product (M $\overset{+}{\cdot}$ 277), the latter being equivalent to free AG excreted. As was found for humans on continuous AG therapy, the amount of N-AcAG excreted decreased as the N-OHAG appeared (Fig. 2). The ratio of m/z 274 to m/z 277 (N-AcAG:AG excreted) was found to be 40:100 on initial dosing, but 20:100 on subsequent dosing.

It appears, therefore, that the C57BL/6 mouse might be a useful model for the human metabolism of AG and AG-related compounds. Work is presently underway with the nude mouse to determine whether the metabolism is species- or strain-dependent and whether a better tumour model for these compounds could be set up using the nude mouse. In summary, the derivatization procedure described serves to stabilize N-OHAG sufficiently for extraction and identification as well as allowing the fraction of metabolites excreted in the free and acetylated form to be determined. Finally, it is proposed that derivatization with d_6-acetic anhydride could be employed generally in the field of nitrogen oxidation so as to confer stability on these relatively unstable hydroxylamine metabolites.

REFERENCES

Baker, M. H., Foster, A. B., Harland, S. J. & Jarman, M. (1981), *Br. J. Pharmacol.*, **74**, 243P.

Coombes, R. C., Foster, A. B., Harland, S. J., Jarman, M. & Nice, E. C. (1982), *Br. J. Cancer*, **46**, 340.

Foster, A. B., Griggs, L. J., Howe, I., Jarman, M., Leung, C-S., Manson, D. & Rowlands, M. G. (1984), *Drug metab. dispos.*, **12**, 511.

Jarman, M., Foster, A. B., Goss, P., Griggs, L. J., Howe, I. & Coombes, R. C. (1983), *Biomed. mass spectrom.*, **10**, 620.

Chapter 15

DIETARY CONTROL OF BACTERIAL NITRO-GROUP METABOLISM

A. K. Mallett and **I. R. Rowland**, The British Industrial Biological Research Association, Carshalton, Surrey SM5 4DS, UK

1. Male Sprague-Dawley rats, MF1 mice and DSN hamsters, were fed a purified fibre-free diet alone or containing 50 g/kg i-carrageenan.
2. Carrageenan treatment significantly decreased the total bacterial population in the rat and mouse caecum, but was without effect on the hamster flora.
3. Carrageenan administration significantly decreased the *in vitro* nitro-reductase activity of caecal contents from all three species of rodent towards nitrobenzoic acid.
4. Dietary carrageenan abolished the ability of the rat caecal microflora to reduce a number of nitrosubstrates, yet was without effect on the metabolism of dinitrotoluene or metronidazole *in vitro*.
5. Diet related modification of caecal microbial enzyme activity may limit bacterial conversion of nitro compounds to toxic products.

INTRODUCTION

Physico-chemical conditions within the mammalian large intestine, in particular the absence of molecular oxygen and a redox potential below -100 mV (Schröder & Johansson, 1973), promote the bacterial reduction of alternative electron acceptors such as nitrocompounds. Bacterial reduction of the nitro group may initially give an anion radical following a single electron transfer, and proceed through the nitroso and hydroxylamino intermediates to yield ultimately the fully reduced amine (Mitchard, 1971; Mason, 1979). Such a theoretical scheme is complicated by the existence of two or more distinct bacterial nitroreductase enzymes, each specific for a given class of nitrocompound (Peterson *et al.*, 1979; Rosenkranz *et al.*, 1982).

Many aromatic and heterocyclic nitro compounds possess toxic, carcinogenic or mutagenic properties after reduction (McCalla & Voutsinos, 1974; Lindmark & Muller, 1976; Reddy *et al.*, 1976) and the resulting amines may be further metabolized by mammalian mixed function oxidases to electrophilic species (Hlavica, 1982). The microflora may therefore antagonize some aspects of mammalian detoxification processes and generate active species in the gut lumen.

The activity of several bacterial biotransformation enzymes may be modified by dietary components, particularly indigestible plant cell-wall material, which can predispose the host to toxic sequelae associated with the increased generation of bacterial products (Wise & Gilburt, 1982; de Bethizy *et al.*, 1983). Such changes may reflect microbial enzyme induction *per se* or be associated with a gross reorganization of the bacterial species composition of the flora with fortuitous alterations in metabolic capacity. Irrespective of the precise change involved, the net effect may greatly alter the biological lifetime of xenobiotics within the tissues of the host. For example, de Bethizy *et al.* (1983) have reported that rats fed a purified diet containing 5% or 10% citrus pectin exhibit a two- or threefold increase caecal bacterial nitroreductase activity which is associated with increased activation and covalent binding of 2,6-dinitrotoluene in vivo.

We have been interested in the effects of a number of plant cell-wall components related to pectin on the metabolic activity of the rodent caecal microflora. These materials, which may be considered as unconventional sources of dietary fibre (Furda, 1983), altered certain reductive or hydrolytic functions associated with the caecal bacteria, yet generally were without effect on the ability of the flora to reduce the model substrate *p*-nitrobenzoic acid under anaerobic conditions *in vitro* (Mallett *et al.*, 1984). One notable exception, however, was carrageenan which dramatically decreased the nitroreductase activity of the caecal microflora (Mallett *et al.*, 1984). We report in this paper the effects of carrageenan on the ability of the gut microflora to reduce a number of model substrates *in vitro*.

EXPERIMENTAL

Male conventional microflora Sprague-Dawley rats, MF1 mice and DSN hamsters (3–4 weeks old) were fed either a purified, fibre-free diet (Wise & Gilburt, 1982) alone, or this diet incorporating 5% (w/w) food grade *i*-carrageenan (Sigma Chemical Co., C1138) for 4–6 weeks. The animals were lightly anaesthetized with ether, killed by cervical dislocation and the contents of caecum weighed and suspended in 25 ml deoxygenated 0.1 M-potassium phosphate buffer. Nitroreductase activity was determined *in vitro* under anaerobic conditions using *p*-nitrobenzoic acid (PNBA) as substrate (Wise & Gilburt, 1982). Microbial reduction of *p*-nitrophenol (PNP) (Dawson *et al.*, 1969) and nitrofurantoin (NF) (Wang *et al.*, 1974) was determined by published procedures using a substrate concentration of 1 mM. Reduction of metronidazole (MNZ) was assayed by the method of Wise & Gilburt (1982). The reductive metabolism of 2,4-dinitrotoluene (2,4-DNT) was determined by g.l.c. (Carlo Erba Fractovap 4160 with nitrogen-phosphorous alkali bead detector, 25 M × 0.3 mm glass capillary column coated with the non-polar gum phase SE25, film thickness 0.3 μm) using 2,6-DNT as internal standard. Total numbers of bacteria present in the caecal contents were determined from a direct microscopic count (Holdeman & Moore, 1972). The data were analysed by one-way analysis of variance using the Minitab Statistical Package (Minitab Inc., Pennsylvania, USA).

RESULTS

Carrageenan treatment was associated with highly significant decreases in the total bacterial population of the rat and mouse caecum, yet was without significant effect on the hamster microflora (Table 1). Total caecal microbial nitroreductase activity (with *p*-nitrobenzoic acid as substrate) was signific-

Table 1
Influence of dietary i-carrageenan on total caecal bacterial numbers and caecal bacterial reduction of *p*-nitrobenzoic acid

		Nitroreductase activity	
	Bacterial count ($\log_{10}$/caecum)	μmol/hr/caecum	μmol/hr/10^{11} bacteria
Rat (6)			
Fibre-free diet	11.06 ± 0.23	0.51 ± 0.07	0.49 ± 0.28
5% Carrageenan	$10.41 \pm 0.14^{***}$	$0.03 \pm 0.02^{***}$	$0.12 \pm 0.06^{**}$
Mouse (8)			
Fibre-free diet	10.58 ± 0.15	0.08 ± 0.20	0.69 ± 0.30
5% Carrageenan	$10.17 \pm 0.08^{***}$	$0.02 \pm 0.01^{***}$	$0.36 \pm 0.02^{*}$
Hamster (8)			
Fibre-free diet	11.34 ± 0.32	1.50 ± 0.20	0.88 ± 0.82
5% Carrageenan	11.38 ± 0.09	$0.31 \pm 0.10^{***}$	$0.13 \pm 0.08^{*}$

The data are given as means ± SD, *n* in parenthesis.
Statistical analysis relative to the fibre-free diet group: $^{*}P < 0.05$; $^{**}P < 0.01$; $^{***}P < 0.001$.

antly decreased by carrageenan administration in all three rodent species (Table 1), with the rat showing the greatest loss in activity (decreased by approximately 95%). Nitroreductase activities per 10^{11} bacteria (approximating the total bacterial content of 1 g caecal contents) were also significantly decreased, indicating a direct effect of carrageenan on microbial metabolism in addition to that associated with the decrease in caecal bacterial numbers.

Caecal contents incubations from rats fed the purified control diet reduced a number of aromatic and heterocyclic substrates (Table 2), with the greatest activity occurring when 2,4-DNT, MNZ and NF were substrate. 2,4-DNT was reduced to yield two intermediates, namely 4-amino-2-nitrotoluene (4A2NT) and 2-amino-4-nitrotoluene (2A4NT) which were then further reduced (presumably to 2,4-diaminotoluene, although this was not identified in the incubation mixture). Dietary carrageenan abolished the ability of the rat caecal microflora to reduce PNP, 2A4NT, 4A2NT and NF, but was without effect on 2,4-DNT metabolism. MNZ reduction showed pronounced interindividual variation after carrageenan administration, although the bacteria retained appreciable activity towards this substrate.

DISCUSSION

Dietary administration of carrageenan, a non-digestible plant cell-wall component consisting of galactose and anhydrogalactose copolymers, significantly decreased the total activity of rodent caecal contents towards PNBA *in vitro*, and abolished the activity of the rat caecal microflora towards PNP, 2A4NT, 4A2NT and NF. Carrageenan treatment has also been shown to decrease the activity of certain other reductive and hydrolytic functions

Table 2

Influence of dietary ɪ-carrageenan on rat caecal microbial nitroreductase activities

Substrate	Nitroreductase activity (μmol/hr/10^{11} bacteria)	
	Fibre-free diet	5% Carrageenan
p-Nitrophenol (PNP)	4.17 ± 1.81	—
2,4-Dinitrotoluene (DNT)	9.08 ± 3.50	9.76 ± 2.59
2-Amino-4-nitrotoluene (2A4NT)	0.33 ± 0.20	—
4-Amino-2-nitrotoluene (4A2NT)	0.42 ± 0.27	—
Metronidazole (MNZ)	20.83 ± 6.94	63.53 ± 89.29
Nitrofurantoin (NF)	33.33 ± 10.21	—

The data are given as mean $\pm$ SD, $n = 6$.
Some activities were not detected (—) after extended incubation period (up to 150 min).

associated with the rat hindgut microflora, yet nitroreductase activity is particularly susceptible to this effect (Mallett *et al.*, 1984). However, reduction of certain substrates (notably 2,4-DNT and MNZ) was unaltered by this dietary influence. Such observations suggest that carrageenan may modulate bacterial nitroreductase functions in the gut lumen. The precise nature of the change in the gastrointestinal flora is unresolved however, and may reflect the direct repression of bacterial nitroreductase enzymes *per se*, or represent a fortuitous event associated with a gross reorganization of the species composition of the hindgut flora. Preliminary observations suggest that all of the major taxonomic groups comprising the rat gut flora are decreased by dietary carrageenan (Mallett & Rowland, unpublished observations). Furthermore, it is unclear why bacterial reduction of nitro substrates should be decreased over and above other reductive processes associated with the gut microflora (Mallett *et al.*, 1984). Nevertheless, the evidence suggests that carrageenan has the potential to limit bacterial conversion of nitro compounds to toxic products, and provides an example of dietary control of foreign compound metabolism.

ACKNOWLEDGEMENTS

We thank Carol Bearne for excellent technical assistance and E. Bailey for the g.l.c. analyses. This work was funded by the UK Ministry of Agriculture Fisheries and Food, to whom we express thanks.

REFERENCES

Dawson, R. M. C., Elliott, D. C., Elliott, W. H. & Jones, K. M. (1969), *Data for Biochemical Research*, 2nd Ed. Oxford University Press, Oxford, p. 451.

de Bethizy, J. D., Sherril, J. M., Rickert, D. E. & Hamm, T. E. (1983), *Toxicol. Appl. Pharmacol.*, **69**, 369.

Furda, I. (1983), *Unconventional Sources of Dietary Fiber*, ACS Symposium Series No. 214, American Chemical Society, Washington, DC.

Hlavica, P. (1982), *CRC Crit. Rev. Biochem.*, **12**, 39.

Holdeman, L. V. & Moore, W. E. C. (1972), *Anaerobe Laboratory Manual*, 3rd ed., VPI Anaerobe Laboratory, Blacksburg, Virginia.

Lindmark, D. G. & Muller, M. (1976), *Antimicrob. Agents. Chemother.*, **10**, 476.

Mallett, A. K., Wise, A. & Rowland, I. R. (1984), *Fd. Chem. Toxic.*, **22**, 415.

Mason, R. P. (1979), in *Reviews in Biochemical Toxicology*, Vol. 1 (ed. Hodgson, E., Bend, J. R. & Philpot, R. M. Elsevier/North Holland Inc. Amsterdam.

McCalla, D. R. & Voutsinos, D. (1974), *Mut. Res.*, **26**, 3.

Mitchard, M. (1971), *Xenobiotica*, **1**, 469.

Peterson, F. J., Mason, R. P., Horsepin, J. & Holtzman, J. L. (1979), *J. Biol. Chem.*, **254**, 4009.

Reddy, B. G., Pohl, R. L. & Krishna, G. (1976), *Biochem. Pharmac.*, **25**, 1119.

Rosenkranz, E. J., McCoy, E. C., Mermelstein, R. & Rosenkranz, H. S. (1982), *Carcinogenesis*, **3**, 121.

Schröder, H. & Johansson, A. K. (1973), *Xenobiotica*, **3**, 233.

Wang, C. Y., Behrens, B. C., Ichikawa, M. & Bryan, G. T. (1974), *Biochem. Pharmac.*, **23**, 3395.

Wise, A. & Gilburt, D. J. (1982), *Drug-Nutr. Interact.*, **1**, 229.

Part 4

OXIDATION OF AMIDES AND CARBAMATES

Chapter 16

ENZYMATIC N-HYDROXYLATION OF AROMATIC AMIDES†

P. D. Lotlikar, Fels Research Institute, Temple University Medical School, Philadelphia, PA 19140, USA

1. Studies on ring and N-hydroxylation of 2-acetylaminofluorene (AAF) and several other aromatic amides including acetaminophen, 4-acetylamino-biphenyl, 2-acetylaminonaphthalene, 2-acetylaminophenanthrene, 4-acetylaminostilbene, N-acetylbenzidine, N,N-diacetylbenzidine and phenacetin by liver microsomes from various species are reviewed.
2. Effects of various pretreatments including 3-methylcholanthrene, phenobarbital and AAF on N- and ring-hydroxylation of AAF by liver microsomes from several species are also discussed.
3. Several lines of evidence for involvement of cytochrome P-450 in AAF N- and ring-hydroxylation are discussed.
4. Catalytic activities of several forms of purified cytochrome P-450 from various mammalian species are examined for AAF N- and ring-hydroxylations in reconstituted systems.
5. Inhibition studies of AAF oxidations in the presence of various polyclonal and monoclonal antibodies prepared against several purified forms of cytochrome P-450 are also reviewed.

†Supported by grants CA-10604 and CA-12227 from the National Cancer Institute, US Department of Health and Human Services, and grant SIG-6 from the American Cancer Society.

INTRODUCTION

Epidemiological evidence has indicated that several aromatic amines including 4-aminobiphenyl, 2-naphthylamine and benzidine have been responsible for the induction of bladder cancer among industrial workers (Scott, 1962).

·It is known that aromatic amides need to be metabolized before they exhibit toxic and carcinogenic properties in any given species (Miller, 1970). Several reviews on the metabolism, toxicity and carcinogenicity of aromatic amines and amides have appeared during the last two decades (Clayson & Garner, 1976; Irving, 1979; Kiese, 1966; Kriek, 1974; Lotlikar *et al.*, 1978; Lotlikar & Hong, 1981; Miller & Miller, 1969; Weisburger & Weisburger, 1973). Among many aromatic amides investigated, several laboratories have used 2-acetylaminofluorene (AAF)[1] as a model compound in studying the mechanism of carcinogenesis by aromatic amides (Fig. 1). AAF is readily ring-hydroxylated at several positions including 1-, 3-, 5-, 7- and 9. The ring-hydroxylation reaction is mediated via a liver microsomal enzyme system requiring NADPH and O_2. These monophenols are usually excreted in the bile and urine as glucuronides and sulphates (Clayson & Garner, 1976, and references cited therein). Since the discovery of N-HO-AAF glucuronide in the urine of rats fed AAF (Cramer *et al.*, 1960b), there is now unequivocal evidence that *N*-hydroxylation is an activation step, whereas ring-hydroxylation is an inactivation step in the carcinogenesis by AAF and several other aromatic amines and amides (Miller & Miller, 1969).

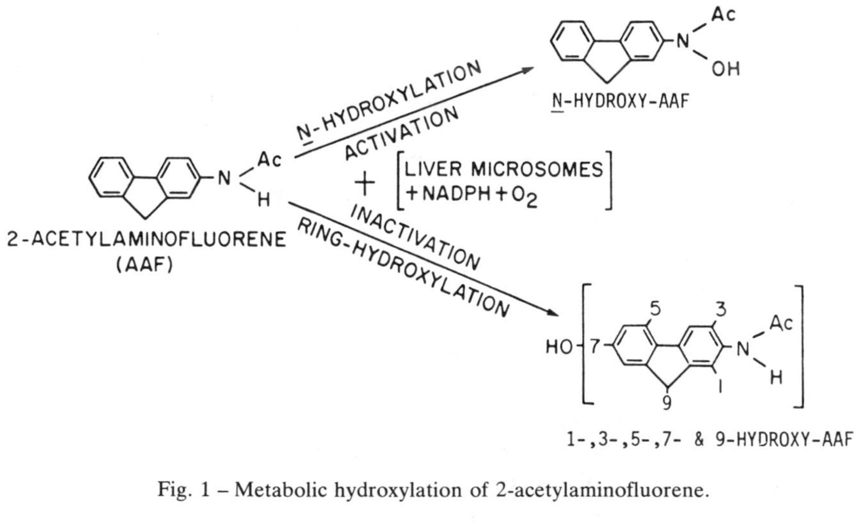

Fig. 1 – Metabolic hydroxylation of 2-acetylaminofluorene.

---

[1]Abbreviations: C-OH, ring-hydroxy; AAF, 2-acetylaminofluorene; N-HO-AAF, N-hydroxy-AAF; MC, 3-methylcholanthrene; PB, phenobarbital; TCDD, 2,3,7,8-tetra-chlorodibenzo-*p*-dioxin; BNF, *β*-naphthoflavone; PCB, polychlorinated biphenyls.

ENZYMATIC *N*-HYDROXYLATION OF AROMATIC AMIDES

Irving (1962) was the first to demonstrate that *N*-hydroxylation of AAF is mediated via a hepatic microsomal mixed function oxidase system. Pioneering studies on *N*- and ring-hydroxylations of AAF and various other aromatic amides by liver microsomes from several species are summarized in Tables 1 and 2.

Table 1

N- and ring-hydroxylation of AAF by liver microsomes

Species	Oxidation		References
	N-	C-	
Guinea pig, hamster, mouse, rabbit and rat		+	Booth & Boyland, 1957 Seal & Gutmann, 1959 Cramer *et al.*, 1960a Benkert *et al.*, 1975
Rabbit	+		Irving, 1962
Cat, chicken, dog, hamster, rabbit and rat	+	+	Irving, 1964 Lotlikar *et al.*, 1967 Matsushima *et al.*, 1972 Thorgeirsson *et al.*, 1973
Guinea pig	+		Gutmann & Bell, 1977 Razzouk *et al.*, 1980
Monkey	+ +	+	Thorgeirsson *et al.*, 1978 Razzouk *et al.*, 1980
Human	+	+	Enomoto & Sato, 1967 Dybig *et al.*, 1979 McManus *et al.*, 1983

ROLE OF CYTOCHROME P-450 IN AAF *N*-HYDROXYLATION

Several lines of evidence have demonstrated that cytochrome P-450 is involved in oxidation of AAF by hepatic microsomes (Table 3).

Inducers of cytochrome P-450 and hydroxylations

Earlier studies from the Millers' laboratory indicated that MC-pretreatment of various animals demonstrated differential effect of AAF *N*- and ring-hydroxylations by their liver microsomes (Table 4). Thus, hamster and rabbit

Table 2

N- and ring-hydroxylation of carcinogenic aromatic amides by liver microsomes

Compound	Species	Oxidation		References
		N-	C-	
Acetaminophen	Mouse	+		Thorgeirsson *et al.*, 1975
4-Acetylaminobiphenyl	Rabbit	+		Booth & Boyland, 1964
	Dog	+		Brill & Radomski, 1971
7-Fluoro-AAF	Rabbit	+		Irving, 1964
	Hamster	+		Lotlikar *et al.*, 1967
2-Acetylaminonaphthalene	Rat		+	Booth & Boyland, 1957
	Dog and rabbit	+		Brill & Radomski, 1971
2-Acetylaminophenanthrene	Hamster	+		Lotlikar *et al.*, 1967
4-Acetylaminostilbene	Rat	+	+	Baldwin & Smith, 1965
	Hamster	+	+	Gammans *et al.*, 1977
N-Acetylbenzidine	Rabbit	+		Booth & Byland, 1964
N,N-Diacetylbenzidine	Hamster, rat and mouse	+	+	Morton *et al.*, 1979
Phenacetin	Hamster	+		Hinson & Mitchell, 1976

Table 3

Evidence for involvement of cytochrome P-450 in AAF N- and ring hydroxylation by liver microsomes

Evidence	Species	References
CO Inhibition	Hamster, mouse and rat	Thorgeirsson et al., 1973 Lotlikar & Zaleski, 1974 Gutmann & Bell, 1977 Malejka-Giganti et al., 1978
Substrate binding	Guinea pig, hamster and rat	Gutmann & Bell, 1977 Malejka-Giganti et al., 1978 Lotlikar et al., 1978
NADPH-cytochrome c reductase antibody inhibition	Hamster	Thorgeirsson et al., 1973
Cytochrome P-450 antibody inhibition	Rat	Hara et al., 1981 Thorgeirsson et al., 1983 Pandey et al., 1983

Table 4

Ring- and N-hydroxylation of AAF by liver microsomes from control and 3-Methylcholanthrene (MC) pretreated animals[a]

Species	MC	nmoles formed/g liver/20 min			
		N-HO-AAF	3-OH-AAF	5-HO-AAF	7-HO-AAF
Rat	−	n.d.[b]	24 ± 5	20 ± 4	57 ± 12
	+	37 ± 8	480 ± 100	395 ± 90	580 ± 190
Hamster	−	69 ± 18	47 ± 7	140 ± 19	640 ± 60
	+	950 ± 145	90 ± 21	186 ± 27	590 ± 130
Mouse	−	37 ± 6	31 ± 17	37 ± 5	120 ± 11
	+	236 ± 53	175 ± 33	220 ± 35	550 ± 210
Rabbit	−	16 ± 7	n.d.	n.d.	160 ± 85
	+	77 ± 15	n.d.	n.d.	200 ± 45
Guinea pig	−	n.d.	n.d.	43 ± 15	1150 ± 320
	+	n.d.	n.d.	73 ± 20	1800 ± 300

[a]Data from Lotlikar et al. (1967). Weaning male animals injected with MC (2–10 mg/ 100 g wt) 24 hr before assay for hydroxylase activity.
[b]n.d. = not detected.

Table 5

Effects of various pretreatments on AAF *N*- and ring hydroxylation by
liver microsomes

Pretreatment	Species	References
MC	Guinea pig, hamster, mouse, rat and rabbit	Cramer *et al.*, 1960a Lotlikar *et al.*, 1967 Matsushima *et al.*, 1972 Atlas *et al.*, 1977 Gutmann & Bell, 1977 Malejka-Giganti *et al.*, 1978 Razzouk *et al.*, 1980 Hara *et al.*, 1981 Astrom *et al.*, 1983
Phenobarbital (PB)	Hamster, mouse and rat	Matsushima *et al.*, 1972 Thorgeirsson *et al.*, 1975 Hara *et al.*, 1981 Lotlikar & Hong, 1981 Razzouk *et al.*, 1982a Astrom *et al.*, 1983
β-Naphthoflavone (BNF)	Mouse	Thorgeirsson *et al.*, 1975
Polychlorinated biphenyls (PCB)	Rat	Hara *et al.*, 1981
TCDD	Mouse	Thorgeirsson *et al.*, 1975

Table 6

Pretreatment of AAF on *N*- and ring-hydroxylation of
AAF by liver microsomes

Species	References
Rat	Malejka-Giganti *et al.*, 1978 Malejka-Giganti & Ritter, 1980 Sato *et al.*, 1981 Razzouk & Roberfroid, 1982 Razzouk *et al.*, 1982a Astrom *et al.*, 1983
Hamster and mouse	Razzouk *et al.*, 1982b Razzouk & roberfroid, 1982

showed preferential increases in N-hydroxylation, whereas rat and mouse gave increases in both N- and ring-oxidations.

In addition to MC, various other inducers have been employed for the study of microsomal oxidation of AAF (Table 5).

However, induction studies with AAF treatment are of great interest (Table 6). All laboratories have shown increased AAF N-hydroxylation with liver microsomes isolated from animals pretreated with AAF. However, there is some controversy as to whether or not this increase in AAF N-hydroxylation is accompanied by an increase in cytochrome P-450 and an induction of a new cytochrome P-450 (Astrom et al., 1983; Malejka-Giganti et al., 1978; Malejka-Giganti & Ritter, 1980). This discrepancy may be resolved by purification and reconstitution of cytochrome P-450 systems and by immunological studies.

Solubilization, purification and reconstitution of the cytochrome P-450 enzyme system

Pioneering studies from Coon's laboratory in solubilization, resolution and reconstitution of cytochrome P-450 enzyme system from hepatic microsomes indicated that the enzyme system is composed of three factors: cytochrome P-450, NADPH-cytochrome P-450 reductase and a lipid factor identified as phosphatidylcholine. All three factors are required for the oxidation of physiological and xenobiotic compounds (see Lu & West, 1980).

Earlier studies from our laboratory utilized bacterial protease treatment followed by Triton X-100 solubilization and ammonium sulphate precipitation for the isolation of cytochrome P-450 fractions from hamsters (Lotlikar et al., 1974), rats (Lotlikar & Zaleski, 1975) and mice (Lotlikar & Wang, 1982). On comparison of differences in AAF N- and ring-hydroxylations between control and MC-treated hamster liver microsomes, our reconstitution experiments indicated that even though both reductase and P-450 fractions are required for AAF oxidations, the increase in N-hydroxylation resided in the cytochrome P-450 fraction (Table 7). Similar data were obtained in reconstitution experiments with rat (Lotlikar & Zaleski, 1975) and mouse (Lotlikar & Wang, 1982) hepatic microsomal systems. Lipid requirement could not be demonstrated in all of these studies because of incomplete removal of Triton X-100 from cytochrome P-450 fractions. Lipid requirement, however, has been demonstrated in reconstitution experiments where a cholate solubilized cytochrome P-450 fraction as been used (Lotlikar & Hong, 1981).

Several laboratories have now purified various hepatic microsomal cytochromes P-450 by affinity chromatography methods. These highly purified forms of cytochrome P-450 isolated from various animals pretreated with different inducers show differential rates of N- and ring-hydroxylations of AAF in reconstituted systems (Table 8).

Table 7

Metabolism of AAF by reconstituted hamster liver microsomal system[a]

Fractions added	nmoles of hydroxy-AAF formed/20 min	
	N-	Ring-
P-450 (C) or P-450 (MC)	<0.8	1.2
Reductase (C) or reductase (MC)	<0.8	0.8
P-450 (C) + reductase (C)	4.8	7.3
P-450 (C) + reductase (C) + lipid (C)	6.1	5.4
P-450 (C) + reductase (MC)	4.8	4.7
P-450 (MC) + reductase (MC)	30.6	12.3
P-450 (MC) + reductase (C)	33.4	13.6
P-450 (MC) + reductase (MC) + lipid (MC)	36.9	15.2

[a]Data from Lotlikar *et al.*, (1974). Where indicated, fractions containing 3.2 nmoles of cytochrome P-450, reductase equivalent to 610 nmole/min of NADPH-cytochrome *c* reduced, and 2 mg of lipid were added.
C = control   MC = methylcholanthrene pretreated.

Table 8

AAF *N*- and ring-hydroxylation by purified forms of P-450 from liver microsomes from various species

Species	P-450	Oxidation		References
		N-	C-	
Rabbit	Form 2 (PB-induced)	0	0	Johnson *et al.*, 1980
	Form 3 (neonate, constitutive)	0	+	
	Form 4 (TCDD-induced)	+	+	
	Form 6 (TCDD-induced in neonate)	0	+	
Rat	Form D (MC-induced)	+	+	Pandey *et al.*, 1983 Lotlikar *et al.*, 1984
	MC-P-450	+	+	Thorgeirsson *et al.*, 1983
Hamster	Form C (MC-induced)	+	+	Pandey *et al.*, this volume

Immunological studies

Hydroxylation of AAF has been examined with either rat liver microsomes or highly purified forms of P-450 in the presence of polyclonal or monoclonal antibodies raised against purified forms of P-450 (Hara *et al.*, 1981; Thorgeirsson *et al.*, 1983; Pandey *et al.*, 1983). Inhibition of oxidation has indicated involvement of a particular form of P-450 against which an antibody was raised. Thus, complete inhibition of AAF N- and ring-oxidations mediated by purified MC-450 in the presence of antibody raised against BNF-P-450 (same as MC-P-450), but no inhibition in the presence of antibody raised against PB-450 implies that AAF N- and ring-oxidations are catalysed by the same isozyme (Fig. 2).

Similar results are obtained with monoclonal antibodies specific to MC-induced rat liver P-450 (Fig. 3). In these studies, monoclonal antibodies inhibited both AAF N- and ring-hydroxylations, except 9-hydroxylation, mediated by either microsomes or highly purified rat liver MC-450. These inhibition patterns indicate that the formation of 7-OH-, 5-OH-, 1-OH-, and N-OH-AAF was catalysed by the same unique isoenzyme or at least isoenzymes of common antigenic determinants.

Such immunological experiments with monoclonal antibodies with microsomes and various purified P-450 forms would enable us to determine if there is any enzyme specificity in N- and ring-hydroxylations of various aromatic amides.

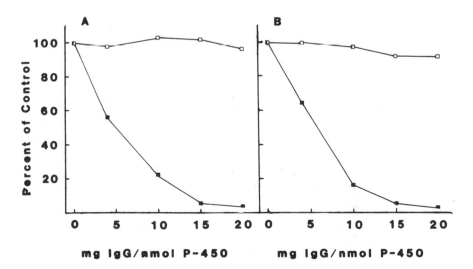

Fig. 2 – Effect of varying concentrations of various IgG to rat P-450 on AAF N- and ring-hydroxylation by MC-rat-P-450 in a reconstituted system. (A) ring-hydroxy-AAF; (B) N-hydroxy-AAF. □—□ goat IgG against PB-rat P-450 B$_2$; ■—■ goat IgG against BNF rat P-450.
    Data of Pandey *et al.* (1983).

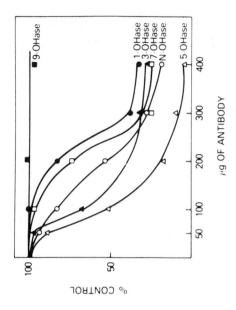

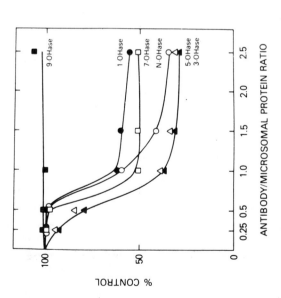

Fig. 3 – Monoclonal antibody inhibition of AAF hydroxylation catalysed either by
MC-liver microsomes (left panel) or by purified rat liver MC-450 (right panel).
Data of Thorgeirsson _et al._ (1983).

REFERENCES

Astrom, A., Meijer, J. & DePierre, J. W. (1983), *Cancer Res.*, **43**, 342.

Atlas, S. A., Boobis, A. R., Felton, J. S., Thorgeirsson, S. S. & Nebert, D. W. (1977), *J. Biol. Chem.*, **252**, 4712.

Baldwin, R. W. & Smith, W. R. (1965), *Brit. J. Cancer*, **19**, 433.

Benkert, K., Fries, W., Kiese, M. & Lenk, W. (1975), *Biochem. Pharmacol.*, **24**, 1375.

Booth, J. & Boyland, E. (1957), *Biochem. J.*, **66**, 73.

Booth, J. & Boyland, E. (1964), *Biochem. J.*, **91**, 362.

Brill, E. & Radomski, J. L. (1971), *Xenobiotica*, **1**, 347.

Clayson, D. B. & Garner, R. C. (1976), in *Chemical Carcinogens*, American Chemical Society Monograph No. 173 (ed. Searle, C. E.), American Chemical Society, Washington DC, p. 366.

Cramer, J. W., Miller, J. A. & Miller, E. C. (1960a), *J. Biol. Chem.*, **235**, 250.

Cramer, J. W., Miller, J. A. & Miller, E. C. (1960b), *J. Biol. Chem.*, **235**, 885.

Dybig, E., von Bahr, C., Aune, T., Glaumann, H., Levitt, D. S. & Thorgeirsson, S. S. (1979), *Cancer Res.*, **39**, 4206.

Enomoto, M. & Sato, K. (1967), *Life Sci.*, **6**, 881.

Gammans, R. E., Sehon, R. D., Anders, M. W. & Hanna, P. E. (1977), *Drug Metab. Dispos.*, **5**, 310.

Gutmann, H. R. & Bell, P. (1977), *Biochim. Biophys. Acta*, **498**, 229.

Hara, E., Kawajiri, K., Gotoh, O. & Tagashira, Y. (1981), *Cancer Res.*, **41**, 253.

Hinson, J. W. & Mitchell, J. R. (1976), *Drug Metab. Dispos.*, **4**, 430.

Irving, C. C. (1962), *Biochim. Biophys. Acta*, **65**, 564.

Irving, C. C. (1964), *J. Biol. Chem.*, **239**, 1589.

Irving, C. C. (1979), in *Carcinogens, Identification and Mechanism of Action* (eds. Griffin, A. C. & Shaw, C. R.), Raven Press, New York, p. 211.

Johnson, E. F., Levitt, D. S., Muller-Eberhard, U. & Thorgeirsson, S. S. (1980), *Cancer Res.*, **40**, 4456.

Kiese, M. (1966), *Pharmacol. Rev.*, **18**, 1091.

Kriek, E. (1974), *Biochim. Biophys. Acta*, **355**, 177.

Lotlikar, P. D., Enomoto, M., Miller, J. A. & Miller, E. C. (1967), *Proc. Soc. Exp. Biol. Med.*, **125**, 341.

Lotlikar, P. D. & Hong, Y. S. (1981), in *Carcinogenic and Mutagenic N-Substituted Aryl Compounds*, J. Natl. Cancer Inst. Monograph, No. 58, p. 101.

Lotlikar, P. D., Hong, Y. S. & Baldy, W. J. Jr. (1978), in *Biological Oxidation of Nitrogen* (ed. Gorrod, J. W.), Amsterdam: Elsevier, p. 185.

Lotlikar, P. D., Luha, L. & Zaleski, K. (1974), *Biochem. Biophys. Res. Commun.*, **59**, 1349.

Lotlikar, P. D., Pandey, R. N., Clearfield, M. S. & Paik, S. M. (1984), *Toxicology Lett.*, **21**, 111.

Lotlikar, P. D. & Wang, T. F. (1982), *Toxicology Lett.*, **11**, 173.

Lotlikar, P. D. & Zaleski, K. (1974), *Biochem. J.*, **144**, 427.

Lotlikar, P. D. & Zaleski, K. (1975), *Biochem. J.*, **150**, 561.

Lu, A. Y. H. & West, S. B. (1980), *Pharmacol. Rev.*, **31**, 277.

Malejka-Giganti, D., McIver, R. C., Glasebrook, A. L. & Gutmann, H. R. (1978), *Biochem. Pharmacol.*, **27**, 61.

Malejka-Giganti, D. & Ritter, C. L. (1980), *Res. Commun. Chem. Pathol. Pharmacol.*, **28**, 53.

Matsushima, T., Grantham, P. H., Weisburger, E. K. & Weisburger, J. H. (1972), *Biochem. Pharmacol.*, **21**, 2043.

McManus, M. E., Minchin, R. F., Sanderson, N. D., Wirth, P. J. & Thorgeirsson, S. S. (1983), *Carcinogenesis*, **4**, 693.

Miller, J. A. (1970), *Cancer Res.*, **30**, 559.

Miller, J. A. & Miller, E. C. (1969), *Prog. Exp. Tumor Res.*, **11**, 273.

Morton, K. C., King, C. M. & Baetcke, K. P. (1979), *Cancer Res.*, **39**, 3107.

Pandey, R. N., Clearfield, M. S., Paik, S. M. & Lotlikar, P. D. (1983), *Cancer Lett.*, **19**, 263.

Pandey, R. N., Clearfield, M. S., Paik, S. M. & Lotlikar, P. D., this volume.

Razzouk, C., Batardy-Gregoire, M. & Roberfroid, M. (1982a), *Mol. Pharmacol.*, **21**, 449.

Razzouk, C., Battardy-Gregoire, M. & Roberfroid, M. (1982b), *Carcinogenesis*, **3**, 1325.

Razzouk, C., Mercier, M. & Roberfroid, M. (1980), *Xenobiotica*, **10**, 565.

Razzouk, C. & Roberfroid, M. B. (1982), *Chem-biol. Interact.*, **41**, 251.

Sato, M., Yoshida, T., Suzuki, Y., Degawa, M. & Hashimoto, Y. (1981), *Carcinogenesis*, **2**, 571.

Scott, T. S. (1962), *Carcinogenic and Chronic Toxic Hazards of Aromatic Amines*, Elsevier, Amsterdam.

Seal, U. S. & Gutmann, H. R. (1959), *J. Biol. Chem.*, **234**, 648.

Thorgeirsson, S. S., Felton, J. S. & Nebert, D. W. (1975), *Mol. Pharmacol.*, **11**, 159.

Thorgeirsson, S. S., Jollow, D. J., Sasame, H., Green, I. & Mitchell, J. R. (1973), *Mol. Pharmacol.*, **9**, 398.

Thorgeirsson, S. S., Sakai, S. & Adamson, R. H. (1978), *J. Natl. Cancer Inst.*, **60**, 365.

Thorgeirsson, S. S., Sanderson, N., Park, S. S. & Gelboin, H. V. (1983), *Carcinogenesis*, **4**, 639.

Weisburger, J. H. & Weisburger, E. K. (1973), *Pharmacol. Rev.*, **25**, 1.

Chapter 17

N-HYDROXYLATION OF THE CARCINOGEN 2-ACETYLAMINOFLUORENE IN A RECONSTITUTED SYSTEM WITH PURIFIED CYTOCHROME P-450 FROM LIVER MICROSOMES OF 3-METHYLCHOLANTHRENE PRETREATED HAMSTERS†

R. N. Pandey,‡ **M. S. Clearfield, S. M. Paik** and **P. D. Lotlikar,** Fels Research Institute, Temple University School of Medicine, Philadelphia, PA 19140, USA

1. Various liver microsomal cytochromes P-450 and NADPH-cytochrome P-450 reductase have been purified from 3-methylcholanthrene pretreated hamsters.
2. Reconstitution studies for the carcinogen 2-acetylaminofluorene *N*- and ring-hydroxylations indicate that cytochrome P-450 C is the most active followed by P-450 D, P-450 B and P-450 A.

---

†Supported by grants CA-10604 and CA-12227 from the National Cancer Institute, US Department of Health and Human Services and grant SIG-6 from the American Cancer Society.
‡Present address: Department of Pathology, Northwestern University Medical School, Chicago, IL 60611, USA

3. On the basis of overall data, it appears that cytochromes P-450 with molecular weights of 47,000 and 54,000 daltons are primarily responsible for the oxidative metabolism of the carcinogen.

INTRODUCTION

2-Acetylaminofluorene (AAF), a potent carcinogenic aromatic amide, undergoes activation and inactivation via *N*- and ring-hydroxylation steps respectively (Miller, 1970, 1978). Both ring- and *N*-hydroxylations occur in the hepatic endoplasmic reticulum (Cramer *et al.*, 1960; Irving, 1964) and are mediated via cytochrome P-450 enzyme system (Gutmann & Bell, 1977; Lotlikar & Zaleski, 1974; Thorgeirsson *et al.*, 1973). Pretreatment of hamsters with 3-methylcholanthrene (MC) produces a several-fold increase in hepatic microsomal *N*-hydroxylation activity without appreciably increasing ring-hydroxylation activity (Lotlikar *et al.*, 1967). Previous reconstitution studies have shown that the increase in *N*-hydroxylation activity resided in the cytochrome P-450 fraction (Lotlikar *et al.*, 1974; Lotlikar, 1981). We have now purified various cytochromes P-450 from hepatic microsomes from MC-pretreated hamsters and examined their catalytic activity for AAF *N*- and ring-hydroxylation in a reconstituted system.

EXPERIMENTAL

Young male Syrian golden hamsters (60–80 g body wt) were injected i.p. with MC (10 mg/100 g body wt) suspended in corn oil 24 hr before sacrifice. Liver microsomes were prepared as described previously (Lotlikar *et al.*, 1974). Cytochrome P-450 content, NADPH-cytochrome P-450 reductase assayed as NADPH-cytochrome *c* reductase, cytochrome b_5, NADH-ferricyanide reductase and protein content were determined by published methods as described previously (Lotlikar *et al.*, 1984). Sodium dodecylsulphate (SDS)-polyacrylamide slab gel electrophoresis was carried out according to Laemmli (1970) with a 7.5% separating gel and 3% stacking gel in a discontinuous buffer system.

RESULTS AND DISCUSSION

Data on purification of various cytochromes P-450 from MC-pretreated hamsters are summarized in Table 1. The recovery of total P-450 fraction obtained after affinity chromatography was about 50%. Among the four isolated P-450 fractions, the specific activity of P-450 C was the highest.

The NADPH-P-450 reductase obtained from pooled fractions from *n*-octylaminosepharose 4B column after P-450 fraction was purified by biospecific affinity chromatography on 2′,5′-ADP agarose (Guengerich, 1977; Yasukochi & Masters, 1976). This reductase preparation with NADPH-

Table 1

Purification of various cytochromes P-450 from liver microsomes of
MC-pretreated hamsters

Fractions	Specific content of cytochrome P-450 nmol/mg protein (% recovery)
Whole microsomes	1.94 (100)
Cholate solubilization	2.1 (90)
105,000 g supernatant	2.2 (82)
n-Octylaminosepharose 4B (P$_3$)	5.3 (50)
P-450 A	8.3 (8.7)
P-450 B	6.0 (2.7)
P-450 C	13.7 (7.1)
P-450 D	11.6 (2.9)

The liver microsomal cytochrome P-450 purification procedure was that described by
Guengerich (1977, 1978) and used by our laboratory (Lotlikar *et al.*, 1984). These four
purified fractions showed no detectable levels of cytochrome b_5, NADPH-P-450 reductase or
NADH-ferricyanide reductase activity.

Table 2

2-Acetylaminofluorene N- and ring-hydroxylation by various
purified cytochromes P-450 from liver microsomes of
MC-pretreated hamsters

P-450 Fraction	nmol of hydroxy-AAF formed/30 min	
	N-	Ring-
Whole microsomes	2.7 (100)	2.9 (100)
Total P-450 (P$_3$)	2.8 (104)	3.1 (107)
P-450 A	0.2 (7)	0.2 (7)
P-450 B	0.6 (22)	0.5 (17)
P-450 C	1.8 (67)	1.8 (62)
P-450 D	0.7 (26)	0.8 (28)

The incubation medium contained 100 nmol of [9-^{14}C]AAF, 2 mM-NADPH
and microsomes or P$_3$ or P-450 fraction containing 0.1 nmol of P-450 in a
total volume of 1.0 ml. With P$_3$ or purified P-450 fractions, reductase fraction
containing 400 nmol/min of NADPH-cytochrome c reductase and 0.05 mg of
dilauroyl phosphatidylcholine were also added. After incubation of duplicate
samples in air for 30 min at 37°C, 50 μg of N-NO-AAF in 0.05 ml acetone
was added as a carrier. Reaction was terminated by the addition of 4 ml of
cold 1 M-Na acetate buffer, pH 6.0. Ring- and N-hydroxy-AAF metabolites
were extracted, separated and quantitated by radioactivity measurements as
described previously (Lotlikar *et al.*, 1974). Data are expressed in terms of
nmol of N- or ring-hydroxy-AAF formed/30 min. Numbers in parentheses
represent data expressed in % of values obtained with whole microsomes.

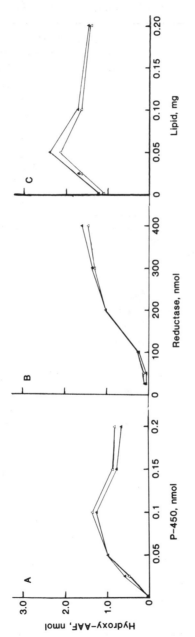

Fig. 1 – AAF N- & ring-hydroxylations as a function of P-450, reductase or lipid concentration. (A) Amount of NADPH-cytochrome P-450 reductase equivalent to 400 nmol NADPH-cytochrome c reduced/min and 0.05 mg dilauroylphosphatidyl-choline (lipid) were kept constant and P-450 C was variable. (B) 0.1 nmol P-450 C and 0.05 mg lipid were kept constant and reductase was variable. (C) 0.1 nmol P-450 C and 400 nmol reductase were kept constant and lipid was variable. △—△ and ▲—▲ represent *N*-OH-AAF and ring-HO-AAF respectively.

cytochrome c reductase activity of 53.2 μmol/mg protein/min was used for reconstitution studies.

In our earlier reconstitution studies for AAF oxidation, crude total P-450 fraction from MC-treated hamsters was used (Lotlikar *et al.*, 1974; Lotlikar, 1981). In the present study four purified P-450 fractions are examined for their capacity to oxidize AAF (Table 2). Total P-450 fraction (P_3) had about the same capacity as whole microsomes to metabolize AAF. Among four purified P-450 fractions, P-450 C is the most active followed by P-450 D, P-450 B and P-450 A. Ratios of N- to ring-hydroxylation with all P-450 fractions were similar to those obtained with whole microsomes. This ratio is several-fold higher than that obtained with P-450 fractions isolated from MC-pretreated rats (Lotlikar *et al.*, 1974; Thorgeirsson *et al.*, 1983).

Kinetic studies have been performed with P-450 C with regard to cytochrome P-450, reductase and phospholipid (Fig. 1). Linear response was seen with P-450 C levels up to 0.05 nmol. In contrast to rat P-450 fraction (Lotlikar *et al.*, 1984), hamster P-450 C showed inhibitory effect at higher P-450 levels (Fig. 1A). Reductase kinetics appeared to indicate a saturation effect in the presence of 400 nmol of reductase (Fig. 1B). The addition of lipid, dilauroyl phosphatidylcholine gave optimum stimulation for N- and ring-hydroxylations at 0.05 mg level (Fig. 1C). At higher concentrations, the lipid produced inhibitory effects.

Gel electrophoresis of microsomes, total P-450 fraction (P_3) and four purified P-450 fractions (data not shown) indicate that whereas P-450 C has two predominant protein bands of 47,000 and 54,000 daltons, P-450 D shows a predominant band of 47,000 daltons.

On the basis of the overall data, it appears that cytochromes P-450 with molecular weights of 47,000 and 54,000 daltons are primarily responsible for the oxidative metabolism of AAF. However, immunological studies will enable us to obtain definitive information about contribution of these proteins in AAF oxidation.

REFERENCES

Cramer, J. W., Miller, J. A. & Miller, E. C. (1960), *J. Biol. Chem.*, **235**, 885.

Guengerich, F. P. (1977), *J. Biol. Chem.*, **252**, 3970.

Guengerich, F. P. (1978), *J. Biol. Chem.*, **253**, 7931.

Gutmann, H. R. & Bell, P. (1977), *Biochim. Biophys. Acta*, **498**, 229.

Irving, C. C. (1964), *J. Biol. Chem.*, **239**, 1589.

Laemmli, U. K. (1970), *Nature*, **227**, 680.

Lotlikar, P. D. (1981), *J. Cancer Res. Clin. Oncol.*, **99**, 125.

Lotlikar, P. D., Enomoto, M., Miller, J. A. & Miller, E. C. (1967), *Proc. Soc. Exp. Biol. Med.*, **125**, 341.

Lotlikar, P. D., Luha, L. & Zaleski, K. (1974), *Biochem. Biophys. Res. Commun.*, **59**, 1349.

Lotlikar, P. D., Pandey, R. N., Clearfield, M. S. & Paik, S. M. (1984), *Toxicol. Lett.*, **21**, 111.

Lotlikar, P. D. & Zaleski, K. (1974), *Biochem. J.*, **144**, 427.

Miller, E. C. (1978), *Cancer Res.*, **38**, 1479.

Miller, J. A. (1970), *Cancer Res.*, **30**, 559.

Thorgeirsson, S. S., Jollow, D. J., Sasame, H., Green, I. & Mitchell, J. R. (1973), *Mol. Pharmacol.*, **9**, 398.

Thorgeirsson, S. S., Sanderson, N., Park, S. S. & Gelboin, H. V. (1983), *Carcinogenesis*, **4**, 639.

Yasukochi, Y. & Masters, B. S. S. (1976), *J. Biol. Chem.*, **251**, 5337.

Chapter 18

METABOLISM OF 2-ACETYLAMINOFLUORENE IN THE PERFUSED MALE RAT LIVER†

P. D. Lotlikar, **K. Desai**, **G. Takahashi**,‡ **H. Shah** and **S. Weinhouse**, Fels Research Institute, Temple University School of Medicine, Philadelphia, PA 19140, USA

1. The effect of pretreatment with 3-methylcholanthrene (MC) or phenobarbital (PB) on the metabolism of 2-acetylaminofluorene in the perfused male rat liver has been examined.
2. MC treatment increased biliary excretion rate of AAF metabolites severalfold.
3. Total biliary excretion in MC-treated rats was twice as much as that in untreated rats. This increase was observed in the glucuronide and non-extractable polar fractions.
4. Excretion of N-HO-AAF glucuronide was twice as much in the bile of MC-treated rats compared to controls; whereas its presence was much less in livers of treated animals than that in controls.
5. PB treatment showed no appreciable differences in the overall metabolism compared to controls.

†Supported by grants CA-10604 and CA-12227 from the National Cancer Institute, US Department of Health and Human Services and grant S1G-6 from the American Cancer Society.
‡Present address: Department of Pathology, Chest Disease Research Institute, Kyoto University, Kyoto, Japan.

INTRODUCTION

Liver plays a central role in the metabolism of foreign compounds. It is known that N-hydroxylation is an activation step in the carcinogenesis of 2-acetylaminofluorene (AAF) and several other aromatic amines and amides (Miller, 1970). Liver tumour induction by AAF is inhibited on simultaneous administration of either 3-methylcholanthrene (MC) (Miller et al., 1958) or phenobarbital (PB) (Peraino et al., 1971). In acute and chronic studies with AAF, it is shown that MC treatment decreases the urinary excretion of N-HO-AAF in the rat (Lotlikar et al., 1967; Miller et al., 1960). However, MC pretreatment of rats gives increased AAF N-hydroxylation with hepatic microsomal systems (Lotlikar et al., 1967, 1978). Since the perfused liver, in contrast to liver homogenate and slices, has a distinct advantage in that it retains the intracellular and extracellular integrity of the in vivo tissue (Bartosek et al., 1973), it was of interest to examine the effect of MC and PB pretreatment on the metabolism of AAF in the perfused rat liver.

EXPERIMENTAL

Male Sprague-Dawley rats (200 g body wt) were used for the study. One group of animals served as controls and was injected i.p. with 0.5 ml sesame oil 24 hr before sacrifice. The second group was injected i.p. with 5 mg MC suspended in 0.5 ml sesame oil 24 hr before liver perfusion. The third group was injected subcutaneously with PB (8 mg PB/100 g body wt/day for 5 days).

Liver perfusion technique was as described by Takahashi et al. (1977). The perfusion medium was Krebs–Henseleit–Ringer bicarbonate at pH 7.4. The procedure for erythrocyte-free perfusion was that of Scholz et al. (1969). After 15 minute perfusion with buffer only, labelled-AAF, 1.5 μmol AAF containing 10 μCi [9-^{14}C]AAF dissolved in 0.2 ml ethanol was added to the perfusion medium, and circulation was continued for 90 minutes. At intervals of 5 minutes, portions of bile were collected for 1 minute and directly assayed for radioactivity; portions collected over each 4-minute period were pooled for separation and quantitation of various metabolites.

Macromolecules from liver tissue were precipitated with ethanol as described by Grantham et al. (1968). Bile, perfusate, and liver samples were then subjected to a series of ethyl ether extractions and enzymatic and acid hydrolysis to determine amounts of metabolites present as free, glucuronide and sulphate conjugates (Mohan et al., 1976). Ether unextractable radioactivity is designated as polar metabolites. Hydroxylated metabolites were separated and quantitated as described previously (Lotlikar et al., 1974).

RESULTS AND DISCUSSION

Biliary excretion of AAF metabolites is shown in Fig. 1. There was a 5 minute lag period in biliary excretion in all animals (Fig. 1A). In controls, the rate of

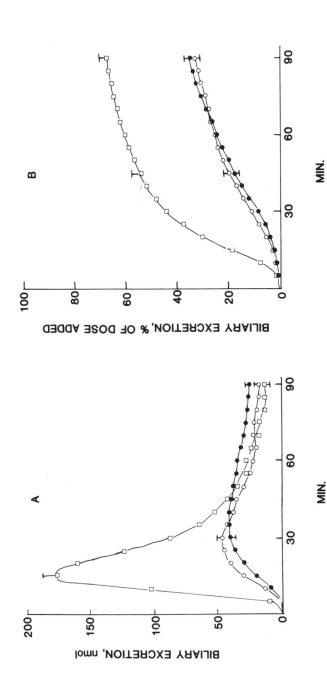

Fig. 1 – Biliary excretion of [9-^{14}C]AAF metabolites in the perfused rat liver.
    A. Time course of appearance of radioactivity in the bile with the ordinate given
in nmol excreted over each 5-minute interval;
    B. Summation of biliary excretion as percentage of dose added. Vertical bars
represent S.E. of data of 4 rats.

●—● controls; ○—○ PB-treated and □—□ MC-treated.

Table 1

Effect of MC and PB treatment on metabolism of AAF in perfused rat liver[a]

Source	Treatment	nmol of radioactivity recovered as					
		Total	AAF	Glucuronide	Sulphate	Polar
Bile	—	637 ± 46	13 ± 2	157 ± 10	165 ± 26	293 ± 9
	MC	1031 ± 144	21 ± 3	326 ± 17	159 ± 24	438 ± 37
	PB	555 ± 101	16 ± 3	128 ± 31	138 ± 44	214 ± 60
Perfusate	—	231 ± 21	54 ± 23	42 ± 60	19 ± 3	62 ± 14
	MC	202 ± 65	9 ± 2	20 ± 1	28 ± 31	149 ± 49
	PB	191 ± 30	31 ± 10	25 ± 5	29 ± 2	50 ± 11
Liver	—	612 ± 58	131 ± 11	73 ± 9	194 ± 26	196 ± 30
	MC	250 ± 23	23 ± 4	24 ± 7	42 ± 29	134 ± 4
	PB	440 ± 41	252 ± 38	20 ± 6	156 ± 23	108 ± 11

[a]AAF, 1.5 μmol containing 10 μCi[9-^{14}C]AAF dissolved in 0.2 ml ethanol was suspended in 75 ml of perfusion medium. Perfusion was continued for 90 min. Four animals were used in each group.

biliary excretion reached optimum levels at about 30–40 minute periods. In PB-treated animals, the optimum rate of excretion was at about 30 minute. However, in MC-treated animals, the rate of biliary excretion reached optimum levels in about 15 minutes and then fell rapidly. The rate of biliary excretion in MC-treated rats was severalfold higher than that in control and PB-treated animals. Total biliary excretion in MC-treated animals was almost twice that in controls (Fig. 1B). Total biliary excretion in control rats is similar to that reported by Miyata *et al.* (1972) in perfusion studies and by Irving *et al.* (1967) in *in vivo* experiments.

Analysis of various metabolites excreted in the bile, perfusate and liver are shown in Table 1. Comparison of biliary excretion in the control and MC-treated rats indicates that glucuronide and non-extractable polar fraction could account for the increase in MC-treated rats over the controls. There were also differences in the excretory patterns in the perfusate of these two groups. Excretion of AAF was much smaller in the bile of all groups.

About 7–8% of the total glucuronide fraction present in the bile was recovered as N-HO-AAF (Table 2). Biliary N-HO-AAF excretion was twice as much in the MC-treated animals than that in controls. However, the presence of N-HO-AAF was threefold higher in the livers of controls than that in MC-treated rats. The total N-HO-AAF recovered as glucuronide was about the same in both groups. Total biliary excretion and N-HO-AAF glucuronide excretion in control animals are similar to those reported by Miyata *et al.* (1972).

Excretion of a large amount of non-extractable polar metabolites in the liver perfusion system described here may be due to the presence of glutathione conjugates (Lotlikar *et al.*, 1965; Meerman *et al.*, 1980, 1982; Mulder, 1982).

The present results and our previous data (Lotlikar *et al.*, 1967, 1978) indicate the MC inhibitory effect on liver tumour induction by AAF in male rats may be due to maintenance and not formation of a lower concentration of N-HO-AAF in liver cells.

Table 2

Formation of N-HO-AAF during metabolism of AAF in
perfused rat liver[a]

Treatment	nmol recovered as glucuronide of N-HO-AAF in		
	Bile	Perfusate	Liver
—	12.2 ± 1.8	1.3 ± 0.4	10.8 ± 2.9
MC	23.5 ± 2.7	1.3 ± 0.2	2.9 ± 1.0
PB	7.5 ± 1.9	1.1 ± 0.4	1.4 ± 0.8

[a]Perfusion conditions as for Table 1 (see text).

REFERENCES

Bartosek, I., Guaitani, A. & Miller, L. (eds.) (1973), *Isolated Liver Perfusion and Its Application*, Raven Press, New York.

Grantham, P. H., Mohan, L., Yamamoto, R. S., Weisburger, E. K. & Weisburger, J. H. (1968), *Toxicol. App. Pharmacol.*, **13**, 118.

Irving, C. C., Wiseman, R. Jr. & Hill, J. T. (1967), *Cancer Res.*, **27**, 2309.

Lotlikar, P. D., Enomoto, M., Miller, J. A. & Miller, E. C. (1967), *Proc. Soc. Exp. Biol. Med.*, **125**, 341.

Lotlikar, P. D., Hong, Y. S. & Baldy, W. J. (1978), *Toxicol. Lett.*, **2**, 135.

Lotlikar, P. D., Luha, L. & Zaleski, K. (1974), *Biochem. Biophys. Res. Commun.*, **59**, 1349.

Lotlikar, P. D., Miller, E. C., Miller, J. A. & Margreth, A. (1965), *Cancer Res.*, **25**, 1743.

Meerman, J. H. N., Beland, F. A., Ketterer, B., Srai, S. K. S., Bruins, A. D. & Mulder, G. J. (1982), *Chem.-biol. Interact.*, **329**, 149.

Meerman, J. H. N., van Doorn, A. B. D. & Mulder, G. J. (1980), *Cancer Res.*, **40**, 3772.

Miller, E. C., Miller, J. A., Brown, R. R. & McDonald, J. C. (1958), *Cancer Res.*, **18**, 469.

Miller, J. A. (1970), *Cancer Res.*, **30**, 559.

Miller, J. A., Cramer, J. W. & Miller, E. C. (1960), *Cancer Res.*, **20**, 950.

Miyata, K., Noguchi, Y. & Enomoto, M. (1972), *Japan J. Exp. Med.*, **42**, 483.

Mohan, L. C., Grantham, P. H., Weisburger, E. K., Weisburger, J. H. & Iodine, J. B. (1976), *J. Natl. Cancer Inst.*, **56**, 763.

Mulder, G. J., Unruh, L. E., Evans, F. E., Ketterer, B. & Kadlubar, F. F. (1982), *Chem.-biol. Interact.*, **39**, 111.

Peraino, C., Fry, R. J. M. & Staffeldt, E. (1971), *Cancer Res.*, **31**, 1506.

Scholz, R., Thurman, R. G., Williams, J. R., Chance, B. & Bucher, T. (1969), *J. Biol. Chem.*, **244**, 2317.

Takahashi, G., Shah, H. & Weinhouse, S. (1977), *Cancer Res.*, **37**, 369.

Chapter 19

N-HYDROXYLATION OF BARBITURATES *IN VITRO*?

Alison Seago[†] and **J. W. Gorrod**, Pharmacy Department, Chelsea College (University of London), Manresa Road, London SW3 6LX, UK

1. Aroclor-induced Syrian hamster hepatic washed microsomes were used to re-examine the *in vitro* metabolism of [^{14}C]pentobarbitone and [^{14}C]phenobarbitone.
2. Supplementing the NADPH-regenerating system at 20-minute intervals in the presence of 1mM-EDTA increased the metabolism so that [^{14}C]pentobarbitone (0.5 μmole) was virtually totally metabolized.
3. Using the supplemented system only 8% metabolism of [^{14}C]phenobarbitone in a 120 minute incubation was observed.
4. N-Hydroxypentobarbitone and N-hydroxyphenobarbitone were synthesized but could not be detected as microsomal metabolites in the NADPH supplemented system.

INTRODUCTION

Previously published studies on *in vivo* barbiturate metabolism in humans reported the formation of an N-hydroxylated metabolite (Tang *et al.*, 1975, 1977a,b). This novel metabolite was subsequently characterized as the N-glucosyl derivative (Tang *et al.*, 1978, 1979). Since the formation of N-hydroxylated metabolites often leads to more toxic and/or carcinogenic

---
†Present address: Institute of Cancer Research, CRC Laboratory, Clifton Avenue, Belmont, Sutton, Surrey SM2 5PX, UK.

products we decided to re-examine the *in vitro* metabolism of pentobarbitone and phenobarbitone to determine whether or not *N*-hydroxylation could occur.

EXPERIMENTAL

N-Hydroxypentobarbitone and *N*-hydroxyphenobarbitone were synthesized by reacting the correctly substituted malonyl chloride with benzyloxyurea. The free *N*-hydroxylated product was liberated by hydrogenation of the *N*-benzyloxybarbiturate. The reference compounds were used to develop extraction methods and thin-layer chromatography systems for examining the metabolic products.

Metabolism studies were carried out using Aroclor-induced Syrian hamster hepatic washed microsomes, equivalent to 0.5 g original liver/ml microsomes, with an NADPH-regenerating system. 2[^{14}C]Labelled substrates were used throughout. Incubations were carried out at 37 °C and the reaction terminated by cooling on ice. Metabolites were extracted with ethyl acetate over 1 g NaCl, the extract reduced to dryness under nitrogen, the residue taken up in 100 μl methanol and analysed by t.l.c. (silica gel GF$_{254}$). Chromatograms were routinely run in benzene:acetic acid:methanol (88:10:2) (system A). Once optimum metabolism conditions were established, *n*-octanol, saturated with water (system B), was used as a second system in two-dimensional chromatography to separate *N*-hydroxypentobarbitone from 3'-oxopento-

Table 1

The effect of adding further cofactors on the *in vitro* metabolism of pentobarbitone, 0.5 μmole, using Aroclor-induced Syrian hamster liver microsomes in the presence of EDTA (1 nM).

	R_f^a	20 min[b]	40 min[b]	60 min[b]	80 min[b]	100 min[b]
		\multicolumn{5}{c}{Incubation time}			
	0.02	0.72	1.66	2.18	2.92	2.71
Pen B	0.09	64.09	84.54	87.66	85.34	82.78
	0.15	0.80	1.63	2.01	2.21	2.56
Pen C	0.19	1.00	3.01	5.31	7.29	9.84
	0.23	0.29	0.64	0.69	0.65	0.83
	0.28	0.90	0.78	0.61	0.75	0.75
Pen A	0.42	32.21	7.73	1.53	0.81	0.53

Pen A = Pentobarbitone; Pen B = 3'-Hydroxypentobarbitone, Pen C = 3'-Oxopentobarbitone.
[a]R_f values in TLC solvent system A. Radioactivity corresponding to R_f 0.02, 0.15, 0.23 and 0.28 were unidentified metabolies.
[b]Results expressed as radioactivity (% of total) associated with each spot on the TLC plate.

barbitone. Metabolites were located by autoradiography; active areas on the chromatogram were then removed and subjected to liquid scintillation counting.

Jacobson *et al.* (1973) showed that addition of 1 mM-EDTA to the microsomal incubate could increase the amount of metabolism with tissue preparations. This was confirmed using [^{14}C]pentobarbitone, and therefore 1 mM-EDTA was routinely incorporated into incubation mixtures.

RESULTS AND DISCUSSION

The effect of substrate concentration was determined over the range 0.054 μmole to 5.0 μmole. It became apparent that the higher concentrations

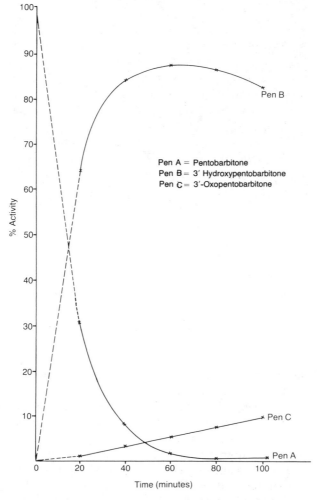

Fig. 1 – Increased microsomal metabolism of pentobarbitone (0.5 μmole), by the addition of cofactors to the incubation system at timed intervals.

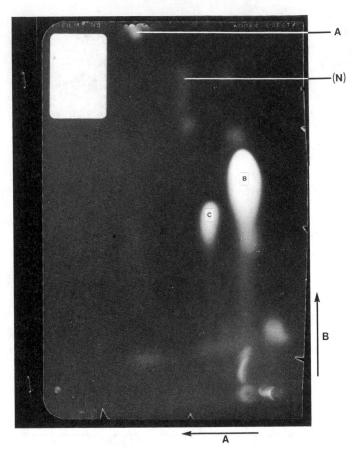

System A: Benzene : Acetic acid : Methanol
           88   10   2

System B: n-Octanol, saturated with water.

Fig. 2 – Autoradiograph showing the two-dimensional chromatographic separation of the metabolites of pentobarbitone formed in the 100-minute incubate, to which additional cofactors had been added. A = pentobarbitone, B = 3′-hydroxy-pentobarbitone, C = 3′-oxopentobarbitone, N = R_f of N-hydroxypentobarbitone.

were giving rise to substrate inhibition and a concentration of 0.5 μmole was chosen for all subsequent studies (approximately 50% conversion in 30 minutes). Increasing the period of incubation from 20 to 100 minutes did not result in an increase in metabolism. Therefore a method of supplementing with further cofactors as described by Jacobson *et al.* (1973) was modified and utilized for these experiments. The standard NADPH-regenerating system used at time zero was again added in 50 μl phosphate buffer (0.01 mM; pH 7.4) at 20 minutes and at 20-minute intervals thereafter. Using this method almost all the substrate ([^{14}C]pentobarbitone) was consumed within

100 minutes. These results are shown in Table 1 and Fig. 1. Initially an increase in 3'-hydroxypentobarbitone (Pen B) was detected; this was seen to fall off gradually after 60 minutes, concomitant with an increase in 3'-oxopentobarbitone (Pen C). After 40 minutes the appearance of a third metabolite (R_f 0.15) was observed; this co-chromatographed with 5-ethyl-5-(1-methylbutyric acid) barbituric acid. Consumption of [^{14}C]pentobarbitone (Pen A) in this system was rapid and almost 100%.

Two-dimensional thin-layer chromatography of the 100-minute incubation extract is shown in Fig. 2. The chromatogram was first run in System A, then System B. As can be seen, very little substrate ([^{14}C]pentobarbitone) remains (A) with the major metabolites 3'-hydroxypentobarbitone (B) and 3'-oxopentobarbitone (C) being present. Any N-hydroxypentobarbitone formed should chromatograph to the area marked (N) but none could be detected using this method (detection limit of 10×10^{-9} M).

Similar metabolic studies were carried out using [^{14}C]phenobarbitone as the substrate. However, no metabolism could be observed in the normal *in vitro* system, confirming the results of Crayford & Hutson (1980) and Peters *et al.* (1973). Using the supplemented microsomal system 8% of substrate (0.5 μmole) was metabolized over 120 minutes.

In conclusion, using the supplemented microsomal system developed in this study, the overall metabolism of [^{14}C]pentobarbitone and [^{14}C]phenobarbitone could be increased, however, the N-hydroxylated metabolites were not detected, at a level of 10×10^{-9} M. Therefore, it would appear that N-hydroxylation is not an *in vitro* metabolic route for the barbiturates tested, using Syrian hamster tissue.

REFERENCES

Crayford, J. V. & Hutson, D. H. (1980), *Food and Cosmetics Toxicology*, **18**, 503.

Jacobson, M., Levin, W., Lu, A. Y. H., Conney, A. H. & Kuntzman, R. (1973), *Drug metab. dispos.*, **1**, 766.

Peters, R. A., Shorthouse, M., Thorne, C. J. R., Ward, P. F. V. & Huskisson, N. S. (1973), *Biochem. Pharmacol.*, **22**, 2633.

Tang, B. K., Inaba, T. & Kalow, W. (1975), *Drug metab. dispos.*, **3**, 479.

Tang, B. K., Inaba, T. & Kalow, W. (1977a), *Drug metab. dispos.*, **5**, 71.

Tang, B. K., Inaba, T. & Kalow, W. (1977b), *Fed. Proc.*, **36**, 966.

Tang, B. K., Kalow, W. & Grey, A. A. (1978), *Res. commun. chem. path. pharmacol.*, **21**, 45.

Tang, B. K., Kalow, W. & Grey, A. A. (1979), *Drug metab. dispos.*, **7**, 315.

Chapter 20

THE FAILURE TO DETECT N-HYDROXYLATION AS A METABOLIC PATHWAY FOR A SERIES OF CARBAMATES

Joanne T. Marsden[†] and **J. W. Gorrod**, Department of Pharmacy, Chelsea College (University of London), Manresa Road, London SW3 6LX, UK

1. A series of aliphatic, aromatic carbamates and N-methylcarbamates and their N-hydroxy derivatives were prepared.
2. Methods for the detection of N-hydroxy derivatives based on thin layer chromatography and colorimetry were established.
3. [^{14}C]Urethane metabolism was studied using radiochemical methods.
4. In no case was N-hydroxylation observed as a metabolic pathway for carbamates, *in vitro* or *in vivo*.

INTRODUCTION

Aliphatic and aromatic carbamates are a group of compounds possessing a wide range of pharmacological activity. They have been used to treat disorders of the nervous system in humans, and in agriculture as pesticides.

There are conflicting views regarding the N-hydroxylation of carbamates in metabolism studies and the role of this pathway in the carcinogenic activity of some carbamates. N-Hydroxylation has been reported to be a metabolic path-

[†]Present address: Department of Chemical Pathology, Kings College School of Medicine and Dentistry, Kings College London, Denmark Hill, London SE5 8RX, UK.

way for urethane, a known carcinogen, and other aliphatic carbamates in several animal species (Boyland & Nery, 1965; Nery, 1968). However, there have been no further reports confirming N-hydroxylation as a metabolic pathway for carbamates, although several workers have looked for this pathway in rodents (Mirvish, 1966; Tatsumi et al., 1980; O'Flaherty & Sichak, 1983).

The present study was designed to ascertain the importance of N-hydroxylation as a metabolic pathway for a series of aliphatic and aromatic carbamates.

EXPERIMENTAL

Urethane was obtained from BDH Ltd. n-Butyl carbamate and phenyl carbamate were purchased from Kodak Ltd and Aldrich Chemical Co. Ltd, respectively. Benzyl carbamate was obtained from Koch-Light Labs., Ltd. Methyl carbamate, methyl N-hydroxycarbamate, n-propyl carbamate, n-propyl N-hydroxycarbamate, benzyl N-hydroxycarbamate, N-hydroxyurethane, p-tolyl N-methylcarbamate, phenyl N-methylcarbamate, carbaryl and [ethyl-1-^{14}C]urethane were gifts. 1-Naphthyl, 2-naphthyl and p-tolyl carbamate were prepared as described by Lambrech (1960). n-Butyl, 1-naphthyl, 2-naphthyl and p-tolyl N-hydroxycarbamates were prepared according to the method of Dorough & Casida (1964). Phenyl, 1-naphthyl and p-tolyl N-hydroxy-N-methylcarbamates were prepared similarly using N-methylhydroxylamine hydrochloride. Carbamates and N-hydroxycarbamates were characterized by melting point determinations, infrared, ultraviolet, nuclear magnetic resonance and mass spectrometry.

Thin-layer chromatography (t.l.c.) was carried out on glass plates (20 × 20 cm) coated with a layer (0.25 mm thickness) of silica gel GF$_{254}$. The solvent systems and detection reagents used are shown in Table 1. Colorimetric analysis was carried out using the methods described by Boyland & Nery (1964) and Nery (1966). The metabolism of [ethyl-1-^{14}C]urethane was studied in vitro using liver fractions from five different animal species and in vivo in the rat as described by Marsden (1983).

The animals used in this study were all male, New Zealand White rabbits, albino Dunkin–Hartley guinea-pigs, Syrian hamsters, albino Wistar rats and albino LACA mice.

The preparation of liver fractions (homogenate, 9000 g and microsomes), and incubations using 5 μmol of substrate were essentially the same as described previously (Gorrod et al., 1975). Fluoride (300 μmol) was added to some incubates as this was shown to inhibit the breakdown of N-hydroxyurethane (Marsden, 1983). The doses and vehicles used to administer carbamates by intraperitoneal injection to various animal species is shown in Table 2. When more than one animal was used in the study, urine was pooled prior to analysis. For in vitro studies, animals were also pretreated with inducing agents prior to tissue preparation. Phenobarbitone (80 mg/kg

Table 1

Typical thin layer chromatographic data for the carbamates and their
N-hydroxy derivatives: Solvent systems and detection reagents

Compound			Solvent system $R_F \times 100$			Detection reagent			
R	R′	R″	S1	S2	S3	D1	D2	D3	D4
Ethyl	H	H	49	60	68	—	—	—	red
Ethyl	OH	H	40	17	36	black	purple	reddish	red
Phenyl	H	H	55	56	69	—	—	—	red
Phenyl	OH	H	43	14	37	black	purple	reddish	red
Phenyl	H	CH_3	64	83	84	—	—	—	—
Phenyl	OH	CH_3	60	35	71	black	purple	—	—

Key:
S1 = Petroleum ether (b.p. 40–60 °C): acetone (70:30)
S2 = Chloroform: methanol (97:3)
S3 = Chloroform: acetic acid (90:10)
D1 = Tollens reagent; D2 = Ferric chloride; D3 = Sodium amminoprusside;
D4 = *p*-Dimethylaminocinnamaldehyde

Carbamate molecule = $R-O-\underset{\underset{O}{\|}}{C}-N\begin{smallmatrix}R' \\ R''\end{smallmatrix}$

— = No colour detected

body weight) was administered by intraperitoneal injection for three consecutive days prior to tissue preparation. 3-Methylcholanthrene (40 mg/kg body weight) was administered by a single intraperitoneal injection 48 hours prior to tissue preparation.

The ability of the liver fraction preparations to carry out cytochrome P-450 dependent and flavin-containing monooxygenase reactions was determined by measuring *N*-oxidation products of aniline, *N,N*-dimethylaniline and 2-acetylaminofluorene (AAF) (Marsden, 1983).

RESULTS AND DISCUSSION

The limit of detection of *N*-hydroxycarbamates in microsomal incubates and urine was established using t.l.c. visualization reagents and colorimetric analysis. These methods were sensitive enough to detect comparable levels of *N*-hydroxylation as reported by Boyland & Nery (1965) and Nery (1968). The results using t.l.c. and detection reagents showed that the limit of detection of *N*-hydroxylation ranged from 0.1–0.4% conversion using 5 μmole of substrate (~1–2 μg per microsomal incubate). Colorimetric analysis was more sensitive than chromatographic analysis for detecting aliphatic *N*-hydroxycarbamates in microsomal incubates. The results showed that this

Table 2

Administration of carbamates by intraperitoneal injection to various
animal species

Carbamate administered	Dose (g/kg body weight)	Animal and number dosed
		Mouse
Urethane	1.0[a]	10
n-Butyl carbamate	0.4[b]	10
Phenyl N-methylcarbamate	0.05[b]	4
Urethane	1.0[a]	4
		Hamster
Urethane	1.0[a]	2
n-Butyl carbamate	0.4[b]	2
Phenyl N-methylcarbamate	0.05[b]	2
		Rat
Urethane	1.0[a]	1
Phenyl carbamate	0.05[b]	2
Phenyl N-methylcarbamate	0.05[b]	2
		Guinea Pig
Urethane	1.0[a]	1
		Rabbit
Urethane	1.0[a]	1

[a]Administered in isotonic saline.
[b]Administered in 2-methoxyethanol.

method had a lower limit of detection of 1 μg per microsomal incubate for all
of the aliphatic N-hydroxycarbamates. The limit of detection of
N-hydroxycarbamates in urine was established using t.l.c. and detection reagents as 0.5 μg per ml of urine for all the N-hydroxycarbamates studies.
Radioactivity studies using [ethyl-1-^{14}C]urethane showed that the limit of
detection of N-hydroxyurethane in a microsomal incubate was equivalent to
0.1–0.2 μg of N-hydroxyurethane per microsomal incubate. Other workers
(Tatsumi et al., 1980; O'Flaherty & Sichak, 1983) have also established limits
of detection of N-hydroxycarbamates lower than the levels measured *in vivo*
by Boyland & Nery (1965) and *in vitro* by Nery (1968). Despite the availability of these sensitive methods, N-hydroxylation was not detected as a
metabolic pathway for any of the carbamates in this study.* The addition of
fluoride to incubates and the pretreatment of animals with inducing agents
did not result in the formation of N-hydroxycarbamates. Further studies were

---

*In all these studies the viability of tissue preparations to carry out N-oxidative reactions of aniline,
N,N-dimethylaniline and AAF was established.

carried out to look at the rate of metabolism or breakdown of these *N*-hydroxycarbamates under identical incubation conditions.

The results using 0.5 μmol of *N*-hydroxycarbamate showed that aromatic *N*-hydroxycarbamates were metabolized more rapidly than aliphatic *N*-hydroxycarbamates, although the rate of breakdown did not exceed 25% in a 10-minute incubation period. Since fluoride was additionally added to carbamate incubates to prevent breakdown, the failure to detect the *N*-hydroxycarbamates was unlikely to be due to their further metabolism or chemical breakdown.

In conclusion, it seems more likely, from these results and the work of Mirvish (1966), Tatsumi *et al.* (1980) and O'Flaherty & Sichak (1983) that *N*-hydroxylation is not a metabolic pathway for carbamates.

REFERENCES

Boyland, E. & Nery, R. (1964), *Analyst*, **89**, 520.
Boyland, E. & Nery, R. (1965), *Biochem. J.*, **94**, 198.
Dorough, H. W. & Casida, J. E. (1964), *J. Agr. Food Chem.*, **12**, 294.
Gorrod, J. W., Temple, D. J. & Beckett, A. H. (1975), *Xenobiotica*, **5**, 453.
Lambrech, J. A. (1960), Union Carbide Chemical Company.
Marsden, J. T. (1983), PhD Thesis, University of London.
Mirvish, S. S. (1966), *Biochem. Biophys. Acta.*, **117**, 1.
Nery, R. (1966), *Analyst.*, **91**, 388.
Nery, R. (1968), *Biochem. J.*, **106**, 1.
O'Flaherty, E. J. & Sichak, S. P. (1983), *Toxicol. Appl. Pharmacol.*, **68**, 354.
Tatsumi, K., Yoshimura, H., Kawazoe, Y., Horiuchi, T. & Koga, H. (1980), *Chem. Pharm. Bull.*, **28**, 351.

Chapter 21

A COMPARISON OF THE LIGHT REVERSAL PROPERTIES OF CO-INHIBITION OF N-DEALKYLATION WITH THOSE OF OTHER MIXED FUNCTION OXIDASES OF HEPATIC MICROSOMES

S. Hamill,[†] **H. Schleyer,**[†] **P. D. Lotlikar**[‡] and **D. Cooper**[†]
†The Harrison Department of Surgical Research and the Department of Pharmacology, School of Medicine, University of Pennsylvania, Philadelphia, PA 19104, USA
‡The Fels Research Institute, Temple University, School of Medicine, Philadelphia, PA 19140, USA

1. Light reversal of CO inhibition of the N-dealkylation of benzphetamine (BP), ethylmorphine (EM), dimethylaniline (DMA), dimethylaniline N-oxide (DMAO), 7-ethoxycoumarin (EOC), diethylnitrosamine (DEN) and dimethylnitrosamine (DMN), N-hydroxylation of 2-acetylamino-fluorene (AAF) and the ring-hydroxylations of testosterone (T) have been studied using a kinetic technique.

2. All reactions are inhibited by CO/O_2 mixtures (Warburg's K ranged from 0.5–13).

3. BP, EM and DMA dealkylation are poorly reversed by wavelengths of light around 445 nm. EOC de-ethylation is readily reversed by 445 nm light as is T hydroxylation by 450 nm light. DMAO demethylation was

reversed about 80% by 455 nm light. There was no light reversal at any wavelength studied of CO inhibition of AAF N-hydroxylation or DEN and DMN dealkylation.

4. Hepatic microsomes contain several membrane-bound, P-450 dependent, mixed function oxidases (MFO) that are distinguishable from one another by the biophysical properties of light reversal of CO inhibition.

5. These differences in light reversal of CO inhibition yield a method for more exact definition of the properties of the multiple MFO activities of highly purified P-450 preparations and permit a comparison of these characteristics with those of intact microsomes.

INTRODUCTION

From the work of many laboratories over the past two decades (Ryan *et al.*, 1979, 1980; Coon, 1981; Lu & West, 1980; Nebert & Negishi, 1982; Lau & Strobel, 1982; Conney, 1982), evidence has accumulated that several highly purified forms of cytochrome P-450 can be prepared from the livers of animals pretreated with various inducers such as phenobarbital, 3-methylcholanthrene, polychlorinated biphenyls, isosafrole, pregnenolone, 16α-carbonitrile, ethanol, benzoflavone and cholestyramine. Although it has been possible chemically to isolate and assess the purity of these isoenzymes by conventional protein purification procedures, the problem remains that none of these purified preparations have single substrate specificity (Conney, 1982). The P-450 preparation obtained thus far that has the highest single substrate specificity is that prepared by Lau & Strobel (1982) which catalyses 7-ethoxycoumarin de-ethylation. The main difficulty complicating the assignment of catalytic activity to the various purified P-450s is that all of these haems have essentially the same spectral characteristics with wide, non-distinct Soret bands. The spectrum of the oxidized form has broad absorption maxima ranging for the different enzymes from 414 to 416 nm, while the reduced Soret band has a lower extinction coefficient and absorption maxima in the range between 410 to 414 nm. Likewise, the absorption of the Soret band of the CO complex of P-450 $(Fe^{2+}).CO$ is broad, with absorption maxima in the spectral region between 440 and 455 nm.

The protein components of the different P-450 isoenzymes can be distinguished by immunochemical techniques employing antibodies to the various purified P-450 isoenzymes. P-450s characterized by such immunochemical techniques, however, also do not have single substrate specificities (Conney, 1982; Thomas *et al.*, 1976, 1979).

PHOTOCHEMICAL ACTION SPECTROSCOPY

Recent studies reported from our laboratory have indicated that the properties of light reversal of CO inhibition of oxidation of various substrates of

P-450 offer a method by which the biophysical properties of P-450 can be better characterized (Cooper *et al.*, 1979; Hamill *et al.*, 1983a,b; Hamill & Cooper, 1984).

This approach is based on a well-known biophysical property: the CO complex of CO-combining haems is most readily dissociated by the wavelengths of light absorbed by the CO complex. Application of this technique, termed photochemical action spectroscopy, to the study of P-450 dependent, mixed function oxidase systems yields distinctive information about microsomal enzymes that cannot be obtained by any other method presently available. Inhibition by CO indicates that the catalyst functioning in a mixed function-oxidation most likely is a heavy metal-containing enzyme, probably a haem. The light reversal maxima and line shape of the light reversal spectrum (Photochemical Action Spectrum) define the absorption spectrum of the pigment and help distinguish it from similar microsomal pigments. The degree of light reversibility is also characteristic for different groups of substrates (Hamill *et al.*, 1983b; Hamill & Cooper, 1984).

A survey of the existing literature of the light reversal of CO inhibition (Cooper *et al.*, 1979) and recent detailed studies on light reversal of CO inhibition in our laboratory (Hamill *et al.*, 1983b; Hamill & Cooper, 1984), show that it is possible to distinguish several P-450 isoenzymes by their properties of light reversal of CO inhibition of hepatic microsomal mixed-function oxidation of six groups of substrates. Further investigation of these properties with additional filters and microsomal systems induced with other agents will probably demonstrate other differences in the biophysical properties of cytochrome P-450.

RESULTS

The biophysical characteristics of the light reversal of CO inhibition of six types of substrates are summarized in Table 1.

CO inhibits five of the mixed-function oxidations of the substrate groups to about the same degree. Dimethylaniline N-oxide demethylation, however, is significantly less sensitive to CO inhibition. The light reversal maximum and the degree of light reversal are the properties that distinguish between the six substrate groups. The degree of light reversal varies from total reversal with steroids and arylhydrocarbons as substrates to no reversal with N-hydroxylation and nitrosamine dealkylations. The line shape of the action spectra of these substrate groups varies. A broad low light reversal spectrum with a maximum at 445 nm that is not in accord with the absorption spectrum of P-450(Fe^{2+}).CO is constantly found with substrates that are N-dealkylated. On the other hand, the light reversal spectrum for the hydroxylation of steroids and benzo[a]pyrene is in good agreement with the absorption spectrum of the CO complex of P-450.

Table 1

Biophysical properties of the light reversal of CO inhibition of mixed function oxidations catalysed by hepatic microsomes

Group	Reaction	Substrate	CO Inhibit.[a] Warburg K	Light Rev. nm	Gamma[b]	References
I	Aliphatic Hydroxylation	Testosterone	1.0–2.0	450	1.0	Estabrook et al., 1963 Cooper et al., 1965 Cooper et al., 1968
	Arylhydrocarbon Hydroxylation	Benzo[a]pyrene				Cooper et al., 1976
II	O-Dealkylation	Ethoxycoumarin	0.5–1.0	445	0.5–0.8	Hamill et al., 1983b Hamill et al.,1983b
	H$_2$O$_2$ Alcohol Oxidation	Ethanol				Fabrey & Lieber, 1979
III	N-Dealkylation	Benzphetamine Ethylmorphine	0.5–1.0	445	0.1–0.3	Hamill et al., 1983b
IV	N-Hydroxylation	2-Acetylamino-fluorene	2.0–3.0	—	0.0	Lotlikar et al., 1983
V	Nitrosamine Dealkylation	Dimethyl-nitrosamine	0.2–2.0			Lotlikar et al., (unpublished)
		Diethyl-nitrosamine	0.3–0.6	—	0.0	
VI	N-Oxide Dealkylation	Dimethylaniline-N-oxide	13.0	445	0.7–0.8	Hamill & Cooper, 1984

[a] $K = n/(1 - n) \times CO/O_2$, where n = rate in CO–O_2, N_2 mixture. K is Warburg's partition constant, which expresses the degree of CO inhibition.
[b] Gamma $= V$ (L,CO) $- V$ (D,CO)/(D,O$_2$) $- V$(D,O$_2$) $\times 100$. Gamma is the ammount by which the various wavelengths and light intensities reverse CO inhibition expressed as a percentage (see Cooper et al., 1979).

DISCUSSION

The cause of these differences of light reversibility seen with these various substrates cannot be explained at the present time. We have previously suggested the following possibilities (Cooper *et al.*, 1979; Hamill *et al.*, 1983b): (1) copper enzymes function in the reactions (CO complexes of copper enzymes are known to be poorly dissociated by light) (Warburg, 1949); (2) several isoenzymes function with different light reversal maxima in the oxidation of a single substrate; (3) heavy metals other than iron and copper combine with CO (selenium, molybdenum, cobalt, etc.) as the metal prosthetic group of the apoenzyme; (4) metal ions such as copper or other CO combining metals not combined with haem act in conjunction with the haem haemoprotein P-450; (5) the substrate interacts with P-450 and produces a conformation change in the cytochrome that alters the light reversal properties; (6) combination of one or more of these possibilities.

Further work is required to determine the mechanism responsible for these observed differences in the light reversibility of the CO inhibition of different mixed function oxidase substrates. What is of importance at the present is that light reversal provides a method of measuring those properties of P-450 that are characteristic for six groups of substrates with different chemical properties. Therefore, by measuring the degree of light reversibility, the line shape, and the absorption maxima of the CO complexes, the catalytic activities of the different P-450s can be assigned to the various substrates.

REFERENCES

Conney, A. H. (1982), *Cancer Res.*, **42**, 4918.

Coon, M. J. (1981), *Drug. Metab. Dispos.*, **9**, 1.

Cooper, D. Y., Levin, S. S., Narasimhulu, S., Rosenthal, O. & Estabrook, R. W. (1965), *Science*, **147**, 400.

Cooper, D. Y., Narasimhulu, S., Rosenthal, O. & Estabrook, R. W. (1968), in *Advances in Chemistry Series*, American Chemical Society, Washington, DC, Vol. 77, p. 220.

Cooper, D. Y., Schleyer, H., Levin, S. S., Eisenhardt, R. H., Novack, B. G. & Rosenthal, O. (1979), *Drug. Metab. Rev.*, **10**, 153.

Cooper, D. Y., Schleyer, H., Rosenthal, O., Levin, W., Lu, A. Y. H., Kuntzman, R. & Conney, A. H. (1976), *Eur. J. Biochem.*, **74**, 69.

Estabrook, R. W., Cooper, D. Y. & Rosenthal, O. (1963), *Biochem. Z.*, **338**, 741.

Fabrey, T. L. & Lieber, C. S. (1979), *Alcoholism*, **3**, 219.

Hamill, S. J., Cooper, D. Y., Schleyer, H. & Rosenthal, O. (1983a), *Arch. Biochem. Biophys.*, **224**, 614.

Hamill, S. J., Cooper, D. Y., Schleyer, H. & Rosenthal, O. (1983b), *Arch. Biochem. Biophys.*, **224**, 625.

Hamill, S. & Cooper, D. (1984), *Xenobiotica*, **14**, 139.

Lau, P. P. & Strobel, H. W. (1982), *J. Biol. Chem.*, **257**, 5257.

Lotlikar, P. D., Hong, Y. S. & Cooper, D. Y. (1983), *Toxicology Lett.*, **19**, 51.

Lotlikar, R., Knight, R. & Cooper, D. Y., unpublished.

Lu, A. Y. H. & West, S. B. (1980), *Pharmacol. Rev.*, **31**, 277.

Nebert, D. W. & Negishi, M. (1982), *Pharmacol.*, **31**, 2311.

Rosenthal, O. & Cooper, D. Y. (1967), in *Methods in Enzymology*, Vol. 10 (eds. Estabrook, R. W. & Pullman, M. E.), Academic Press, New York, pp. 616–628.

Ryan, D. E., Thomas, P. E., Korzeniowski, D. & Levin, W. (1979), *J. biol. Chem.*, **254**, 1365.

Ryan, D. E., Thomas, P. E. & Levin, W. (1980), *J. biol. Chem.*, **255**, 7941.

Thomas, P. E., Korzeniowski, D., Ryan, D. E. & Levin, W. (1979), *Arch. Biochem. Biophys.*, **192**, 524.

Thomas, P. E., Lu, A. Y. H., Ryan, D. E., West, S. B., Kwalek, J. & Levin, W. (1976), *Mol. Pharmacol.*, **12**, 746.

Ullrich, V. & Webber, P. (1972), *Hoppe-Seylers Z. Physiol. Chem.*, **353**, 1171.

Warburg, O. (1949), *Heavy Metal Prosthetic Groups and Enzyme Action*, Oxford Univ. Press (Clarendon), London/New York.

Part 5

HETEROAROMATIC
N-OXYGENATION

Chapter 22

OXIDATION OF TERTIARY HETEROAROMATIC AMINES

L. A. Damani, Department of Pharmacy, University of Manchester, Manchester M13 9PL, UK

1. Compounds containing aromatic nitrogen heterocyclic rings as part structures are widely distributed in nature, and many find use as drugs.
2. Metabolic reaction at a tertiary heteroaromatic ring nitrogen leads to a
   greater formal positive charge on the nitrogen, i.e. quaternization. All
   such reactions may be technically defined as 'N-oxidations'.
3. Examples of such 'biological quaternization reactions' include not only
   N-oxygenation (i.e. N-oxide formation), but also N-methylation and
   N-glucuronidation.
4. Chemical quaternization reactions occur with ease in most ring systems;
   this review examines to what extent these chemical reactions have biological counterparts, and where they do, their physiological significance and
   enzymic basis is discussed.

INTRODUCTION

Compounds containing *heterocyclic rings* as part-structures are widely distributed in nature, and some of these play an important role in many essential
biochemical processes. In addition, a large proportion of synthetic drug
molecules in use today also contain such ring systems. In view of this widespread exposure, or potential exposure, of mammals to nitrogen heterocycles,
there has been an increased interest in recent years in studying the biotrans-
formaton of model heterocyclic ring systems. The aims of such investigations

have been fourfold. Firstly to elucidate the different types of metabolic reactions that are possible in nitrogen heterocycles; secondly to examine their enzymic basis; thirdly to study the interrelationship and relative importance of each of these, often competing, biotransformation pathways; and lastly to attempt to relate structural and chemical features of specific heterocycles to their metabolism. In this respect, it must be noted that compared to the number of papers on the metabolism of carbocycles, such as benzene, naphthalene and anthracene derivatives, relatively little, and on occasions nothing, is known about the metabolism of their simple heterocyclic isosteres. At present, therefore, it is still premature to derive structure-activity relationships, even in a simple series of compounds, but clearly it is desirable to establish some general principles with model compounds and to employ this information prospectively whenever possible with more complex drug molecules.

The main metabolic reactions of heterocycles are often the same as those undergone by carbocycles. However, the heteroatom (e.g. nitrogen or sulphur) may also serve as an alternative site for metabolic attack, but perhaps more important, the heteroatom may profoundly influence the chemical and biochemical reactivity of different positions in the ring. Metabolic options open to aromatic nitrogen heterocycles include (a) oxidative reactions at ring carbons, (b) metabolic reactions at ring substituents, and (c) metabolic reactions at the electronegative heteroatoms. A more comprehensive discussion of the above reactions has been given elsewhere (Damani & Crooks, 1982; Damani & Case, 1984). The present review will therefore concentrate on metabolic reactions at the heteroatom, leading to the production of water-soluble quaternary ammonium metabolites. Examples of *'biological quaternization reactions'* include *N*-oxygenation (i.e. *N*-oxide formation), *N*-methylation and *N*-glucuronidation. Since all these reactions lead to a greater formal positive charge on the nitrogen, they may be all classified as '*N*-oxidations', i.e. metabolic *N*-oxidation of organic nitrogen compounds does not always produce *N*-oxygenated metabolites (Ziegler, 1984). This is technically correct, and it suits the purpose of the present reviewer to use this more broader definition of *N*-oxidation, to include a discussion of the other, often competing, reactions.

CHEMICAL QUATERNIZATION REACTIONS

Aromatic tertiary heterocyclic amines are less basic than their non-aromatic counterparts. The weakness of some nitrogenous bases has been associated with the proportion of s character in the orbital containing the unshared electrons. In trialkylamines the unshared pair of electrons are in an sp^3 hybrid orbital, in the less basic pyridine in an sp^2 orbital, and in nitriles, which show almost no basicity, in an sp orbital, the proportion of s character therefore increasing from left to right (Fig. 1a). In the case of pyrrole, the lone pair of electrons on the nitrogen atom is needed to make up the sextet of electrons essential for aromaticity, making pyrrole (and related ring systems) a very

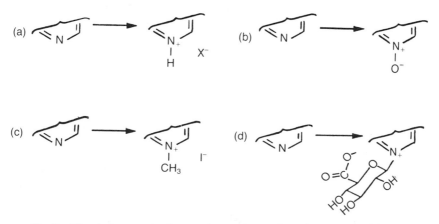

(a)

C
C—N
C

sp^3

C
N

sp^2

— C ≡ N

sp

(b)

1

1.194
1.058
1.052

1.096
N
H

0.881
N
0.941

Fig. 1 – Types of hybridization involved in nitrogen containing molecules (a), and π-electron density distributions in benzene, pyrrole and pyridine (b).

weak base. It is not surprising that such pyrrolic or indolic nitrogens are not readily susceptible to electrophilic enzymic attack (Damani & Case, 1984). The lone pair of electrons on the nitrogen in pyridine (and related ring systems) is not required to make up the sextet of electrons essential for aromaticity, which makes pyridine a much stronger base than pyrrole. In fact, electron density calculations (Fig. 1b) for pyridine show a drift of electrons from ring carbons to the electronegative nitrogen. Therefore, chemical quaternization reactions readily take place at nitrogen in pyridine and other tertiary aromatic nitrogen heterocycles (Fig. 2). These include protonation as a simple case, or N-oxide formation with peroxy acids (Ochiai, 1967; Katritzky & Lagowski, 1971), or N-alkylation with alkyl halides and reactive esters (Menschutkin, 1902; Bergmann et al., 1952), and N-glucuronide formation via a multi-step synthetic route (Aboul-Enein, 1977; Dalgaard, 1983; Shaker, 1984). It would now seem, as subsequent sections of this review will indicate, that these quaternization reactions also occur during the metabolism of nitrogen heterocyclic drugs and other foreign compounds.

(a)

N → N+
|
H X−

(b)

N → N+
|
O−

(c)

N → N+
|
CH3 I−

(d)

N → N+

O=C
O−
HO
HO
OH

Fig. 2 – Chemical quaternization reactions at nitrogen in azaheterocycles, (a) protonation (b) N-oxygenation (c) N-alkylation, and (d) N-glucuronidation.

BIOLOGICAL QUATERNIZATION REACTIONS—ENZYMOLOGY

N-Oxygenation of aromatic nitrogen heterocycles was established as a metabolic route in 1959 by Chaykin & Bloch. They isolated nicotinamide *N*-oxide from the tissues of rats administered radioactive nicotinamide (Fig. 3). From this discovery of heteroaromatic *N*-oxygenation as a metabolic route until 1971, the year of the 1st International Symposium on Biological Oxidation of Nitrogen in Organic Molecules (Bridges, Gorrod & Parke, 1972), a few other xenobiotics had been reported to form heteroaromatic *N*-oxides. Amongst these were 3-acetylpyridine (Neuhoff & Kohler, 1966), nikethamide (Chambon-Mougenot *et al.*, 1970, 4,4'-bipyridyl and quinoline (Gorrod, 1971) (Fig. 4). The data pertaining to the metabolism of these compounds, together with the natural occurrence and toxicology of heterocyclic *N*-oxides was reviewed at the 1st *N*-Oxidation Symposium by Gorrod (1971). The reviewer drew attention to the paucity of information on this type of reaction, and in particular the lack of data on the enzyme systems mediating this reaction, and further postulated (Gorrod, 1973) that heteroaromatic *N*-oxygenation may be mediated via a cytochrome P-450 system.

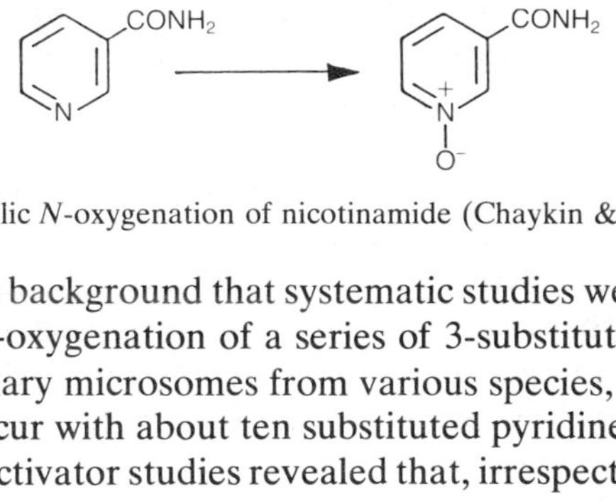

Fig. 3 – Metabolic *N*-oxygenation of nicotinamide (Chaykin & Bloch, 1959).

It was against this background that systematic studies were initiated in 1974 on the metabolic *N*-oxygenation of a series of 3-substituted pyridines. Using hepatic and pulmonary microsomes from various species, *N*-oxygenation was demonstrated to occur with about ten substituted pyridines. Results of induction, inhibitor and activator studies revealed that, irrespective of the pK_a of the compound, the reaction was mediated by a phenobarbitone-inducible cyto-

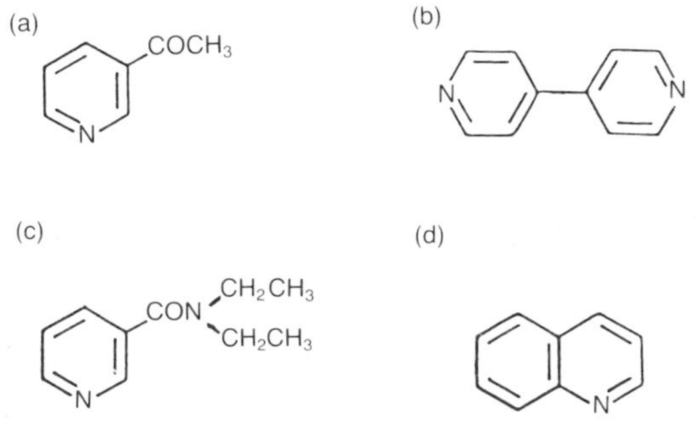

Fig. 4 – Aromatic nitrogen heterocycles, other than nicotinamide, reported to be *N*-oxygenated up to 1971. (a) 3-acetylpyridine, (b) 4,4'-bipyridyl, (c) nikethamide, and (d) quinoline.

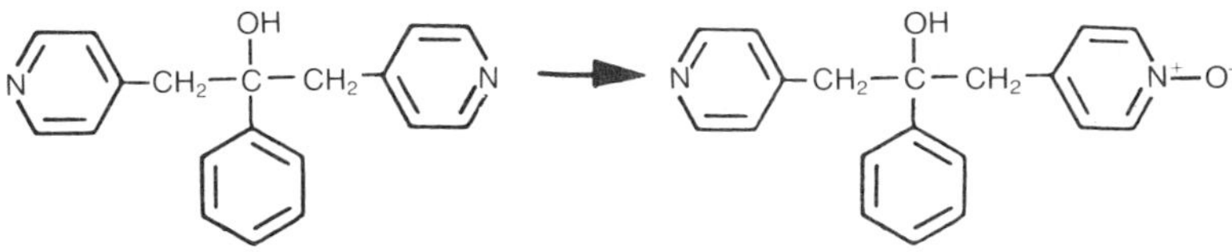

Fig. 5 – Metabolic N-oxygenation of prochiral 2-phenyl-1,3-di(4-pyridyl)-2-propanol to a chiral N-oxide (Schwartz et al., 1978).

chrome P-450 system. Furthermore, pyridyl N-oxides were found as urinary metabolites in six different laboratory animal species, and in the case of pyridine itself, N-oxygenation was a major route of metabolism, accounting for up to 40% of the administered dose in some species. These results were in part presented in 1977 at the 2nd N-Oxidation Symposium (Damani et al., 1978; Gorrod, 1978), and published in full elsewhere (Damani, 1977; Cowan et al., 1978; Gorrod & Damani, 1979a,b; Gorrod & Damani, 1980). These conclusions have to a large extent been confirmed by subsequent studies. For example, the prochiral compound, 2-phenyl-1,3-di(4-pyridyl)-2-propanol (Fig. 5), affords a chiral pyridyl N-oxide as a major urinary metabolite in rat, dog and man (Schwartz et al., 1978). Similarly, 2-methyl-1,2-bis(3-pyridyl)propane-1-one (metyrapone) is extensively metabolized via N-oxygenation in the rat (Usansky & Damani, this volume). The N-oxygention of both the above compounds was demonstrated in vitro, the reaction in each case being inducible by phenobarbitone, indicating cytochrome P-450 involvement. In fact, heterocyclic aromatic amines, e.g. pyridines, purines, pyrimidines, etc., cannot be N-oxygenated by pure hog liver flavin-containing monooxygenase (Ziegler, 1980). On the basis of published data, it would seem that the N-oxygenation of nucleophilic trialkylamines (e.g. trimethylamine) and N,N-dialkylarylamines (e.g. N,N-dimethylaniline) is a flavoprotein reaction (Gorrod, 1973; Ziegler, 1980, 1982, 1984), whereas N-oxygenation of the less nucleophilic heterocyclic aromatic amines is by cytochrome P-450 (see Fig. 6). There appears to be a rough parallel in S-oxygenation of thioethers. For example, dialkylsulphides (e.g. diethylsulphide) are oxidized by the flavoprotein only, whereas the diarylsulphides (e.g. dibenzothiophene) are oxidized by cytochrome P-450 (Houdi & Damani, 1984). In this respect, it is interesting to note that compounds of intermediate nucleophilicity, alkylarylsulphides (e.g. p-tolylethylsulphide), are substrates for both the microsomal mixed function oxidases (Waxman et al., 1982) (Fig. 6). It would be very tempting to suggest that this situation may also exist with intermediate amines, i.e. alkyldiarylamines, but as events over the last decade have shown, such simplistic rules based on pK_a and nucleophilicity are not likely to be either totally predictive, or fully explain the enzymic basis of such reactions (Gorrod, 1978).

N-Methylation and quarternization of a heteroaromatic nitrogen was first observed about a hundred years ago, when His detected N-methylpyridinium hydroxide in the urine of dogs after injection of pyridine acetate (His, 1887)

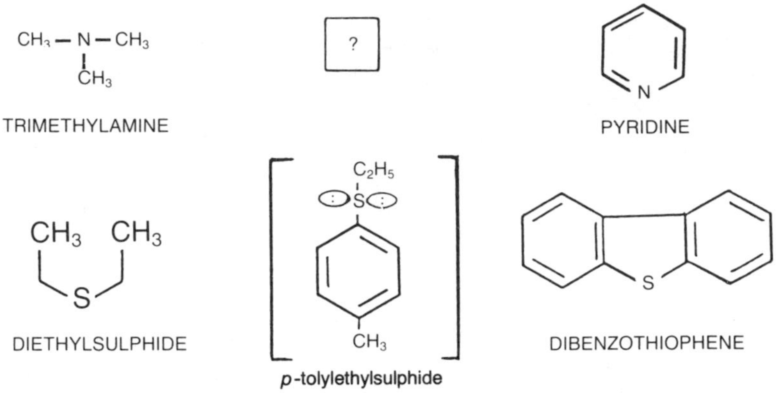

CH₃ — N — CH₃
        |
        CH₃

TRIMETHYLAMINE

?

PYRIDINE

CH₃ CH₃

DIETHYLSULPHIDE

C₂H₅
 |
 S

CH₃
p-tolylethylsulphide

DIBENZOTHIOPHENE

Fig. 6 – Cytochrome P-450 and FAD-containing monooxygenase catalysed *N*- and *S*-oxygenations in compounds of differing nucleophilicities.

(Fig. 7). Subsequent work in many laboratories confirmed this early study and extended this observation to many different animal species (Abderhalden *et al.*, 1909; Totani & Hoshiai, 1910). More recent studies have indicated that this type of reaction is a quantitatively important *in vivo* route of metabolism for pyridine (D'Souza *et al.*, 1980; Shaker *et al.*, 1982; Damani *et al.*, 1982, and references cited therein for early studies on *N*-methylation). *N*-Methylpyridinium ion has also been identified as an *in vitro* metabolite of pyridine using rabbit tissue preparations. Pyridine *N*-methyltransferase activity resides mainly in the pulmonary cytosolic fraction, and to a lesser extent in the hepatic and renal cytosolic fractions of rabbit tissues. The rabbit lung pyridine *N*-methyltransferase has been partially purified by ammonium sulphate precipitation. The enzyme utilizes *S*-adenosyl-L-methionine, but not N^5-methyltetrahydrofolate, as a methyl donor, it has a pH optimum at around 7.7, and a very low K_m and high V_{max} values for pyridine (Damani & Shaker, 1984a; Shaker, 1984). *N*-Methylation of pyridine was also demonstrated by the above investigators with prified rabbit liver 'arylamine *N*-methyltransferase', contrary to previous reports of the inability of this enzyme to *N*-methylate tertiary azaheterocyclic nitrogens (Lyon & Jakoby, 1982). The pyridine *N*-methylating enzyme in rabbit lung has some similarities with the rabbit 'arylamine *N*-methyltransferase' and the rabbit lung 'non-specific *N*-methyltransferase' first described by Axelrod (1962). It is probable that mammalian tissues contain a number of *N*-methyltransferases with circumscribed patterns of specificity for their methyl group acceptor and

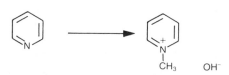

Fig. 7 – *N*-Methylation and quaternization of pyridine in the dog (His, 1887).

utilizing *S*-adenosyl-L-methionine as the methyl group donor. Further work is required to purify the various 'non-specific *N*-methyltransferases' and to determine their substrate specificities, since it is now recognized that such enzymes do function as enzymes of detoxication for a variety of nitrogenous compounds. Although quaternary *N*-glucuronide formation is an important route of metabolism for many tertiary amines (see below), to date no detailed *in vitro* studies have been carried out on this type of *N*-oxidation and therefore, the enzymology of this reaction is still obscure.

BIOLOGICAL QUATERNIZATION REACTIONS—PHYSIOLOGICAL IMPORTANCE

This section examines the quantitative importance of *N*-oxygenation, *N*-methylation and *N*-glucuronidation as *in vivo* routes of metabolism for azaheterocyclic drugs and other foreign compounds. A recent study has attempted a comparison of the role of these three biological quaternization reactions in the simple model compound pyridine (Damani & Shaker, 1984a,b; Shaker, 1984). After administration of a 5 mg/kg or 50 mg/kg dose to several animal species, the pyridyl *N*-glucuronide was not detected. Synthetic *N*-glucuronide was available, and it was estimated that if *N*-glucuronidation does occur at all in these species, it must account for less than 5% of the administered dose. *N*-Methylation and *N*-oxygenation, on the other hand, were important routes of metabolism in all species (Table 1). *N*-Metylation of pyridine was species dependent, and decreased in the order hamster, guinea-pig, rabbit, rat and mouse. In the hamster and guinea-pig, *N*-methylation was also sex-dependent; in both species *N*-methylation occur-

Table 1

Quaternization reactions at nitrogen in pyridine (*N*-methylation, *N*-oxygenation, *N*-glucuronidation)

Species	% Dose excreted in urine (0-72H) as: (dose 5 mg/kg body weight)		
	N-Methylpyridinium ion	Pyridine *N*-oxide	Pyridine *N*-glucuronide
Hamster	49.6 ± 10.5	13.5 ± 3.3	<5
Guinea-pig	31.7 ± 4.3	9.9 ± 7.0	<5
Rabbit	12.6 ± 4.8	13.2 ± 4.8	<5
Rat	8.1 ± 0.3	11.0 ± 1.6	<5
Mouse	6.3 ± 1.0	15.0 ± 3.0	<5

Damani & Shaker (1984 a,b)

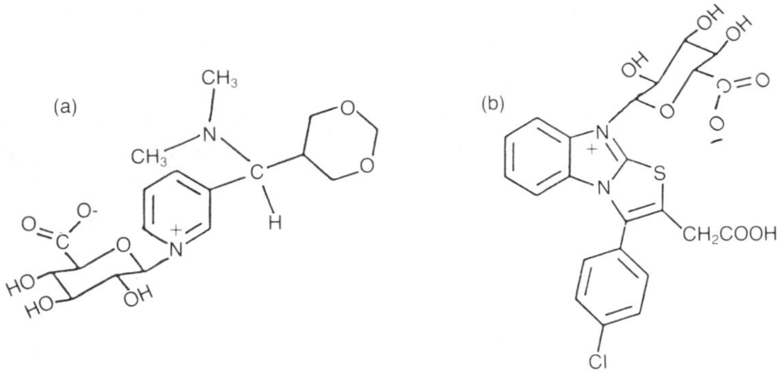

Fig. 8 – Quaternary *N*-glucuronides of LY-108380 (a), and WY-18251 (b) (Rubin *et al.*, 1979, and Janssen *et al.*, 1982, respectively).

ring to a greater extent in males than in females. This reaction was also dose-dependent; whereas at low doses it was the major route of metabolism, at the higher doses (e.g. 50 or 500 mg/kg) the reaction appeared to be a saturable pathway. At these high doses *N*-oxygenation assumes greater importance, and a larger proportion of the dose is excreted as the *N*-oxide. Although *N*-glucuronidation at a pyridyl nitrogen was not observed in the study described above, such a reaction has previously been reported. The potential analgesic agent 1,3-(dimethylamino)(m-dioxan-5-yl)pyridine (LY-108380) affords a pyridyl quaternary *N*-glucuronide (Fig. 8a) as a major urinary metabolite (55% of dose) in man (Rubin *et al.*, 1979). This quaternization reaction is not limited to pyridyl rings, and has been reported for the benzimidazole derivative WY-18251 (Fig. 8b); in this case, the *N*-glucuronide was not detected in rats and dogs, but accounted for about 30% of the dose in rhesus monkey (Janssen *et al.*, 1982). In addition to the azaheterocycles discussed above, quaternary *N*-glucuronides have also been reported in tertiary alicyclic and aliphatic amines, e.g. cyproheptadine and tripelennamine (Lehman & Fenselau, 1982 and references cited therein). In all these instances, it is interesting that this reaction occurs more readily in man and primates than in the experimental animals more commonly used. These interspecies differences should be considered seriously when findings in infrahuman species are used to predict the metabolism in humans of tertiary amino drugs.

   N-Methylation appears to be more important in the common laboratory animals, including the dog, which is widely used in industry for evaluating the metabolism of new drugs. Quaternary *N*-methylated metabolites are difficult to quantify, since they cannot be readily converted back to the parent tertiary amines enzymically or chemically. In addition, such metabolites, in conformity with their ionic properties, are very difficult to extract into organic solvents in the absence of ion pairing agents. Therefore, although such metabolites have been reported for some azaheterocyclic compounds

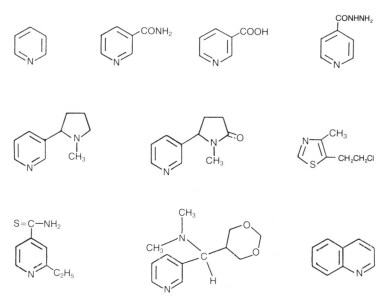

Fig. 9 – Structures of some azaheterocycles reported to be *N*-methylated, but where quantitative data is not available.

(Fig. 9), the quantitative importance of this route of metabolism for the various compounds is not as yet certain. The development in recent years of specific ion-pair extraction techniques and ion-exchange chromatography should make the detection and analysis of such metabolites relatively easy, and it is hoped that future studies will pay more attention to this biotransformation pathway, in view of the importance of heterocyclic compounds in medicinal chemistry.

The number of pyridino-compounds that undergo *N*-oxygenation *in vivo* is now large (Table 2). In some of the earlier studies e.g. with nicotinamide, 4,4'-bipyridyl and cotinine, detailed quantitative data are not available. But in the more recent studies, where specific h.p.l.c. or radiometric methods have been utilized, pyridyl *N*-oxides have been reported as major urinary metabolites. For example, *N*-oxide formation with the new antihypertensive agent pinacidil accounts for 40%, 54% and 54% of the dose in rat, dog and man, respectively (Table 2). *N*-Oxides are also observed as metabolites in other ring systems. For example the triazene derivative *N*,*N*-diallylmelamine (Zins *et al.*, 1968), the benzopyridines quinoline and isoquinoline (Gorrod, 1971), the pyrimidine derivative trimethoprim (Brooks *et al.*, 1973), and the antimalarial agent chloroquine (Essien & Afamefuna, 1982). This list is by no means comprehensive, and other examples of azaheterocyclic *N*-oxygenations will be found elsewhere in this volume, and in the reviews by Damani & Crooks (1982) and Damani & Case (1984).

The metabolism of some 3-pyridyl ketones has attracted attention in recent

Table 2

Examples of pyridino-compounds that undergo N-Oxygenation *in vivo*

Compound	Species [% of dose as N-oxide(s)]	References
3-Acetylpyridine	rat (10%)	Neuhoff & Köhler, 1966
Nicotinamide	rat, mouse, hamster, man	Sparthan & Chaykin, 1969
Nikethamide	guinea-pig (7.5%) rats (4%)	Chambon-Mougenot *et al.*, 1970
4,4'-Bipyridyl	guinea-pig	Gorrod, 1971
Cotinine	monkey	Dagne & Costagnoli, 1972
3,5-Bis(4-pyridyl)-1,2,4-triazole	rat, dog, monkey (major metabolite)	Duggan *et al.*, 1975
Pyridine, 3-chloropyridine, 3-methylpyridine	mouse, rat, hamster, guinea-pig, rabbit and ferret (1-40%)	Damani *et al.*, 1978
2-Phenyl-1,3-di(4-pyridyl) 2-propanol	rat (45%), dog (83%), man (57%)	Schwartz *et al.*, 1978
Rosoxacin	dog (4.0%)	Kullberg *et al.*, 1979
Pinacidil	rat (40%), dog (54%), man (54%)	Eilersten *et al.*, 1982
3-Benzoylpyridine	rat (51.15%), dog (5.1%)	Eyer & Hell, 1983
Metyrapone	rat (85%)	Usansky & Damani, 1985

years. 3-Acetylpyridine had previously been reported to afford 1-(3-pyridyl *N*-oxide) ethanol as a urinary metabolite in the rat (Neuhoff & Köhler, 1966). Studies *in vitro* with hepatic subcellular preparations showed that whereas 3-acetylpyridine itself was a substrate for *N*-oxidation, the keto-reduction product 1-(3-pyridyl) ethanol was not *N*-oxygenated (Damani *et al.*, 1980). From these findings, the authors concluded that the route to 1-(3-pyridyl *N*-oxide) ethanol as a principal urinary metabolite of 3-acetylpyridine is via *N*-oxidation of the parent followed by carbonyl reduction (Fig. 10), and reasoned that lipophilicity may be an important prerequisite for *N*-oxidation. Such a condition is obviously fulfilled by the more lipophilic benzyl alcohol in 3-hydroxy-BP (Fig. 11). The parent compound, 3-benzoylpyridine undergoes *N*-oxidation and keto-reduction. The major metabolite in the rat is 3-hydroxy-BP N-oxide, arising not only as a result of keto-reduction of 3-BP *N*-oxide, but also as a result of *N*-oxidation of 3-hydroxy-BP (Eyer & Hell, 1983). There are strong differences in the urinary metabolic profile of 3-benzoylpyridine between the rat and dog. In the dog, the major pathways of metabolism appear to be *N*-methylation and *O*-glucuronidation, after an ini-

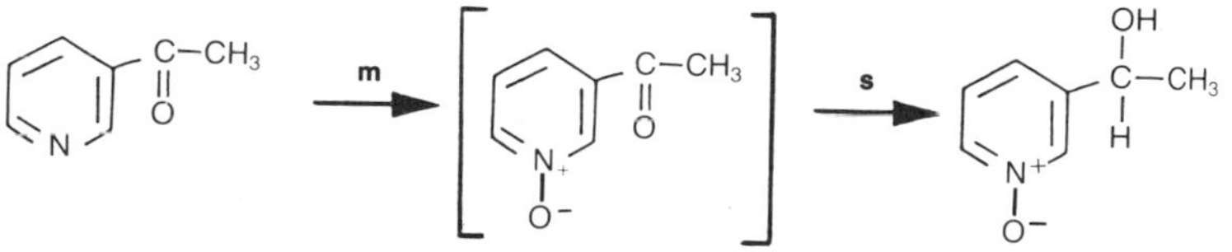

Fig. 10 – The metabolic route to 1-(3-pyridyl *N*-oxide) ethanol from 3-acetylpyridine. m = microsomal reaction, s = cytosolic reaction (Damani *et al.*, 1980).

tial keto-reduction to 3-hydroxy-BP. This work on 3-benzoylpyridine appears to support current work from the authors laboratory on the metabolism and pharmacokinetics of metyrapone (Usansky & Damani, this volume). The major route of metabolism for metyrapone in the rat appears to be keto-reduction, followed by *N*-oxidation at one or the other pyridyl nitrogens to afford two isomeric metyrapol mono *N*-oxides as urinary metabolites, together accounting for about 75% of the administered dose.

In summary, this review has described the enzymic basis and physiological significance of *'biologial quaternization reactions'* (*N*-oxygenation, *N*-methylation and *N*-glucuronidation). It is quite clear that these reactions can be important, depending on the compound, the dose and species used, and on what other metabolic vulnerable sites or substituents are available for reaction on the nitrogen heterocycle. In this respect it is interesting to quote this rather prophetic statement made over a decade ago by Gorrod (1971) '. . . in the many cases where the total metabolic pathways of drugs are not known, it may be that heteroaromatic *N*-oxidation has occurred and that the products with their extreme water solubility have been missed . . .'

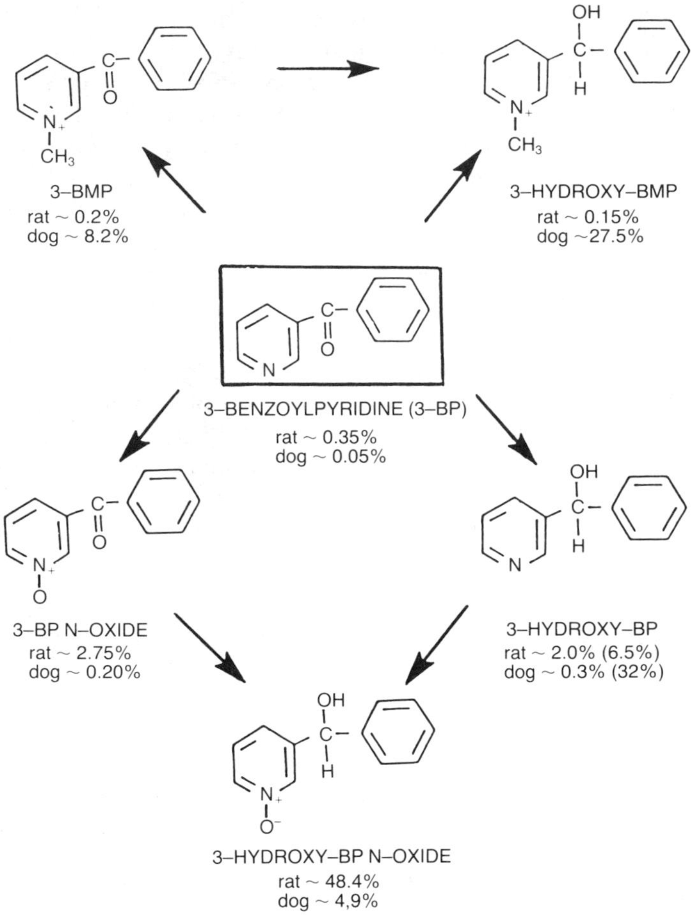

Fig. 11 – The metabolism of 3-benzoylpyridine (3-BP) in rats and dogs (Eyer & Hell, 1983). Values in parenthesis for 3-hydroxy-BP are amounts excreted as O-glucuronide, mean percentage of dose in urine after i.p. and i.m. administration.

ACKNOWLEDGEMENTS

The author thanks his colleagues, J. I. Usansky, M. S. Shaker and A. A. Houdi for permission to use as yet unpublished data in compilation of this review, and Mrs E. A. Bailey for her help in typing this manuscript.

REFERENCES

Abderhalden, E., Brahm, C. & Schittenhelm, A. (1909), *Hoppe-Seyler's Z. Physiol. Chem.*, **59**, 32.
Aboul-Enein, H. Y. (1977), *J. Carbohyd. Nucleosides Nucleotides*, **4**, 77.
Axelrod, J. (1962), *J. Pharmac. exp. Ther.*, **138**, 28.

Bergmann, E. D., Crane, F. E., Jr. & Fuoss, R. M. (1952), *J. Am. Chem. Soc.*, **74**, 5979.

Bridges, J. W., Gorrod, J. W. & Parke, D. V. (eds) (1972), *Biological Oxidation of Nitrogen in Organic Molecules*, Taylor & Francis, London.

Brooks, M. A., De Silva, J. A. F. & D'Arconte, L. (1973), *J. Pharm. Sci.*, **62**, 1395.

Chambon-Mougenot, R., Riotte, M., Bringuier, J. & Chambon, P. (1970), *Cr. r. Seanc. Soc. Biol.*, **164**, 856.

Chaykin, S. & Bloch, K. (1959), *Biochim. biophys. Acta*, **31**, 213.

Cowan, D. A., Damani, L. A. & Gorrod, J. W. (1978), *Biomed. Mass Spectrom.*, **5**, 551.

Dagne, E. & Castagnoli, N., Jr. (1972), *J. med. Chem.*, **15**, 840.

Dalgaard, L. (1983), *Acta Chemica Scandinavica B.*, **37**, 923.

Damani, L. A. (1977), PhD Thesis, University of London, UK.

Damani, L. A., Bryan, J., Cowan, D. A. & Gorrod, J. W. (1980), *Xenobiotica*, **10**, 645.

Damani, L. A. & Case, D. E. (1984), in *Comprehensive Heterocyclic Chemistry*, Volume 1 (ed. Meth-Cohn, O.), Pergamon Press, Oxford, p. 223.

Damani, L. A., Disley, L. G. & Gorrod, J. W. (1978), in *Biological Oxidation of Nitrogen* (ed. Gorrod, J. W.), Elsevier/North Holland Biomedical Press, Amsterdam, p. 157.

Damani, L. A. & Crooks, P. A. (1982) in *Metabolic Basis of Detoxication* (cds. Jakoby, W. B., Bend, J. R. & Caldwell, J.), Academic Press, New York, p. 69.

Damani, L. A., Crooks, P. A., Shaker, M. S., Caldwell, J., D'Souza, J. & Smith, R. L. (1982), *Xenobiotica*, **12**, 527.

Damani, L. A. & Shaker, M. S. (1984a), unpublished observations.

Damani, L. A. & Shaker, M. S. (1984b), *J. Pharm. Pharmacol.*, **36** (supplement), 63P.

D'Souza, J., Caldwell, J. & Smith, R. L. (1980), *Xenobiotica*, **10**, 151.

Duggan, D. E., Noll, R. M., Baer, J. E., Novello, F. C. & Baldwin, J. J. (1975), *J. med. Chem.*, **18**, 900. Also personal communication (1976).

Eilersten, E., Magnussen, M. P., Peterson, H. J., Rastrup-Andersen, N., Sorensen, H. & Arrigoni-Martelli, E. (1982), *Xenobiotica*, **12**, 187.

Essien, E. E. & Afamefuna, G. C. (1982), *Clin. Chem.*, **28**, 1148.

Eyer, P. & Hell, W. (1983), *Xenobiotica*, **13**, 649.

Gorrod, J. W. (1971), *Xenobiotica*, **1**, 349.

Gorrod, J. W. (1973), *Chem-biol. Interact.*, **7**, 289.

Gorrod, J. W. (1978) in *Biological Oxidation of Nitrogen* (ed. Gorrod, J. W.), Elsevier/North Holland Biomedical Press, Amsterdam, p. 201.

Gorrod, J. W. & Damani, L. A. (1979a), *Xenobiotica*, **9**, 209.

Gorrod, J. W. & Damani, L. A. (1979b), *Xenobiotica*, **9**, 219.

Gorrod, J. W. & Damani, L. A. (1980), *Eur. J. Drug Metab. Pharmacokinet.*, **5**, 53.

His, W. (1887), *rch. exp. Path. Pharmak.*, **22**, 253.

Houdi, A. A. & Damani, L. A. (1984), *J. Pharm. Pharmacol.*, **36**, (supplement), 62P.

Janssen, F. W., Kirkman, S. K., Fenselau, C., Stogniew, M., Hofmann, B. R., Young, E. M. & Ruelius, H. W. (1982), *Drug Metab. Dispos.*, **10**, 599.

Katritzky, A. R. & Lagowski, J. M. (1971), *Chemistry of Heterocyclic N-Oxides*, Academic Press, London.

Kullberg, M. P., Koss, R., O'Neil, S. & Edelson, J. (1979), *J. Chromatogr.*, **173**, 155.

Lehman, J. P. & Fenselau, C. (1982), *Drug Metab. Dispos.*, **10**, 446.

Lyons, E. S. & Jakoby, W. B. (1982), *J. biol. Chem.*, **257**, 7531.

Menchutkin, J. (1902), *J. Russ. Phys. Chem. Soc.*, **34**, 411.

Neuhoff, V. & Köhler, F. (1966), *Naunyn-Schmiedebergs Archs. exp. Path. Pharmac.*, **254**, 301.

Ochiai, E. (1967), *Aromatic Amine Oxides*, Elsevier/North Holland Biomedical Press, Amsterdam.

Rubin, A., Dhahir, P. H., Crabtree, R. E. & Henry, D. P. (1979), *Drug Metab. Dispos.*, **7**, 149.

Schwartz, M. A., Williams, T. H., Kolis, S. J., Postma, E. & Sasso, G. J. (1978), *Drug Metab. Dispos.*, **6**, 647.

Shaker, M. S. (1984), PhD Thesis. University of Manchester, UK.

Shaker, M. S., Crooks, P. A. & Damani, L. A. (1982), *J. Chromatogr.*, **237**, 489.

Sparthan, M. & Chaykin, S. (1969), *Analyt. Biochem.*, **31**, 286.

Totani, G. & Hoshiai, Z. (1910), *Hoppe-Seyler's Z. physiol. Chem.*, **68**, 83.

Usansky, J. I. & Damani, L. A. (1985), this volume.

Waxman, D. J., Light, D. R. & Walsh, C. (1982), *Biochemistry*, **21**, 2499.

Ziegler, D. M. (1980), in *Enzymatic Basis of Detoxication*, (ed. Jakoby, W. B.) Volume 1, Academic Press, New York, p. 201.

Ziegler, D. M. (1982), in *Metabolic Basis of Detoxication* (eds. Jakoby, W. B., Bend, J. R. & Caldwell, J.), Academic Press, New York, p. 171.

Ziegler, D. M. (1984), in *Drug Metabolism and Drum Toxicity* (eds. Mitchell, J. R. & Horning, M. G.), Raven Press, New York, p. 33.

Zins, G. R., Emment, D. E. & Walk, R. A. (1968), *J. Pharmac. exp. Ther.*, **159**, 146.

Chapter 23

AMINE–IMINE TAUTOMERISM AS A DETERMINANT OF THE SITE OF BIOLOGICAL N-OXIDATION

J. W. Gorrod, Department of Pharmacy, Chelsea College (University of London, Manresa Road, London SW3 6LX, UK

1. Amino heteroaromatic structures occur widely as drugs, environmental contaminants and natural products.
2. They are capable of being converted into hydroxylamines by oxidation of the *exo*-amino group or amine oxides by *endo*-nitrogen oxidation.
3. Hydroxylamines may be toxic, whereas amine oxides are usually innocuous.
4. It is now proposed that amine–imine tautomerism is a determinant of the site of biological N-oxidation.
5. Metabolic N-oxidation of amino heteroaromatic compounds appear to involve cytochrome P-450 isozymes and not the flavoprotein-containing monooxygenase.

AMINO AZAHETEROCYCLES CONVERTED TO INNOCUOUS AMINE OXIDES

Many compounds possessing an amino group adjacent to a nitrogen hetero atom in an aromatic system are used in clinical medicine (Fig. 1). These compounds have a wide range of pharmacological activity and include the naturally occurring vitamin thiamine and the nucleic acid bases adenine, guanine and cytosine. Metabolic studies have rarely accounted for the major

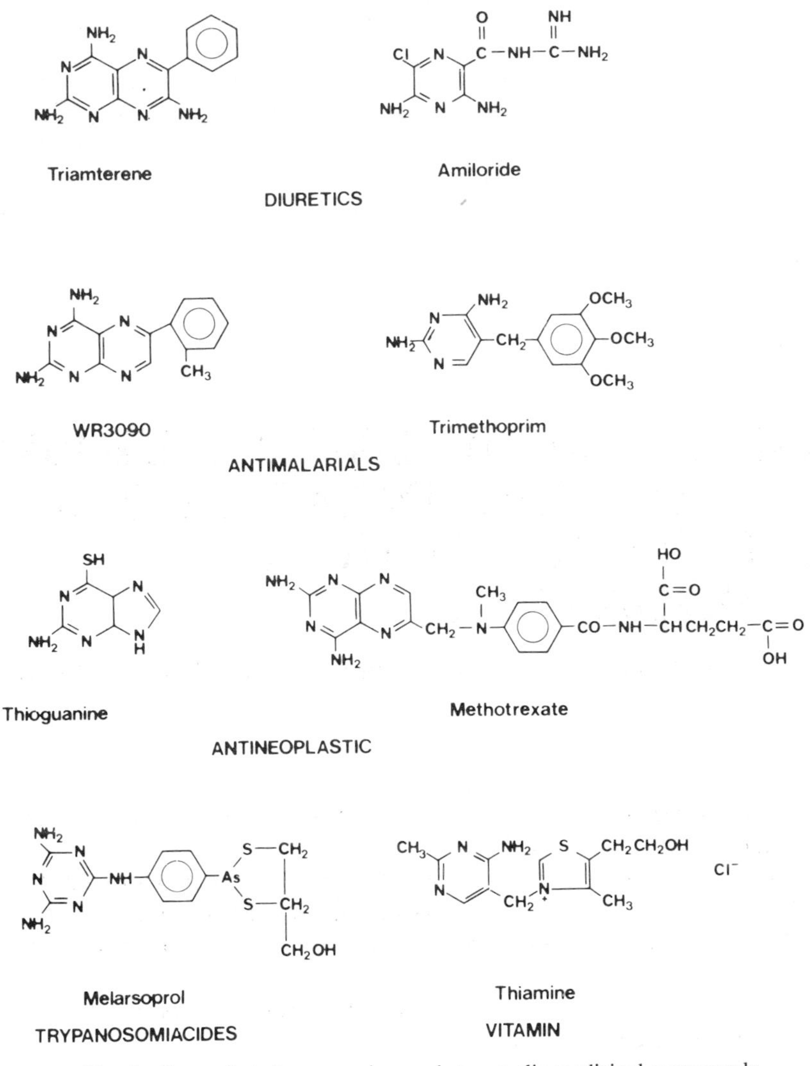

Triamterene

Amiloride

DIURETICS

WR3090

Trimethoprim

ANTIMALARIALS

Thioguanine

Methotrexate

ANTINEOPLASTIC

Melarsoprol

Thiamine

TRYPANOSOMIACIDES

VITAMIN

Fig. 1 – Examples of some amino azaheterocyclic medicinal compounds.

portion of the administered dose, yet certain metabolic patterns are emerging.

Early studies with trimethoprim (Schwartz *et al.*, 1970) showed that this antibacterial compound was converted to the 1-*N*-oxide in rats, dog and man. The metabolism of trimethoprim was also studied by Meshi & Sato (1972), who reported that the rat excreted 7% of a dose as the *N*-oxide. Whilst the position of *N*-oxidation was not established, these authors indicated that they thought it was the 1-*N*-oxide which was formed. Further studies showed that *N*-oxidation at the 3-position also occurred in man and that these pathways accounted for between 3–5% of the dose (Sigel, 1973; Sigel & Brent, 1975). A species variation in the site of *N*-oxidation of trimethoprim was observed by

Brooks *et al.*, (1973) who developed a differential pulse polarographic technique for measuring the *N*-oxides. These authors found that whereas man excretes about equal amounts of both the 1- and 3-*N*-oxide, the rat excreted predominantly the 1-isomer and the dog predominantly the 3-isomer (Fig. 2). Other studies by Hubbel *et al.* (1978) have confirmed that trimethoprim is converted by the rat to both the 1- and 3-*N*-oxide. The former isomer accounting for 5.5% of an oral dose, whereas the 3-isomer was present in urine only to the extent of 0.4%.

These latter workers (Hubbel *et al.*, 1978) also reported their results on the *in vivo* metabolism of pyrimethamine and metoprine by rats. In the case of pyrimethamine, a major metabolite (34%) was the 3-*N*-oxide of a compound which had been hydroxylated on the α-carbon of the β-ethyl group. The 1- and 3-*N*-oxides were also present in urine as free compounds (1-NO = 4%, 3-NO = 1.4%) and glucuronic acid conjugates (1-NO = 1.4%, 3-NO = 1%) (Fig. 2). Metoprine was also metabolized in the rat by *endo N*-oxidation to form the 1-*N*-oxide (which accounted for 60% of the dose), and the 1-*N*-oxide glucuronic acid conjugate a further 7% of the orally administered dose. This compound was also metabolized by mice to give both the 1- and 3-*N*-oxides, the 1-isomer predominating in both sexes and males excreting more than female mice (Hubbel *et al.*, 1980) (Fig. 2).

Even prior to this work, Zins (1965) had shown that the vasodilator, diallylmelamine (Fig. 2), was converted to a nuclear *N*-oxidized compound in which *N*-oxidation had taken place at the nitrogen *para* to the diallylamine group. In this case, *N*-oxidation produced an active hypotensive metabolite. In a later paper (Zins *et al.*, 1968), a further *N*-oxidized mono-*N*-deallylated metabolite was recognized and a species difference in *N*-oxidation was established. Thus, whilst the rat and dog excreted 4–5% of the dose as the *N*-oxide, humans only excreted 0.3–0.4% and were therefore unresponsive to the hypotensive effect of the drug. It is interesting that in this paper the authors consider the *N*-oxide metabolite as existing in the *N*-hydroxyimine tautomeric form (Fig. 4).

A further case in which the metabolism of an amino heteroaromatic compound has been investigated is the cocciostatic compound Aprinocid [6-amino-9-(2-chloro- 6-fluorobenzyl)purine]. Wolf *et al.* (1978), showed that microsomal preparations from chicken and dog liver convert Aprinocid to the 1-*N*-oxide, (Fig. 2). This *N*-oxide was also found in the urine of chickens, rats and mice, even though *in vitro* experiments using hepatic preparations from the latter two species indicated that *N*-debenzylation was the major pathway. From these examples, one comes to the general conclusion that: (a) where an amino group is present in a molecule adjacent to a ring nitrogen, the *N*-oxidation takes place on the ring nitrogen, (b) the presence of the *exo*-amino group appears to be essential to allow nuclear *N*-oxidation to proceed (although this may be too great a generalization as many *N*-oxidations may have been overlooked (Gorrod, 1971), (c) this biological *N*-oxidation process does not produce genotoxic metabolites.

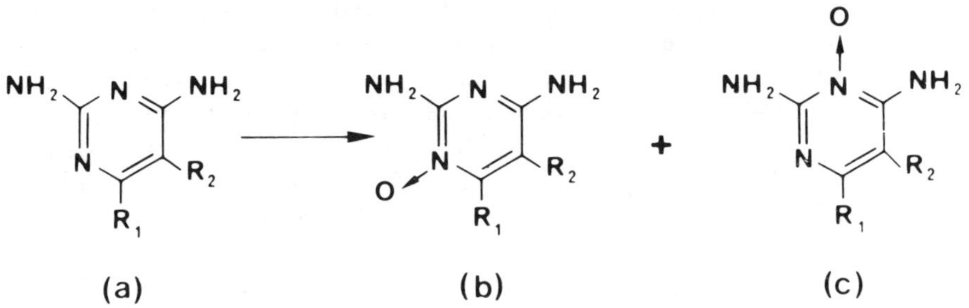

(a)   (b)   (c)

		R_1	R_2
1.	Trimethoprim (a)	H	3,4,5,-trimethoxybenzyl
2.	Metoprine (a)	CH_3	3,4-dichlorophenyl
3.	Pyrimethamine (a)	C_2H_5	4-chlorophenyl

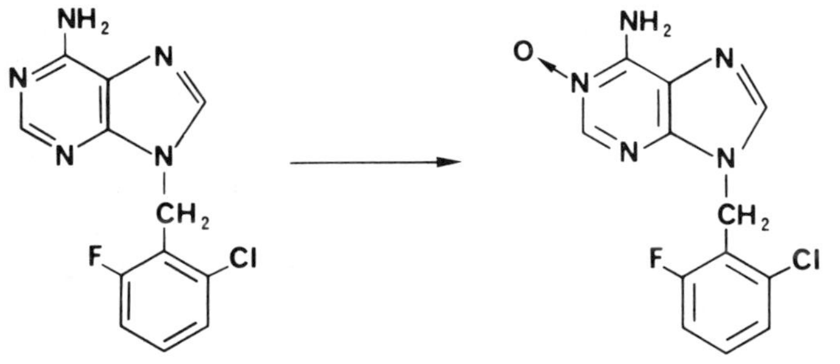

Aprinocid

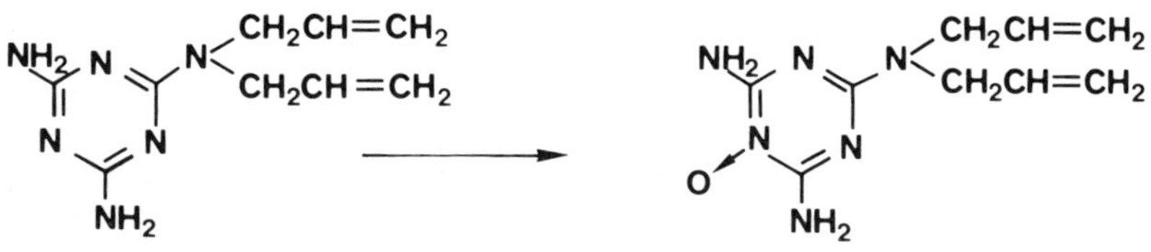

Diallylmelamine

Fig. 2 – Amino heterocycles converted to innocuous amine oxides.

AMINO AZAHETEROCYCLES CONVERTED TO GENOTOXIC HYDROXYLAMINES

Over the last few years considerable interest has been shown in mutagenic material produced during the cooking of food or the pyrolysis of certain amino acids (Sugimura *et al.*, 1977), and this has led to the isolation and characterization of several amino aromatic heterocyclic compounds as the mutagenic agents. Of these products 3-amino-1-methyl-5*H*- pyrido-(4,3*b*)-indole (Trp-P-2) and the corresponding 1,3-dimethyl compound (Trp-P-1) were among the most active mutagens. Further studies have shown that these compounds needed metabolic activation to be mutagenic (Nemoto *et al.*, 1979). In the case of Trp-P-1, the metabolites having greater mutagenicity have been shown to be the 3-*N*-hydroxy and 3-nitro derivatives (Yamazoe *et al.*, 1980a; Hashimoto *et al.*, 1980a) (Fig. 3). Similarly, another mutagen, 2-amino-6-methylpyrido(1,2*a*:3,2*d*)imidazole (Glu-P-1) found in pyroly-sates of L-glutamic acid is converted to the 2-hydroxylamino compound following incubation with fortified rat liver microsomes (Hashimoto *et al.*, 1980b) (Fig. 3).

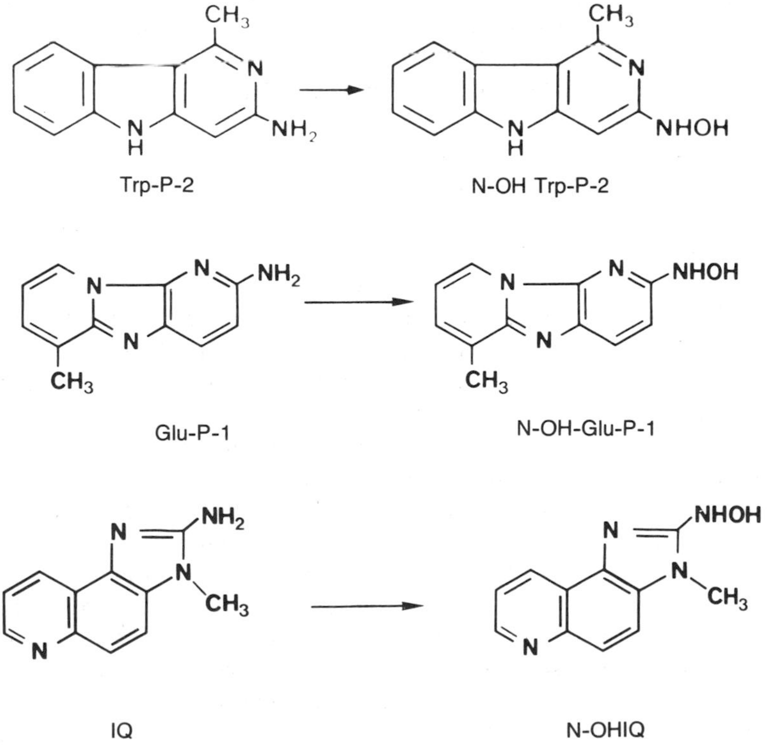

Fig. 3 – Amino heterocycles converted to genotoxic hydroxylamines.

A further aromatic aminoheterocyclic compound which is mutagenic was isolated from broiled sardines. The structure of the compound was shown to be 2-amino-3-methylimidazole[4,5f] quinoline [IQ]. The mutagenic activity of this compound is also dependent upon metabolic activation, which again occurs by oxidation of the *exo*-amino group (Okamoto *et al.*, 1981) (Fig. 3). From the above it appears that in these cases, contrary to the earlier situation, *N*-oxidation occurs on the *exo*-amino group to produce a hydroxylamine which is genotoxic. This means that *N*-oxidation can occur at either nitrogen site in an amino aromatic heterocyclic molecule and clearly some physico-chemical parameter must determine the site of *N*-oxidation and hence the toxicity of the molecule.

AMINE–IMINE TAUTOMERISM

It is possible that the controlling parameter is amine–imine tautomerism, as several pieces of evidence point in this direction. Firstly, there is the observation that an *exo*-amino group is presumably required in order to allow *N*-oxidation of certain aromatic heterocyclic molecules to proceed. This is only required in multi-nitrogen aromatic systems and is not a prerequisite for 'pyridine' *N*-oxidation (Cowan *et al.*, 1978; Gorrod & Damani, 1979a,b). This *exo*-amino group cannot be required only to increase the electron density of the ring nitrogen as the calculated values for pyridine and pyrimidine are remarkably similar, -0.097 and -0.095 respectively (Albert, 1968), and this suggests that the *exo*-amino group is involved in another manner.

However these electron densities may either be irrelevant or inaccurate, as practical experience has shown wide variation in the vulnerability of the constituent nitrogens in pyridine and pyrimidine to chemical oxidizing agents (Katritzky & Lagowski, 1971).

It is of interest that the aromatic amino heterocycles which undergo biological annular *N*-oxidation all have more than one nitrogen in the ring system, whereas both Trp-P-2 and Glu-P-1 which are oxidized on the *exo*-amino group have only one ring nitrogen. Unfortunately, IQ with two ring nitrogens precludes any generalization in this respect.

At present, it is not known which tautomer, if either, leads to *exo-N*-hydroxylation, but studies on the mutagenicity of isomeric aminopyridines in the presence of norharman have shown that only the 3-aminopyridine is active (Sugimura *et al.*, 1982). If this is mutagenically active because of its metabolic conversion to 3-hydroxylaminopyridine, as would be expected by analogy with Trp-P-2 and Glu-P-1, then metabolic *exo-N*-hydroxylation might be associated with the amine tautomer. That the situation is not however so simple, was seen when isomeric methyl-2-aminopyridines were examined for mutagenicity. 3-Methyl-2-aminopyridine, a compound which could presumably tautomerize, was mutagenic (presumably after *exo-N*-oxidation); whereas 4-,5- and 6-methyl-2-aminopyridines were not mutagenic under the same conditions (Sugimura *et al.*, 1982) despite the structures still being theoretically able to tautomerize to an imine.

Aminopyridines are known to exist predominantly in the amino form and in the case of 2-aminopyridine the ratio of amine to imine tautomer was thought to be only 200,000 to 1 (Angyal & Angyal, 1952). This value has been recalculated (Cook *et al.*, 1972) following a redetermination of the basicity of *N*-methylpyridyl-2-oneimine and an even lower proportion of the imine tautomer was thought to be present. These results would seem to be at variance with those obtained using ^{14}N and ^{15}N-n.m.r. spectroscopy (Stefaniak, 1979; Stefaniak *et al.*, 1981) where in acetone solution the imine tautomer apparently constitutes between 2 and 11% of the equilibrium mixture (depending upon the nitrogen and isotope used). It is clear that many factors can influence the amine:imine ratio and in the case of dihydrocytosine it has been shown by Brown & Hewlins (1968) that whereas the amino form predominates in aqueous media, the imine form predominates in chloroform. Thus it is tempting to suggest that the lipoprotein constituent of the enzyme matrix within the endoplasmic reticulum may shift the equilibrium in favour of the imine tautomer. It is recognized that dihydrocytosine is an aliphatic compound and extrapolation to the aromatic systems of current interest may not be warranted. Additional effects may be expected due to the specific binding of substrates to enzyme(s) site(s). As *N*-oxidation would be expected to remove one tautomer from the equilibrium, low imine:amine ratios may not necessarily preclude their involvement in the enzymic reaction.

A further complication arises, in that the amine oxides and the hydroxylamines formed as metablic *N*-oxidation products would also be expected to exist as two tautomeric forms (Fig. 4). Again the overwhelming evidence indicates that the nuclear amine oxide and *exo*-hydroxylamine tautomers predominate (Elguero *et al.*, 1976).

One piece of evidence indicating that tautomerism of *N*-oxides of amino pyrimidines can occur in biological systems is seen with the drug Minoxidil. In this case, the *N*-oxide and amine function are thought to tautomerize *in vivo* to an *N*-hydroxyimine structure which then allows conjugation with glucuronic acid (Thomas & Harpoolian, 1975). However this view may have to be modified in the light of recent findings with this drug (Johnson *et al.*, 1983).

At the present time the enzymology of the above *N*-oxidations have not been absolutely established. Preliminary work on the metabolism of 2,6-diamino pyrimidines (Dr J. P. Hubbel, Wellcome Research Laboratories, Personal Communication) has implicated cytochrome 'P-450' as the terminal oxidase. Similar work on Trp-P-2 (Yamazoe *et al.*, 1980a,b) and Glu-P-1 (Kamataki *et al.*, 1982) indicate that cytochrome P-448 is involved in these *N*-hydroxylations. These conclusions were supported by work using specific antibodies to purified cytochrome P-450 and P-448 (Watanabe *et al.*, 1982).

If these observations are substantiated, then species differences in site of pyrimidine *N*-oxidations as well as differences in *exo*- or *endo-N*-oxidation of amino aromatic heterocyclic molecules may be accounted for by differences in substrate specificity of cytochrome 'P-450' isozymes.

The finding that structurally similar substrates can lead to such dissimilar

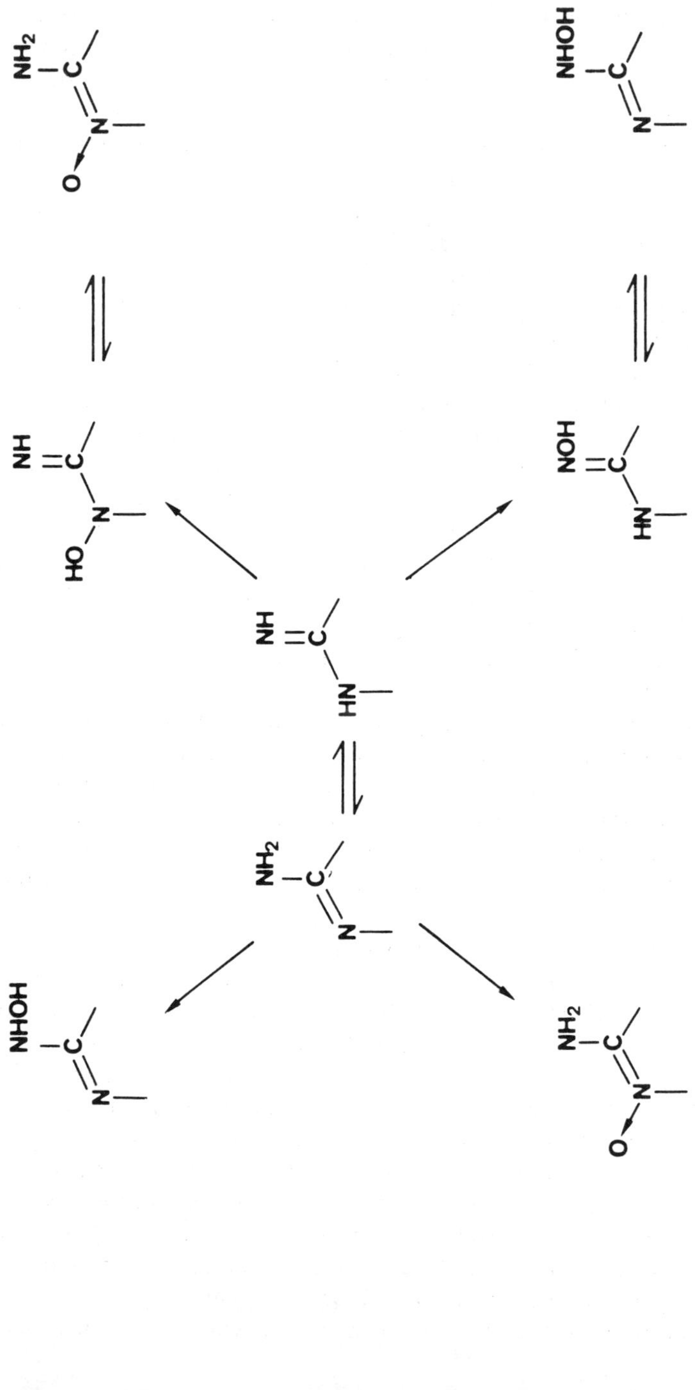

Fig. 4 – Tautomeric forms of amines, amine oxides and hydroxylamines and their possible involvement in the *N*-oxidation of amino azaheterocycles.

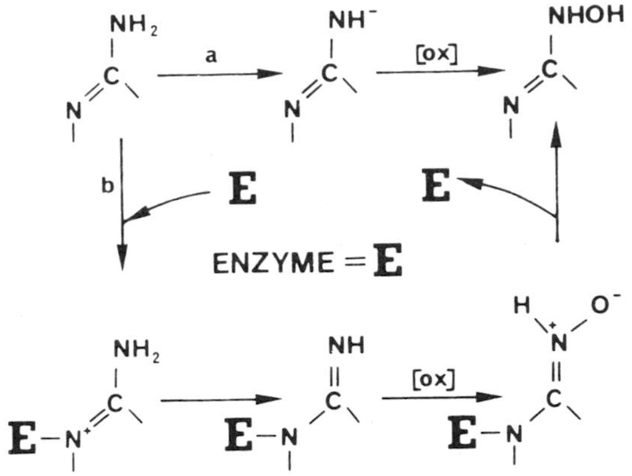

Fig. 5 – Possible mechanisms involved in the oxygenation at the *exo*-nitrogen of amino azaheterocycles.

N-oxidation products poses interesting mechanistic questions. If as is generally thought the ring nitrogen is the most basic, and oxidation normally occurs at this site, with the substrate molecule existing as the amine tautomer, then how can one envisage the *exo*-nitrogen becoming vulnerable to biological *N*-oxidation?

In order to get oxygenation of the *exo*-nitrogen, a different mechanism to that involved in ring *N*-oxidation may be required. One possibility would be to suppose the *exo*-amino group could ionize to a anion (Fig. 5, route a) which was oxidized. This does not seem feasible at physiological pH, but may occur within the enzyme micro environment. An alternative mechanism would involve binding of the substrate to the enzyme via the basic ring nitrogen (Fig. 5, route b) and thereby induce the *exo*-amino group to tautomerize to the imino form which could then be oxidized. Dissociation of the oxidized product from the enzyme would release the *exo*-hydroxylamine.

It is known that organic nitrogen can be oxygenated by a variety of enzymes including certain cytochrome P-450 isozymes, chloroperoxidase and flavin containing monooxygenase, and oxidized by peroxidase and prostaglandin H synthetase. The latter two enzymes giving rise to a radical cation by a one electron oxidation rather than an initial oxygenated product.

It has hitherto been thought that aromatic amines were not usually substrates for the microsomal flavin containing oxygenases (Ziegler *et al.*, 1973) and the specific oxygenation of 2-naphthylamine was due to the presence of imine tautomer (Ziegler *et al.*, 1980). This view has recently been contested by Doerge & Corbett (1984) who, using a model flavin hydroperoxide, conclude that amine nucleophilicity is an important determinant of amine susceptibilty to oxygenation and propose that direct nucleophilic attack of the amine on the flavin hydroperoxide occurs.

At present no evidence suggests that aromatic amino heterocycles are oxidized by the flavin hydroperoxide mechanism and indeed the studies which have been performed indicate that cytochrome P-450 isozymes are involved. In this respect it is of interest that heterocyclic annular nitrogen seems to be oxidized via the phenobarbital inducible species (P-450 or LM2) whereas the *exo*-amino group is oxidized by the methylcholanthrene induced species (P-448 or LM4). Further work is required to examine the site of oxidation and substrate specificity of these and other microsomal cytochromes. At the present time the exact nature of the oxidizing species has not been definitely established for these cytochromes and mechanisms based on the generation of superoxide anion or iron perferryl oxene have been proposed. If the differences in site of oxidation by individual cytochromes is confirmed for aromatic amino heterocycles then a common oxidizing mechanism may not appertain and this difference could account for the different oxidized product.

However if, as seems likely, the oxidizing mechanism is the same for all P-450 isozymes, then differences in substrate orientation produced via substrate–enzyme interaction may occur to yield different *N*-oxidation products.

Experiments designed to establish the enzymology of these processes and the possible role of amine–imine tautomerism in determining the site of enzymic *N*-oxidation are currently in progress.

REFERENCES

Albert, A. A. (1968), *Heterocyclic Chemistry* (2nd edition, pages 64–65), The Athlone Press, London.

Angyal, S. & Angyal, C. L. (1952), *J. Chem. Soc.*, 1461.

Brooks, M. A., de Silva, J. A. F. & D'Arconte, L. (1973), *J. Pharm. Sci.*, **62**, 1395.

Brown, D. M. & Hewlins, M. J. E. (1968), *J.Chem. Soc.*, 2050.

Cook, M. J., Katritzky, A. R., Linda, P. & Tack, R. D. (1972), *J. Chem. Soc. Perkin II*, 1295.

Cowan, D. A., Damani, L. A. & Gorrod, J. W. (1978), *Biomed. Mass Spectrom.*, **5**, 551.

Doerge, D. R. & Corbett, M. D. (1984), *Biochem. Pharmacol.*, **33**, 3615.

Elguero, J., Marzin, C., Katritzky, A. R. & Linda, P. (1976), *Tautomerism of Heterocycles*, Academic Press, New York.

Gorrod, J. W. (1971), *Xenobiotica*, **1**, 349.

Gorrod, J. W. & Damani, L. A. (1979a), *Xenobiotica*, **9**, 209.

Gorrod, J. W. & Damani, L. A. (1979b), *Xenobiotica*, **9**, 219.

Hashimoto, Y., Shudo, K. & Okamato, T. (1980), *Biochem. Biophys. Res. Comm.*, **96**, 355.

Hashimoto, Y., Shudo, K. & Okamato, T. (1980), *Biochem. Biophys. Res. Comm.*, **92**, 971.

Hubbel, J. P., Henning, M. L., Grace, M. E., Nichol, C. A. & Sigel, C. W. (1978), in *Biological Oxidation of Nitrogen* (ed. Gorrod, J. W.) Elsevier, Amsterdam, p. 177.

Hubbel, J. P., Kao, C. J., Sigel, C. W. & Nichol, C. A. (1980), in *Current Chemotherapy and Infectious Disease* (eds Nelson, J. D. & Grassi, C.), American Society for Microbiology, Washington, p. 1620.

Johnson, G. A., Barsuhn, K. J. & McCall, J. M. (1983), *Drug Metab. Disposit.*, **11**, 507.

Kamataki, T., Yamazoe, Y., Mita, S. & Kato, R. (1982), in *Microsomes Drug Oxidation and Drug Toxicity*, (eds Sato, R. & Kato, R.) Japan Scientific Society Press, p. 463.

Katritzky, A. R. & Lagowski, J. M. (1971), *Chemistry of the Heterocyclic N-Oxides*, Academic Press, London.

Meshi, T. & Sato, Y. (1972), *Chem. Pharm. Bull. (Japan)*, **20**, 2079.

Nemoto, N., Kusumi, S., Takayama, S., Nagao, M. & Sugimura, T. (1979), *Chem.-biol. Interact.*, **27**, 191.

Okamoto, T., Shudo, K., Hashimoto, Y., Kosuge, T., Sugimura, T. & Nishimura, S. (1981), *Chem. Pharm. Bull.*, **29**, 590.

Schwartz, D. E., Vetter, W. & Englert, G. (1970), *Arzneim. Forsch.*, **20**, 1867.

Sigel, C. W. (1973), *J. Infec. Dis. Suppl.*, **128**, 567.

Sigel, C. W. & Brent, D. A. (1973), *J. Pharm. Sci.*, **62**, 694.

Stefaniak, R. (1979), *Org. Magn. Resonance*, **12**, 379.

Stefaniak, L., Witanowski, M. & Webb, G. A. (1981), *Polish J. Chem.*, **55**, 1441.

Sugimura, T., Nagao, M. & Wakabayashi, K. (1982), in *Biological Reactive Intermediates II Part B*, p. 1011 (eds Snyder, R., Parke, D. V., Kocsis, J. J., Jollow, D. J., Gibson, C. C. & Witmer, C. M.), Plenum Press, New York.

Sugimura, T., Kawachi, T., Nagao, M., Yahagi, T., Seino, Y., Okamoto, T., Shudo, K., Kosuge, T., Tsuji, K., Wakabayashi, K., Itaka, Y. & Itai, A. (1977), *Proc. Japan. Acad.*, **53**, 58.

Thomas, R. C. & Hartpoolian, H. (1965), *J. Pharm. Sci.*, **64**, 1366.

Watanabe, J., Kawajiri, K., Yonekawa, H., Nagao, M. & Tagashira, Y. (1982), *Biochem. Biophys. Res. Comm.*, **104**, 193.

Wolf, F. J., Steffens, J. J., Alvaro, R. F. & Jacol, T. A. (1978), *Fed. Proc.*, **37**, 814.

Yamazoe, Y., Imai, T., Mita, S., Ishii, K., Kamataki, T. & Kato, R. (1982), in *Microsomes Drug Oxidation and Drug Toxicity* (eds Sato, R. & Kato, R.) Japan Scientific Society Press, p. 525.

Yamazoe, Y., Ishii, K., Kamataki, T., Kato, R. & Sugimara, T. (1980a), *Chem-Biol. Interact.*, **30**, 125.

Yamazoe, Y., Yamaguchi, N., Kamataki, T. & Kato, R. (1980b), *Xenobiotica*, **10**, 483.

Ziegler, D. M., McKee, E. M. & Poulsen, L. L. (1973), *Drug Metab. Disposit.*, **1**, 314.

Ziegler, D. M., Poulsen, L. L. & Duffel, M. W. (1980) in *Microsomes, Drug Oxidations and Chemical Carcinogenesis* (eds Cook, M. J., Conney, A. H.,

Estabrook, R. W., Gelboin, H. V., Gillett, J. R. & O'Brien, P. J.) Academic Press, New York.

Zins, G. R. (1965), *J. Pharmacol. Exptl. Therap.*, **150**, 109.

Zins, G. R., Emmert, D. W. & Walker, R. A. (1968), *J. Pharmacol. Exptl. Therap.*, **159**, 194.

Chapter 24

THE *IN VIVO* METABOLISM OF METYRAPONE IN THE RAT

J. I. Usansky and **L. A. Damani**, Department of Pharmacy, University of Manchester, Manchester, M13 9PL, UK

1. The urinary metabolic profile of metyrapone has been studied following administration of 50 mg/kg by intraperitoneal injection to male Sprague–Dawley rats.
2. Six metabolites have been identified; these include the two isomeric mono *N*-oxides of metyrapone and of metyrapol respectively, the α-pyridone metabolite, and the keto-reduction product metyrapol.
3. Quantitative analysis has shown that the two isomeric metyrapol mono *N*-oxides are the major urinary metabolites, together accounting for about 75% of the administered dose.

INTRODUCTION

Metyrapone is used clinically for the assessment of ACTH secretion from the anterior pituitary in a procedure called the 'Metyrapone Test' (Meikle *et al.*, 1975 and references cited therein), and finds some other clinical uses. In addition, metyrapone is widely used in drug metabolism studies as an inhibitor, and in some instances as an inducer, of cytochrome P-450 monooxygenations. Despite this widespread use of metyrapone, detailed reports on its pharmacokinetics and *in vivo* metabolism in man and in laboratory animal species are lacking. A review of the literature pertaining to the *in vitro* metabolism of metyrapone, together with a brief summary of the enzyme systems mediating various reactions is given elsewhere (Usansky & Damani, Chapter 26 of this volume).

There appear to be only two reports in the literature concerning the urinary excretion profile of metyrapone. Sprunt *et al.* (1968) reported the presence of metyrapone and metyrapol (in the free and conjugated form) in the urine of patients undergoing the 'Metyrapone Test'. A more detailed study by Hannah & Sprunt (1969) essentially confirmed these findings, but the reported metabolites only accounted for a small percentage (<50%) of the total administered dose of metyrapone, although presence in urine of more polar unidentified metabolites was recognized. The present communication describes results of studies carried out to fully characterize the pharmacokinetics and urinary metabolic profile of metyrapone in the rat, as part of a larger study to fully evaluate its disposition in several animal species, including man.

EXPERIMENTAL AND RESULTS

Male Sprague–Dawley rats were administered 50 mg/kg metyrapone by intraperitoneal injection, and urine collected for a period of 72 hours. These urine samples were extracted with dichloromethane, and concentrated extracts subjected to two-dimensional t.l.c. on silica plates, together with authentic metabolites, which were synthesized by chemical methods in our laboratory (Crooks *et al.*, 1981). This revealed the presence in urine of six metabolites, the two isomeric mono *N*-oxides of metyrapone, the two isomeric mono *N*-oxides of metyrapol, the α-pyridone metabolite of metyrapone, and the keto-reduction product metyrapol. Whereas four of these compounds have previously been reported as *in vitro* metabolites, this is the first report identifying *N*-oxides of metyrapol *in vivo*. The materials corresponding to the two isomeric mono *N*-oxides of metyrapol were recovered by preparative t.l.c. and the isolated compounds analysed by electron-impact mass spectrometry. Mass spectra of the two metabolites showed characteristic fragmentation patterns, similar to those previously reported for synthetic metyrapol *N*-oxides (Damani *et al.*, 1981). In each case, a molecular ion was observed at m/z 244, which corresponds to metyrapol plus 16 additional mass units. No $[M\text{-}16]^+$ ion peak indicating cleavage of the *N*-oxide bond was seen, but cleavage of the carbonyl methylene C—C bond occurs, yielding peaks at m/z 120 and m/z 121, and a diagnostic peak at m/z 125 $[C_6H_7NO_2]^+$ for one compound (MPOL-NO I), and a diagnostic peak at m/z 137 $[C_8H_{11}NO]^+$ for the other compound (MPOL-NO II). High resolution mass measurements confirmed these fragmentation patterns as those of the two metyrapol *N*-oxides.

An h.p.l.c. system was developed to enable separation and quantitation of metyrapone and its metabolites using a Partisil 10 ODS column and a mixture of phosphate buffer (pH 7.4, 0.01 M) and acetonitrile as mobile phase. Metabolites were extracted from urine as described above. Metyrapone was

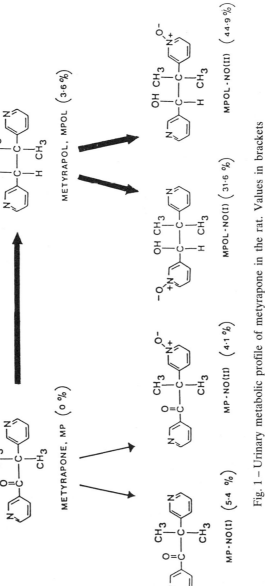

Fig. 1 – Urinary metabolic profile of metyrapone in the rat. Values in brackets indicate mean percentage of dose (50 mg/kg) excreted as each compound (n = 7).

not observed, but peaks corresponding to the retention times for the synthetic metabolites metyrapol, the two metyrapone *N*-oxides, and the two metyrapol *N*-oxides were observed. No increase in peak height ratio of metyrapol or any of the other metabolites was seen after incubation of urine with either β-glucuronidase or acid, indicating the apparent absence of any conjugated metabolites. Titanium trichloride treatment of urine samples however resulted in the disappearance of the peaks corresponding to the metyrapone *N*-oxides and the metyrapol *N*-oxides, with concomitant appearance of metyrapone and metyrapol peaks, accounting for about 4% and 75% of the dose respectively.

DISCUSSION

It can be seen that by a combination of t.l.c., h.p.l.c., mass spectrometry and titanium trichloride reduction, the two isomeric metyrapol *N*-oxides have been identified as new *in vivo* metabolites of metyrapone from the urine of rats administered metyrapone. Quantitative analysis of urine extracts by h.p.l.c. showed that unchanged metyrapone was not excreted, and only 3.6% of the dose was excreted as free metyrapol. The metyrapone *N*-oxides, MP-NO(I) and MP-NO(II), were also minor metabolites, accounting for 5% and 4% of the dose respectively. However, the two metyrapol *N*-oxides MPOL-NO(I) and MPOL-NO(II), accounted for 32% and 45% of the total dose administered (Fig. 1).

Thus, the major urinary metabolites of metyrapone in the rat are the two isomeric metyrapol *N*-oxides. In this respect, it is interesting that our pharmacokinetic studies have shown that metyrapone is almost completely reduced to metyrapol (Usansky *et al.*, 1984). Male Sprague–Dawley rats were given either metyrapone or metyrapol (50 mg/kg) intravenously, and the blood levels of metyrapone and metyrapol measured. The area under the curve (AUC) of metyrapol after administration of either metyrapone or metyrapol were similar ($p > 0.05$), indicating almost complete conversion of metyrapone to metyrapol. The *N*-oxides of metyrapol, which are the major urinary metabolites of metyrapone, must be primarily formed by *N*-oxidation of metyrapol (Fig. 1), which is itself initially produced by the reduction of metyrapone. In contrast, the *N*-oxidation of metyrapone to yield the metyrapone *N*-oxides only accounts for a small percentage of the dose.

REFERENCES

Crooks, P. A., Damani, L. A. & Cowan, D. A. (1981), *J. Pharm. Pharmacol.*, **33**, 309.

Damani, L. A., Crooks, P. A. & Cowan, D. A. (1981), *Biomed. Mass Spectrom.*, **8**, 270.

Hannah, D. M. & Sprunt, J. G. (1969), *J. Pharm. Pharmacol.*, **21**, 877.

Meikle, A. W., West, S. C., Weed, J. A. & Tyler, F. H. (1975), *J. Clin. Endocrinol. Metab.*, **40**, 290.

Sprunt, J. G., Browning, M. C. G. & Hannah, D. M. (1968), *Mem. Soc. Endocrinol.*, **17**, 193.

Usansky, J. I. & Damani, L. A. (1985), this volume.

Usansky, J. I., Damani, L. A. & Houston, J. B. (1984), *J. Pharm. Pharmacol.*, **36** (supplement), 28P.

Chapter 25

THE INFLUENCE OF SEX, SPECIES AND CHEMICAL PRETREATMENT ON THE URINARY EXCRETION OF THE N-OXIDE METABOLITES OF DIETHYLCARBAMAZINE AND METYRAPONE

P. A. F. Dixon, **C. A. Joseph**, **S. E. Okereke** and **M. C. Enwelum**, Department of Pharmacology, University of Ife, Ile-Ife, Nigeria

1. The *in vivo* N-oxidation of diethylcarbamazine, and the N-oxidation and keto-reduction of metyrapone in the rat, cat and rabbit has been studied.
2. There are marked species and sex differences in the metabolism of these compounds.
3. A study of the effect of safrole, metyrapone, cimetidine, ethanol and anthracene treatment appears to suggest the involvement of different sub-populations of cytochrome P-450 in the N-oxidation of these two compounds.

INTRODUCTION

Cytochrome P-450 dependent monooxygenases catalyse the oxidation of a multitude of structurally diverse substrates. These are different forms of cytochrome P-450 and they are inducible to varying degrees by different xenobiotics (Ryan *et al.*, 1982; Guengerich *et al.*, 1982). Kato & Kamataki (1982) have also provided evidence for the involvement of multiple forms of cytochrome P-450 in the occurrence of the sex difference in the oxidative metabolism of drugs.

The *N*-oxidation of diethylcarbamazine (DEC) has been reported by Faulkner & Smith (1971) and confirmed by Joseph & Dixon (1984). In the latter paper we suggest from our *in vivo* data the possible involvement of cytochrome P-450 in this process. Metyrapone (MP) *N*-oxidation has been shown to occur via a phenobarbitone-inducible P-450 (Damani *et al.*, 1981; Usansky & Damani, this volume). This paper reports on the *in vivo* metabolism of these compounds in rat, cat and rabbit and the effect of chemical pretreatment on their metabolism in rat.

EXPERIMENTAL

Drug administration and chemical pretreatment are as described in the foot-notes to Tables 1 and 2. Diethylcarbamazine *N*-oxide was measured by g.l.c. as described before (Joseph & Dixon, 1984). Metyrapone and its metabolites were analysed by the modified h.p.l.c. method of Damani *et al.* (1981) using a

Metyrapone, $R_1=R_2=$nothing
Metyrapone$-$N$-$oxide I, $R_1=$O, $R_2=$nothing
Metyrapone$-$N$-$oxide II, $R_1=$nothing $R_2=$O

Metyrapol

Fig. 1 – Structures of metyrapone and its metabolites.

Table 1

Species and sex differences in the *N*-oxidation of DEC and metyrapone

	% dose excreted in 72-h urine as:			
Species	DEC-N-oxide	Metyrapone N-oxide I	Metyrapone N-oxide II	Metyrapol
Rat (male)	45	2	4	92
Rat (female)	28	20	5	28
Rabbit	0	0	4	48
Cat	0	1	27	8

Dosage: DEC = 20 mg.kg^{-1}, metyrapone = 50 mg.kg^{-1}, Intraperitoneal injection.

Table 2

Effect of drug pretreatment on the *N*-oxidation of DEC and metyrapone in male Wistar rats

Drug	Dose	DEC-*N*-oxide	Metyrapone-*N*-oxide I	Metyrapone-*N*-oxide II
[a]Safrole	25% v/v (0.2 ml)	↓	ND	ND
[a]Metyrapone	50 mg.kg^{-1}	↓	—	—
[a]Cimetidine	100 mg.kg^{-1}	ND	↑	↓
[b]Ethanol	25% v/v (0.2 ml)	↓	↑	↑
[b]Phenobarbitone	30 mg.kg^{-1}	↑	↔	↑
[b]Anthracene	50 mg.kg^{-1}	↑	↔	↔

↓ = inhibition, ↑ = induction, ↔ = no effect, [a] = co-administered with drug, [b] = pretreated for 5 days before drug administration. ND = not determined.

cyanopropyl bonded phase (Varian Micropak CN-10) column (4 mm ×
30 cm) and a mobile phase of isopropanol:dichloromethane:n-hexane (2:1:1,
v/v) at a flow rate of 3 ml min^{-1}.

RESULTS AND DISCUSSION

There was a marked sex difference in the N-oxidation of DEC in the rat, with
the male excreting more than the female, while the cat and rabbit were
apparently deficient in this N-oxidation pathway. With metyrapone (Fig. 1)
the female rat excreted more of N-oxide I than N-oxide II, the reverse was the
case in the male rat. There was also more metyrapol excretion in male compared to female rats. The rabbit failed to excrete N-oxide I, whereas in
contrast the cat excreted both N-oxides (N-oxide II being the major metabolite, Table 1).

Safrole, metyrapone and ethanol decreased the amount of urinary DEC
N-oxide, whereas phenobarbitone and anthracene enhanced the amount
excreted. Phenobarbitone only enhanced MP N-oxide II excretion, whereas
ethanol affected both N-oxide I and II. Anthracene did not influence the
excretion of MP N-oxides. Interestingly, cimetidine apparently inhibited
N-oxide II formation but enhanced N-oxide I formation (Table 2).

In summary, therefore, the following points emerge from the studies
described. (a) There were sex and species differences in the N-oxidation of
DEC and MP, (b) ethanol, which is known to induce a particular form(s) of
cytochrome P-450 (Ingelman-Sundberg & Hagbjork, 1982) induced both
N-oxide I and II of metyrapone but inhibited DEC-N-oxide, (c) metyrapone
and cimetidine treatment (known cytochrome P-450 inhibitors) decrease
DEC N-oxide and metyrapone N-oxide II excretion respectively, (d)
phenobarbitone and anthracene treatment resulted in increased DEC-N-
oxide excretion, but had no effect on metyrapone N-oxide I, phenobarbitone
alone enhanced N-oxide II excretion. These *in vivo* results suggest that different subpopulations of cytochrome P-450 may be involved in the formation of
these three N-oxides.

REFERENCES

Damani, L. A., Crooks, P. A. & Cowan, D. A. (1981), *Biomed. Mass Spec-
trom.*, **8**, 270.
Faulkner, J. & Smith, K. J. A. (1971), *Xenobiotica*, **1**, 321.
Guengerich, F. P., Dannan, G. A., Wright, S. T. & Martin, M. V. (1982),
Xenobiotica, **12**, 710.
Ingelman-Sundberg, M. & Hagbjork, A. L. (1982), *Xenbiotica*, **12**, 673.
Joseph, C. A. & Dixon, P. A. F. (1984), *J. Pharm. Pharmacol.* (in press).
Kato, R. & Kamataki, T. (1982), *Xenobiotica*, **12**, 787.
Ryan, D. E., Thomas, P. E., Reik, L. M. & Levin, W. (1982), *Xenobiotica*,
12, 727.

Chapter 26

METABOLISM OF METYRAPONE BY RAT LIVER SUBCELLULAR PREPARATIONS

J. I. Usansky and **L. A. Damani**, Department of Pharmacy, University of Manchester, Manchester, M13 9PL, UK

1. Incubation of metyrapone with rat hepatic microsomes results in the formation of two isomeric metyrapone mono N-oxides, and in keto-reduction to afford metyrapol.
2. Incubation with the cytosol, on the other hand, affords an α-pyridone metabolite and metyrapol in small amounts.
3. Microsomal N-oxidation is a cytochrome P-450 reaction, whereas microsomal keto-reduction is not. The characteristics of this novel microsomal keto-reductase are described.
4. Cytosolic α-C-oxidation of metyrapone is mediated via a molybdenum hydroxylase, aldehyde oxidase. Cytosolic keto-reduction is via the well characterized NADPH-requiring aromatic-ketone reductases.

INTRODUCTION

Metyrapone is used clinically as a diagnostic agent for the determination of ACTH secretion from the anterior pituitary gland, as a diuretic in some cases of resistant oedema, and in treatment of Cushing's syndrome. Metyrapone is also used in drug metabolism studies as an inhibitor of hepatic microsomal

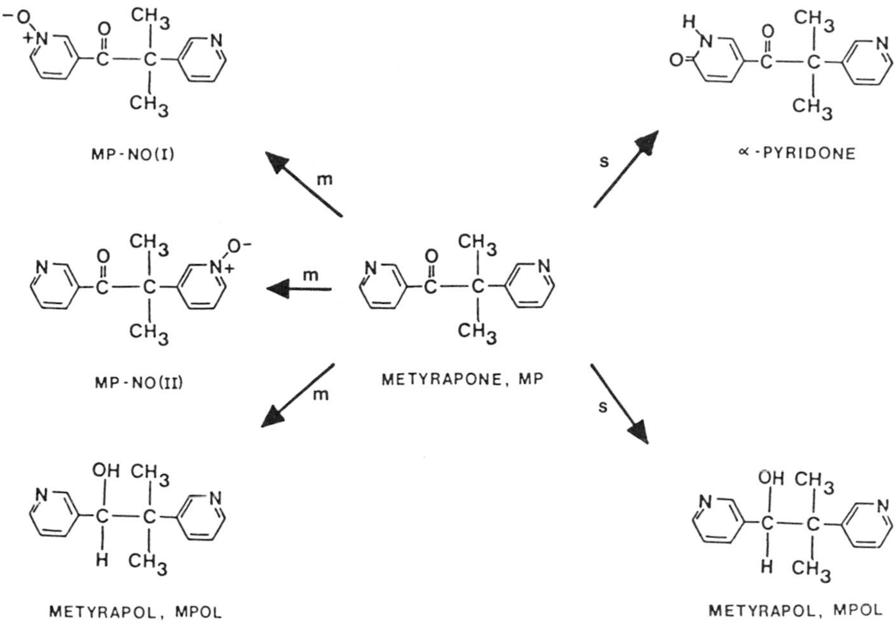

Fig. 1 – Biotransformation of metyrapone *in vitro* with hepatic subcellular preparations. m = reaction catalysed by hepatic microsomal enzymes, s = rection catalysed by hepatic cytosolic enzymes.

cytochrome P-450. The biotransformation of metyrapone has not been fully
investigated, although it has been shown that when metyrapone is incubated
with microsomal preparations three metabolites are produced; the two
isomeric metyrapone mono-*N*-oxides in which either one or the other of the
pyridyl rings becomes *N*-oxidized, and metyrapol, formed by keto-reduction
(De Graeve *et al.*, 1979; Damani *et al.*, 1979; Crooks *et al.*, 1981). When
incubated with cytosolic preparations, an α-pyridone metabolite and
metyrapol are formed (see Fig. 1) (Damani *et al.*, 1981).

Information on the enzyme systems mediating the biotransformation of
metyrapone are fragmentary. The cytosolic metyrapone reductase has been
partially purified and characterized and has been shown to be the well-known
NADPH-requiring keto-reductase which catalyses the reduction of a wide
variety of aromatic and aliphatic ketones (Felsted & Bachur, 1980a,b). Preliminary results on the enzyme catalysing α-pyridone formation indicate that
it is aldehyde oxidase, one of the molybdenum hydroxylase enzymes similar
to xanthine oxidase (Usansky & Damani, 1983). Microsomal *N*-oxidation of
metyrapone is most likely to be mediated by a phenobarbitone-inducible
cytochrome P-450 (De Graeve *et al.*, 1979; Damani *et al.*, 1981). This would
appear to confirm previous reports (Damani, 1977; Gorrod & Damani,
1979a,b) that *N*-oxidation at a tertiary heteroaromatic amino functionality,
such as is present in pyridine and related compounds, is mediated via a P-450
system, rather than the flavin-containing monooxygenase. Microsomal reduc-

tion of metyrapone by rat liver microsomes in the presence of NADH or a NADPH-generating system was first noted by Sprunt et al. (1968). The microsomal reductase was briefly investigated by Kahl (1970) and Kahl et al. (1974) who reported that only 6% of the total homogenate activity could be found in rat cytosol, whereas a mean value of 65% was obtained for the microsomal fraction. It was not, however, ascertained if the metyrapone reductase was linked to the microsomal P-450 system, or whether it was mediated by an independent electron transport system. The present communication describes results of studies carried out in our laboratory further to characterize the rat microsomal enzymes mediating metyrapone N-oxidation and keto-reduction.

EXPERIMENTAL

Incubation conditions, unless otherwise stated, were as follows: 5 μmole metyrapone was added to a cofactor generating system consisting of 2 μmole NADP$^+$, 10 μmole G6P, 1 unit G6PD and 2 mg MgCl$_2$. Enzymatic reactions were initiated by the addition of washed microsomes, from control male Sprague–Dawley rats, equivalent to 0.5 g original tissue per ml. The total incubation volume was 2 ml. Reactions were stopped by the addition of alkali, internal standard was added, and the aqueous phase then extracted with dichloromethane. The extracts were evaporated to dryness, reconstituted into methanol ($\sim$25 μM) and analysed by h.p.l.c. The h.p.l.c. system consisted of a Partisil 10 ODS column (25 cm $\times$ 0.5 cm), using a phosphate buffer (pH 7.4, 0.01 M) and acetonitrile mixture (75/25%, v/v) as mobile phase at a flow rate of 2.5 ml/min.

RESULTS AND DISCUSSION

Initially, studies were carried out on the effect of pretreatment of rats with phenobarbitone (PB) and β-naphthoflavone (BNF). The amount of the two N-oxides produced increased approximately fivefold and 3.5-fold with microsomes from PB-pretreated rats, as compared to controls (Table 1). Upon addition of 10^{-3} M-SKF 525A (a well known inhibitor of cytochrome P-450) to the PB-pretreated microsomes, N-oxidation was inhibited to approximately 33% of the uninhibited PB-pretreated microsomes, but the amount of N-oxidation was still approximately 1.7-fold more than control values. In contrast, microsomal keto-reduction was not enhanced by PB-treatment of rats, and addition of 10^{-3} M-SKF 525A to these microsomes had no further effect. Pretreatment of rats with β-naphthoflavone, which is an inducer of cytochrome P-448, did not result in an increase in either N-oxidation or of keto-reduction. These data would suggest that N-oxidation is mediated via a PB-inducible cytochrome P-450, but keto-reduction is probably mediated via an independent electron transport system.

The effect of incorporating various inhibitors and activators on N-oxidase

Table 1

Effect of inducers, inhibitors and activators on rat hepatic microsomal
N-oxidation and keto-reduction of metyrapone

Inducer/Activator/Inhibitor	Enzyme activity (% of control)*		
	MP-NO (I)	MP-NO (II)	MPOL
Control	100	100	100
PB-pretreatment	600	460	88
BNF-pretreatment	60	76	64
SKF 525A (10^{-3} M)	60	56	101
n-Octylamine (10^{-3} M)	66	54	148
Carbon monoxide (CO/O$_2$/N$_2$:5/1/4)	41	78	97
Cimetidine (10^{-3} M)	92	80	112
Aniline (10^{-3} M)	51	62	114
Naphthylthiourea (10^{-3} M)	0	0	51
Cyanide	97	100	100
Triton (X-100) (0.5%)	0	0	233
Acetone (5%)	66	66	126

*Control activities MP-NO (I) = 354 ± 71 pmole/min/mg protein
MP-NO (II) = 155 ± 19 pmole/min/mg protein
MPOL   = 2.75 ± 0.19 nmol/min/mg protein.

and reductase activity was studied with microsomes from untreated rats
(Table 1). Typical P-450 inhibitors, such as SKF 525A, n-octylamine, carbon
monoxide, cimetidine, aniline and 1-naphthylthiourea, all had a significant
inhibitory effect on microsomal N-oxidation. With the exception of
1-naphthylthiourea, none of the compounds mentioned above had an inhibitory effect on microsomal keto-reduction of metyrapone. In fact, keto-
reduction was slightly enhanced in the presence of n-octylamine, cimetidine
and aniline. Whereas cyanide had no effect on either reactions, addition of
Triton X-100 or acetone to rat microsomes drastically decreased or abolished
N-oxidase activity, but stimulated keto-reductase activity. The inhibitory
effects of certain compounds on the N-oxidation at the two pyridyl nitrogens
appeared to suggest that reactions at these nitrogens may be mediated via
different P-450 isozymes. Further studies are underway to check this possibility. The data above, however, clearly indicated that keto-reduction of
metyrapone was not linked to the microsomal mixed function oxidase electron transport system.

The cofactor requirements for the N-oxidation and keto-reduction of
metyrapone were studied as part of the process of optimizing incubation
conditions. The full NADPH-generating system gave maximal metabolism

Table 2
Cofactor requirements for rat hepatic microsomal N-Oxidation and keto-reduction of metyrapone

Cofactor	Enzyme activity (% of control)*		
	MP-NO (I)	MP-NO (II)	MPOL
Full generating system†	100	100	100
NADPH (5 μmole)	100	100	100
NADH (2 μmole)	20	5	17
NADP$^+$ (2 μmole)	0	0	0
G6P (10 μmole)	0	0	48
G6PD (1 unit)	0	0	9
NADP$^+$ (2 μmole) +G6P (10 μmole)	0	0	85
NADP$^+$ (2 μmole) +G6PD (1 unit)	0	0	0
G6P (10 μmole) +GDP (1 unit)	0	0	57

*control values as in footnote to Table 1.
†as given in text.

and was used as a control value of 100% (see Table 2). NADPH was essential for N-oxidation, and could be supplied either with an NADPH-generating system or pure reduced cofactor. Exclusion of any one of the components of the generating system resulted in complete absence of N-oxidation. Similarly, keto-reduction required reducing equivalents, either in the form of an NADPH-generating system, or as pure reduced cofactor. In contrast to N-oxidation, the effects of exclusion of certain components of the NADPH-generating system on keto-reducton is complex. Keto-reduction appears to be dependent on the presence of G6P, but addition of G6PD results in further increase in activity, although keto-reduction is maximal only with further addition of NADP$^+$. N-Oxidation was found to have a pH optimum of about 7.7, which is similar to that observed for other cytochrome P-450 pyridyl N-oxidations (Damani, 1977). The pH optimum for keto-reduction was about 6.7.

In conclusion, the N-oxidation of metyrapone is catalysed by a PB-inducible cytochrome P-450, as seen by its similar induction, inhibition, cofactor and pH characteristics as other cytochrome P-450 mediated reactions. The microsomal reduction of metyrapone does not appear to be catalysed by cytochrome P-450. Recently, Sawada *et al.* (1981) have solubilized and partially purified a microsomal keto-reductase from guinea-pig liver. This keto-reductase has similar properties to those described in this report for metyrapone reductase, and the purified enzyme has, in fact, been shown to reduce metyrapone to metyrapol.

REFERENCES

Crooks, P. A., Damani, L. A. & Cowan, D. A. (1981), *J. Pharm. Pharmacol.*, **33**, 309.

Damani, L. A. (1977), PhD Thesis, University of London.

Damani, L. A., Crooks, P. A. & Gorrod, J. W. (1979), *J. Pharm. Pharmacol.*, **31** (supplement) 94P.

Damani, L. A., Crooks, P. A. & Cowan, D. A. (1981), *Drug Metab. Disposit.*, **9**, 270

De Graeve, J., Glelen, J. E., Kahl, G. F., Tuttenberg, K. H., Kahl, R. & Maume, B. (1979), *Drug Metab. Disposit.*, **7**, 166.

Felsted, R. L. & Bachur, N. R. (1980a) in *Enzymatic Basis of Detoxication* (ed. Jakoby, W. B.), Academic Press. New York, p. 281.

Felsted, R. L. & Bachur, N. R. (1980b), *Drug Metab. Rev.*, **11**, 1.

Gorrod, J. W. & Damani, L. A. (1979a), *Xenobiotica*, **9**, 209.

Gorrod, J. W. & Damani, L. A. (1979b), *Xenobiotica*, **9**, 219.

Kahl, G. F. (1970), *Naunyn-Schmiedebergs Arch. Pharmak.*, **266**, 61.

Kahl, R., Tuttenberg, K. H., Niedermeier, F. & Kahl, G. F. (1974), *Naunyn-Schmiedebergs Arch. Pharmak.*, **282**, R44.

Sawada, H., Hara, A., Hayashibara, M., Nakayama, T., Usui, S. & Saeki, T. (1981), *J. Biochem.*, **90**, 1077.

Sprunt, J. G., Browning, M. C. K. & Hannah, D. M. (1968), *Mem. Soc. Endocrinol.*, **17**, 193.

Usansky, J. I. & Damani, L. A. (1983), *J. Pharm. Pharmacol.*, **35** (supplement), 72P.

Chapter 27

COMPARATIVE *N*-OXIDATION OF NICOTINE AND COTININE BY HEPATIC MICROSOMES

A. R. Hibberd[†] and **J. W. Gorrod**, Department of Pharmacy, Chelsea College, University of London, Manresa Road, London SW3 6LX, UK

1. Nicotine-1'-oxidation was activated by SKF525A, DPEA and *n*-octy-lamine, but unaffected by DABCO and CO.
2. Cotinine-1-*N*-oxidation was inhibited by SKF525A, DPEA, *n*-octylamine, DABCO and CO.
3. Nicotine-1'-*N*-oxidation was strongly inhibited and cotinine-1-*N*-oxidation only weakly inhibited by MMI; both *N*-oxidations were inhibited by adrenaline.
4. Cotinine-*N*-oxidation but not nicotine-1'-*N*-oxidation, was enhanced by pretreatment of animals with phenobarbitone.
5. The evidence supports the concept that nicotine-1'-*N*-oxidation is mediated via the flavin containing monooxygenase, whilst cotinine-*N*-oxidation occurs via the cytochrome P-450 system.

INTRODUCTION

Nicotine is metabolized by *C*- and *N*-oxidation to yield cotinine and nicotine-1'-*N*-oxide respectively as major metabolites (Gorrod & Jenner, 1975). Cotinine is further metabolized to yield cotinine-1-*N*-oxide *in vivo*

[†]Present address: Hibbro Research, 135 Eign Street, Hereford, HR4 0AJ, UK.

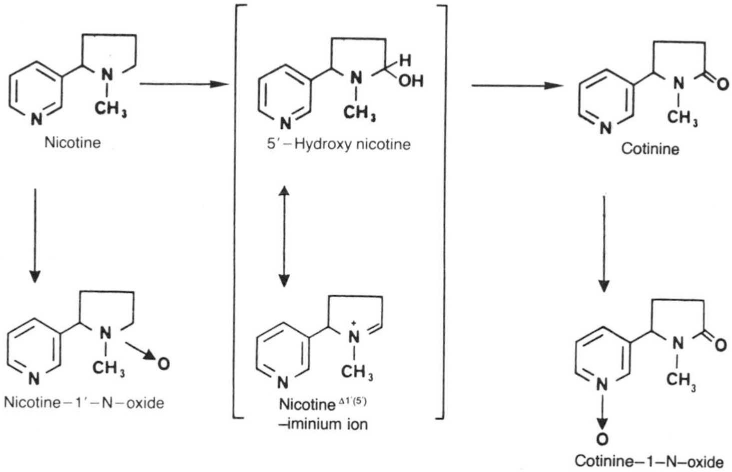

Fig. 1 – *In vitro* metabolic pathways to nicotine-1'-*N*-oxide and cotinine-1-*N*-oxide.

(Dagne & Castagnoli, 1972; Yi *et al.*, 1977; McKennis *et al.*, 1978; Hibberd, 1979) and *in vitro* (Damani, 1977; Hibberd, 1979) (Fig. 1). Considerable evidence indicates that nicotine-1'-*N*-oxidation is not cytochrome P-450 dependent but is mediated by the flavin containing monooxygenase (Gorrod *et al.*, 1971). The enzymology of cotinine-1-*N*-oxidation remained undermined and a comparative study of the two *N*-oxidative pathways appeared desirable.

EXPERIMENTAL

All materials were obtained as previously described (Hibberd, 1979). Resuspended hepatic microsomes were prepared from adult male Syrian hamster and Dunkin–Hartley guinea pig livers by the method of Gorrod *et al.* (1975).

Incubates consisted of microsomes (0.5 g liver/ml) 1.0 ml, substrate (either (−)-nicotine or (−)-cotinine (5 µmoles) in 0.5 ml water, cofactor solution 1 ml (containing NADP 4 µmoles, G6P, 10 µmoles, G6PD 2 units, dissolved in phosphate buffer, pH 7.4, 0.2 M) and potential inhibitors or activators in phosphate buffer, pH 7.4, 0.2 M, 1 ml. Incubations were also performed using phosphate buffer (1 ml) in place of inhibitor or activator solution, to serve as controls. Incubations were carried out for 10 minutes at 37 °C in 25 ml stoppered Erhlenmeyer flasks using a shaking water bath. Enzymatic activity was terminated by ading 5 M-NaOH (0.25 ml).

Nicotine-1'-*N*-oxide was assayed according to a previously described method (Beckett *et al.*, 1971). Cotinine-1-*N*-oxide was extracted from incubates with dichloromethane (4 × 3 ml) after the addition of sodium chloride (2.5 g). Extracted cotinine-1-*N*-oxide was separated by thin layer chromatography and assayed by gas liquid chromatography (Hibberd, 1979).

RESULTS AND DISCUSSION

SKF 525A, DPEA and *n*-octylamine all produced some activation of nicotine-1'-*N*-oxidation but caused inhibition of cotinine-1-*N*-oxidation (Table 1). The activation with SKF 525A may have been due to inhibition of the alternative cytochrome mediated pathway from nicotine to 5'-hydroxy-nicotine via nicotine-$^{\Delta1'(5')}$-iminium ion (Murphy, 1973). Activation by the primary amines, DPEA and *n*-octylamine is thought to be due to stimulation of the flavin containing monooxygenase (Ziegler *et al.*, 1971, 1972, 1973).

Naphthylthiourea did not differentiate between the two processes, although it is a known inhibitor of the flavoprotein monooxygenase (Ziegler *et al.*, 1972). Methimazole (MMI), which is a substrate for the flavoprotein monooxygenase (Poulsen *et al.*, 1974) inhibited nicotine-1'-*N*-oxidation much more than it did cotinine-1-*N*-oxidation at the concentration used. Whilst nicotine-1'-*N*-oxidation was mildly inhibited by KCN, slight activation of cotinine-1-*N*-oxidation was seen. The latter observation agrees with earlier work on the *N*-oxidation of 3-substituted pyridines (Gorrod & Damani, 1979a,b).

Inhibition of both pathways by adrenaline may implicate the involvement of a superoxide species (Uemera *et al.*, 1977; Ohnishi & Lieber, 1978) in the

Table 1

The effect of potential inhibitors and activators, on the *in vitro* production of nicotine-1'-*N*-oxide from nicotine and cotinine-1-*N*-oxide from cotinine, using hepatic microsomes from hamster and guinea-pig

Inhibitor (10⁻⁴ M)	Nic-1'-*N*-oxide formed*	Cot-1-*N*-oxide formed*	Species used
SKF 525A	116	32	H
DPEA	118	0	H
	—	50	GP
n-Octylamine	109	67	GP
Naphthylthiourea	83	67	GP
Potassium cyanide	74	115	H
Metyrapone	75	77	H
Methimazole	16	77	H
Adrenaline	33	32	H
DABCO	100	42	H

*compared to controls = 100
— = Not determined; Nic = Nicotine; Cot = Cotinine; H = Hamster; GP = Guinea-pig.
SKF 525A = 2-diethylaminoethyl-2,2-diphenyl valerate,
DPEA = 2,4-dichloro-6-phenylphenoxyethylamine,
DABCO = diazabicyclo-(2,2,2)-octane.

reactions. Previous work has shown that cotinine-1-N-oxidation is inhibited by superoxide dismutase (Hibberd, 1979) again suggesting the participation of the superoxide anion (O_2^-) in this pathway. DABCO, a known singlet oxygen ($^1O_2$) quencher (Quannes & Wilson, 1968; Scheulen et al., 1975) did not affect the N-oxidation of nicotine but did inhibit the formation of cotinine-1-N-oxide.

Microsomal cotinine-1-N-oxidation is known to be inhibited by CO (Hibberd, 1979) and this is good evidence for the involvement of cytochrome P-450 in the process (Kato, 1966; Parli et al., 1971; Thorgiersson et al., 1973). Cotinine-1-N-oxidation is also much enhanced using hepatic microsomes from hamsters pretreated with phenobarbitone (Hibberd, 1979) again implicating the involvement of cytochrome P-450 in the process. The pH optimum for the enzymic conversion of cotinine to cotinine-1-N-oxide occurs at 7.4 (Hibberd, 1979), whilst that for nicotine oxidation (producing predominantly nicotine-1'-N-oxide) occurs at approximately 8.0 (Booth & Boyland, 1971).

The results obtained support the view that the two tertiary amines are N-oxidized by different microsomal enzyme systems.

Previous studies (Gorrod, 1978; Gorrod & Damani, 1979a,b) comparing the N-oxidation of 3-substituted pyridines with tertiary aromatic amines came to a similar conclusion. Thus, it appears that tertiary heteroaromatic amines are N-oxidized by a cytochrome dependent system, whereas the N-oxidation of alicyclic or aromatic tertiary amines is mediated via the flavoprotein system described by Ziegler et al. (1971, 1972, 1973).

REFERENCES

Beckett, A. H., Gorrod, J. W. & Jenner, P. (1971), *J. Pharm. Pharmac.*, **23**, 55S.

Booth, J. & Boyland, E. (1971), *Biochem. Pharmac.*, **20**, 407.

Dagne, E. & Castagnoli, N. (1972), *J. Med. Chem.*, **15**, 840.

Damani, L. A. (1977), PhD Thesis, University of London.

Gorrod, J. W. (1978), in *Mechanisms of Oxidizing Enzymes* (eds. Singer, T. P. & Ondarza, R. N.) Elsevier/North-Holland, Amsterdam, p. 189.

Gorrod, J. W. & Damani, L. A. (1979a), *Xenobiotica*, **9**, 209.

Gorrod, J. W. & Damani, L. A. (1979b), *Xenobiotica*, **9**, 219.

Gorrod, J. W. & Jenner, P. (1975), in *Essays in Toxicol.*, Vol. 6, (ed. W. J. Hayes (Jr)) Academic Press, London, p. 35.

Gorrod, J. W., Jenner, P., Keysell, G. R. & Beckett, A. H. (1971), *Chem. Biol. Interact.*, **3**, 269.

Gorrod, J. W., Temple, D. J. & Beckett, A. H. (1975), *Xenobiotica*, **5**, 453.

Hibberd, A. R. (1979), PhD Thesis, University of London.

Kato, R. (1966), *J. Biochem. (Tokyo)*, **59**, 574.

McKennis, H., Yi, J. M. & Sprouse, C. T. (1978), in *Biological Oxidation of Nitrogen* (ed. Gorrod, J. W.) Elsevier/North-Holland, Amsterdam.

Murphy, P. J. (1973), *J. Biol. Chem.*, **248**, 2796.

Ohnishi, K. & Lieber, C. S. (1978), *Arch. Biochem. Biophys.*, **191**, 798.

Parli, C. J., Wang, N. L. & McMahon, R. E. (1971), *J. Biol. Chem.*, **246**, 6953.

Poulsen, L. L., Hyslop, R. M. & Ziegler, D. M. (1974), *Biochem. Pharmacol.*, **23**, 3431.

Quannes, C. & Wilson, T. (1968), *J. Amer. Chem. Soc.*, **90**, 6527.

Scheulen, M., Wollenberg, P., Bolt, H. M., Kappus, H. & Remmer, H. (1975), *Biochem. Biophys. Res. Comm.*, **66**, 1396.

Thorgierrson, S. S., Jollow, D. J., Sasame, H. A., Green, I. & Mitchell, J. R. (1973), *Mol. Pharmacol.*, **9**, 398.

Uemera, T., Chiesara, E. & Cova, D. (1977), *Mol. Pharmacol.*, **13**, 196.

Yi, J. M., Sprouse, C. T., Bowman, E. R. & McKennis, H. (Jr) (1977), *Drug Metab. Disposit.*, **5**, 355.

Ziegler, D. M., McKee, E. M. & Poulsen, L. L. (1973), *Drug Metab. Disposit.*, **1**, 314.

Ziegler, D. M. & Mitchell, C. H. (1972), *Arch. Biochem. Biophys.*, **150**, 116.

Ziegler, D. M., Poulsen, L. L. & McKee, E. M. (1971), *Xenobiotica*, **1**, 423.

Part 6

AMIDINE, IMINE, TRIAZENE, HYDRAZINE AND AZO OXIDATION

Chapter 28

THE BIOLOGICAL
N-OXIDATION OF AMIDINES
AND GUANIDINES

B. Clement, Pharmazeutisches Institut, Albert-Ludwigs-Universität, D-7800 Freiburg im Breisgau, FRG

1. The metabolic N-hydroxylation of strong basic amidines and guanidines has been studied for the first time.
2. A great number of drugs and model compounds have been tested for N-oxygenation *in vitro*, using synthetic reference compounds.
3. N-Unsubstituted and N-arylbenzamidines are N-oxygenated to benzamidoximes, a new type of metabolite. N-Alkylbenzamidines and all the guanidines tested are not N-hydroxylated.
4. Evidence is presented for the involvement of cytochrome P-450 in the N-oxygenation of benzamidine. A mechanism for this reaction is suggested.
5. An attempt is made to develop a predictive concept for the metabolism of basic nitrogen-containing functional groups with regard to the two different monooxygenases (cytochrome P-450 and flavin-containing monooxygenase).

INTRODUCTION

This review summarizes recent work on the metabolism of amidines and guanidines. Whereas the N-oxidation of aliphatic amines (see Coutts & Beckett, 1977 for a review) is now a well established metabolic pathway, until

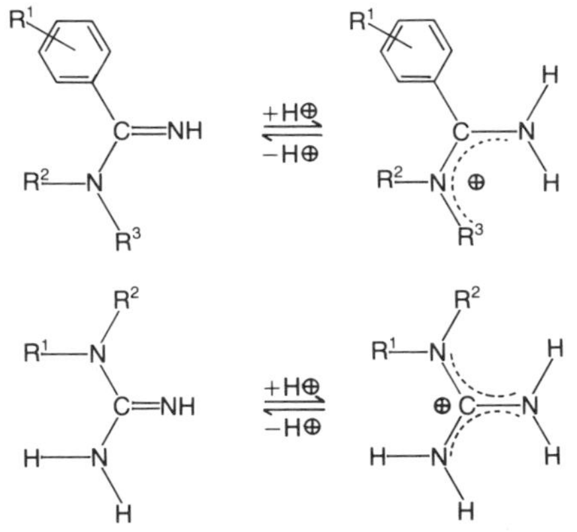

Fig. 1 – Basicity of amidines and guanidines.

recently very little was known about the biotransformation of amidines and guanidines. At the beginning of this study, *N*-oxidation of a typical strong basic amidine or guanidine had not been reported. Of all the known *N*-oxygenations, none of the oxidized nitrogen atoms was part of such a basic functional group.

These amidines and guanidines are the strongest known organic bases, because the cations formed upon protonation at the double-bonded nitrogen are excellently stabilized by resonance (Fig. 1). The compounds studied possess pK_a values between 11.6 and 13.6, are more basic than aliphatic amines, are protonated under physiological conditions and are water soluble (Weast, 1974; Häfelinger, 1975). It was believed at the beginning of the study that these properties of amidines and guanidines might prevent their attack by *N*-oxidative enzymes.

The examination of the pK_a-values and properties of the substrates used in this study demonstrate that only typical basic amidines and guanidines have been tested for *N*-oxidation. In no case is the nitrogen atom a part of an aromatic system, nor does it bear strongly electron-withdrawing groups. The objective of the study was to ascertain whether enzymic attack at such nitrogen centres is possible, despite the basic properties, and to identify new types of *N*-hydroxylated metabolites.

Amidines and guanidines present a family of pharmacologically important compounds, developed mainly for their antiprotozoal (see Dann, 1975 and 1978 for reviews) and antihypertensive (see Schlittler, Druey & Marxer, 1962 for a review) properties.

SUBSTRATES STUDIED AND POTENTIAL METABOLITES

When selecting the substrates for this study, drugs containing these functional groups were considered. If the drugs available were not suitable because of

their complex structure or their physico-chemical properties, model compounds were selected instead.

Because of the known involvement of N-hydroxylated metabolites in the
production of certain reactive or toxic metabolites (Damani, 1982, and references cited therein) it was more interesting to study N-hydroxy compounds
than N-oxides.

1.   Amidines:

Aromatic trypanocide amidines, like pentamidine (Ashley *et al.*, 1942), com-
ply with the criteria which were outlined above. But compounds like pentamidine are all diamidines. So two amidine groups have to be taken into
account. Furthermore, due to their water solubility, extractability of
these compounds into organic solvents is very poor. Therefore, simple benzamidines of type I were chosen as model compounds (Fig. 2). Benzamidines
I and their potential N-hydroxylated metabolites are not as readily hydrolysed
as their aliphatic analogues (De Wolfe, 1975).

By variation of the substituents at the nitrogen centres and in the ring, their
influence on the metabolic attack at nitrogen could be studied. The different
substituted amidines could form the following N-oxidized products.

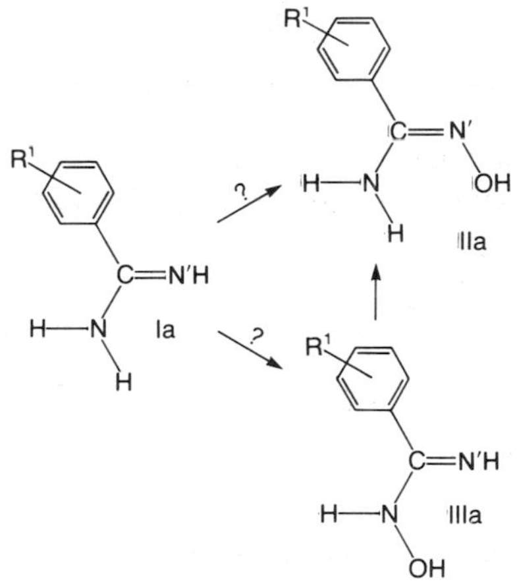

Fig. 2 – N,N-Unsubstituted benzamidines (for R see Table 1).

a)   *Unsubstituted amidines* Ia:

If N-oxidation occurs, the isolation of stable amidoximes IIa would be possible regardless of whether oxygen is transferred to the amide nitrogen N or
the imide nitrogen N' (Fig. 2). Such N-hydroxylated amidines exist in the
oxime structure II and have the Z-configuration. If the N-hydroxyamidines
III are formed initially, they would be expected to tautomerize to the amidox-

imes II (Dignam *et al.*, 1980). The influence of substitution of the aromatic ring was also studied in the case of these *N,N'*-unsubstituted benzamidines.*

b) Monosubstituted amidines Ib:

Beside the amidoximes IIb, the formation and isolation of stable so called *N*-hydroxyamidines IIIb should be possible in case of *N*-oxidation (Fig. 3). IIIb cannot tautomerize to amidoximes. The possible tautomerism of amidines will not be discussed here. It is assumed that the free bases and not the protonated amidines can be *N*-hydroxylated by electrophilic enzymes.

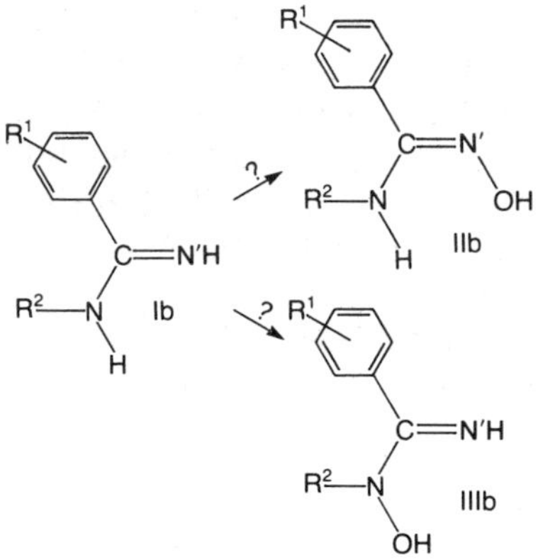

Fig. 3 – *N*-Monosubstituted benzamidines (for R see Table 1).

c) Disubstituted amidines Ic:

Again the isolation of stable amidoximes IIc should be possible (Fig. 4).

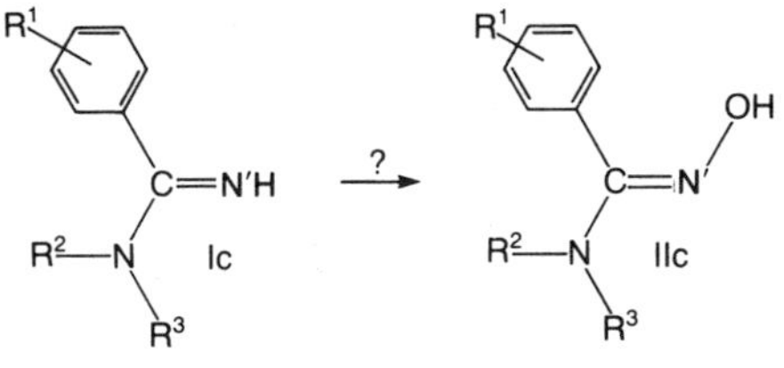

Fig. 4 – *N,N*-Disubstituted benzamidines (for R see Table 1).

With all derivatives of I, amidoximes should always be isolable, if *N*-oxygenation occurs. *N*-Monosubstituted derivatives Ib could also form stable *N*-hydroxyamidines IIIb. Therefore, this study examined not only the possible *N*-oxidation of single-bonded, but also of double-bonded nitrogen.

---

*Although not in complete accordance with the IUPAC rules—for better understanding—throughout primes are used to assign substitution at the nitrogen atoms (system of the *Chemical Abstracts*). So in amidines and related compounds *N* refers to the single bonded and *N'* to the double bonded nitrogen. Accordingly for guanidines and related compounds *N, N'* and *N"* (for the =NH—group) are used.

2.   Guanidines:

Because of the results achieved with amidines, compounds of type IV (Fig. 5) with a NH_2-group were of interest. The unsubstituted guanidine is normally present in the blood and urine of humans and other mammals (Reiter & Horner, 1979), and is therefore not a xenobiotic and was not included in the study.

The guanidines studied can be further divided into N-mono-substituted IV a and b and N,N-disubstituted IVc (Fig. 5). These guanidines, in particular guanoctinum IVb (Schoepke *et al.*, 1965) and debrisoquine IVc (Wenner, 1965), have antihypertensive properties. They were also selected due to their simple structures and the stability of their potential metabolites V.

If N-oxidation occurs at the nitrogen atoms N' or N'' of compound IV, the isolation of these stable N''-hydroxyguanidines V would be possible. Such N-hydroxylated guanidines exist in the oxime structure V and not in the tautomeric form VI (Clement & Kämpchen, in preparation). Therefore, this study was aimed at the identification of compounds V as products of biological N-oxidation.

If N-oxidation occurs at the nitrogen N, another type of N-hydroxylated guanidine VII would be formed (Fig. 6). These compounds VII are very difficult to synthesize, are not very stable and have not yet been included in the studies.

In order to test if N-oxidation is dependent on the pK_a, the less basic N,N''-diphenylguanidine ($pK_a = 10.0$) (Hull, 1930) was chosen as a substrate

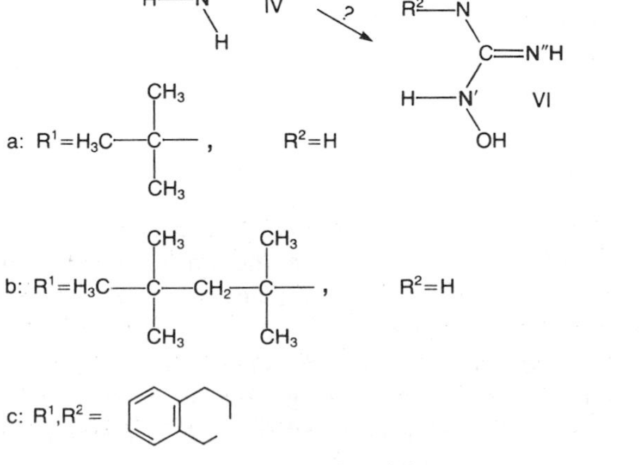

Fig. 5 – Guanidines.

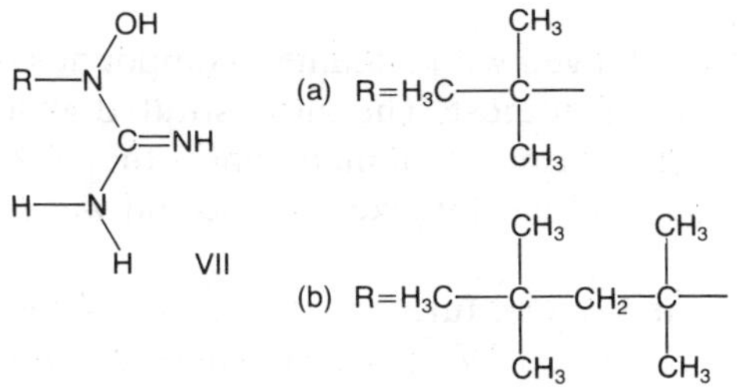

Fig. 6 – *N*-Hydroxyguanidines.

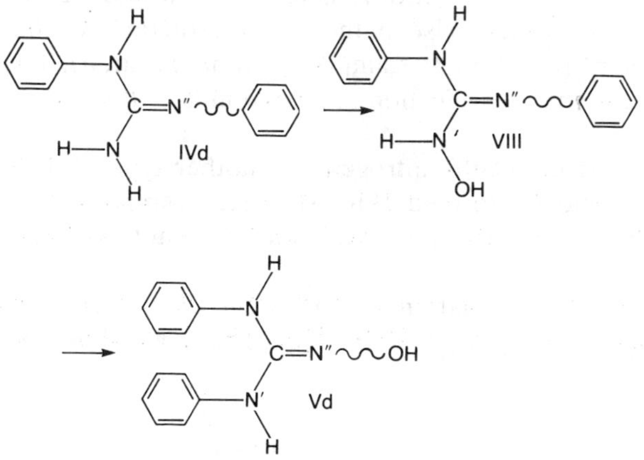

Fig. 7 – *N,N'*-Diphenylguanidine.

(Fig. 7). This guanidine could also be oxidized to the stable *N"*-hydroxy-guanidine Vd. If compound VIII were formed initially it would tautomerize to Vd (Clement & Kämpchen, in preparation).

EXPERIMENTAL

Synthesis

Chemical synthesis utilized methods previously reported with suitable modifications (see Clement, 1983, for amidines Ia and amidines IIa; Clement, 1984, for amidines Ib and c, amidoximes IIb and c and *N*-hydroxyamidine IIIb; guanidines IV and *N"*-hydroxyguanidines V; Clement & Kämpchen, in preparation). Compounds were analysed in the usual way. ^{15}N-n.m.r.-spectra were recorded to elucidate for the first time the preferred tautomeric forms and the configurations at the C=N— double bond (Clement & Kämpchen, in preparation).

Metabolic studies

Preliminary experiments were performed to ascertain that the expected metabolites could be isolated and identified if they are formed. All potential metabolites could be extracted with organic solvents at a pH of 7.4. The basic amidines and guanidines are not extractable at this pH. In this way metabolites can be separated from the substrates. Suitable solvent systems for the thin layer chromatographic separation, and spray reagents (Fe^{3+}) for visualization were developed. The mass spectra of the N-hydroxylated potential metabolites showed molecular ions and typical fragments of diagnostic value (for t.l.c. and mass spectral characteristics of amidoximes IIa see Clement, 1983; for amidoximes IIb and c see Clement, 1984; for N''-hydroxyguanidines Clement, in preparation).

The *in vitro* experiments were performed using mainly 9000 g supernatant fractions of rabbit liver homogenates. The incubations were carried out under aerobic conditions and employed an NADPH-regenerating system.

The metabolites were extracted, separated and identified by comparing their t.l.c. and mass spectral characteristics with the data for the synthetic reference products. Control incubations were carried out with 9000 g supernatant preparations to check the enzyme dependence of the products formed (for details of typical experimental procedures, materials and instruments see Clement, 1983).

A quantitative study (high-performance liquid chromatography) of benzamidoxime formed after incubation of benzamidine, under various conditions was undertaken to elucidate the enzymology of this new type of N-hydroxylation (Clement & Zimmermann, in preparation). Tests with inhibitors and purified flavin-containing monooxygenase were carried out (Clement, 1983).

RESULTS AND DISCUSSION

1. Guanidines:

Neither the guanidines IVa,b,c nor N,N''-diphenylguanidine IVd were N-hydroxylated in detectable amounts to the corresponding N''-hydroxyguanidines. The methods used were capable of detecting around 0.1% metabolism. These functional groups are therefore metabolically stable.

2. Amidines:

N-Hydroxy-N-methylbenzamidine of type IIIb could not be detected after the incubation of N-methylbenzamidine although reference material was available. N-Hydroxy-N-phenylbenzamidine (IIIb) has not yet been synthesized and was therefore not included in the metabolic studies (Clement, 1984). However, benzamidoximes of type II were identified for the first time as metabolites of benzamidines.

The details are summarized in Table 1. The N-hydroxylation could be ver-

Table 1

Metabolism of a series of amidines I by 9000*g* supernatant fractions of rabbit liver homogenates *in vitro*.

Substrates: Amidines I			Metabolites: Amidoximes II		
R¹	R²	R³	isolated + not isolated −		
H	H	H	+		
p−CH₃	H	H	+		
p−OCH₃	H	H	+		
p−NO₂	H	H	−		
Iᵃ m−CH₃	H	H	+	IIᵃ	
m−NO₂	H	H	+		
o−CH₃	H	H	+		
o−NO₂	H	H	−		
Iᵇ H	phenyl	H	+	IIᵇ	
H	CH₃	H	−		
H	CH₃	CH₃	−		
Iᶜ H	−[CH₂]₅−		−	IIᶜ	
H	phenyl	CH₃	−		

ified in the case of benzamidine and nearly all ring-substituted derivatives Ia (Clement, 1983). These benzamidines possess pK_a-values of about 11.6 (Albert *et al.*, 1948) and represent the most basic nitrogens which are accessible to *N*-oxidation. Only the strong electron-withdrawing nitro group in *para* and *ortho* positions prevents this metabolism.

Among the *N*-substituted derivatives only *N*-phenylbenzamidine is slightly *N*-oxidized. *N*-Methylbenzamidine and all *N*,*N*-disubstituted compounds were not metabolized to the corresponding amidoximes (Clement, 1984).

In summary, for optimal metabolism of nitrogen in amidines the 'amino' nitrogen*N* has to be unsubstituted. These results suggest that for*N*-oxygenation of amidines a NH₂ group as part of this functional group must be present. So in contrast to derivatives Ic compounds, Ia can form amidoximes, probably via *N*-oxygenation of a single bonded nitrogen and not by direct *N*-oxygenation of the imino nitrogen *N'* (Fig. 2). The *N*-oxidation of amidines is comparable to the oxidation of primary alkylamines but not to the oxidation of imines to oximes (Christou & Gorrod, this volume). This explains also the difference between the compounds Ib. *N*-Phenylbenzamidine exists exclusively in the amino form with a NH$_2$-structural element. *N*-Methylbenzamidine, however, possesses predominantly an imino type structure and is therefore not *N*-oxygenated (Clement, 1984). Therefore, one explanation for the difference of these two *N*-monosubstituted benzamidines towards oxidative enzymes is provided by their preference for one tautomeric structure (Clement, 1984).

One can further explain the difference by assuming that the *N*-alkyl derivatives are *N*-dealkylated instead of *N*-oxygenated. No *N*-oxygenation occurs if α-hydrogens to the nitrogen atom *N* are available. This suggests a radical mechanism which has been proposed for the different functions of cytochrome P-450 (Guengerich *et al.*, 1982). Studies of the enzymology of the *N*-oxygenation of benzamidine also provided evidence for the involvement of cytochrome P-450. The flavin-containing monooxygenase, however, is capable of direct oxygen transfer (from a peroxyflavin intermediate) (Ziegler, this volume). The results obtained with *N,N*-unsubstituted benzamidines can be explained by the following mechanism (Fig. 8): A P-450 perferryl oxygen (Guengerich *et al.*, 1982) abstracts an electron to yield a cation radical, which can lose a proton (or alternatively a H·) to form an amine radical (or a cation) which is stabilized by resonance. Finally by *N*-oxygenation, *N*-hydroxyamidines are produced which tautomerize to the amidoximes isolated in the metabolic studies. For *N*-phenylbenzamidine (replacement of the hydrogen at *N'* by phenyl) the same mechanism can be operative.

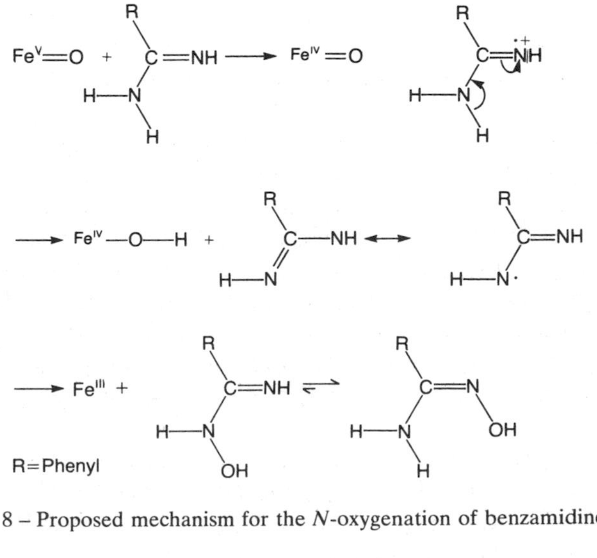

Fig. 8 – Proposed mechanism for the *N*-oxygenation of benzamidines.

I suggest this mechanism also for the *N*-oxygenation of amidino-like, but non-basic, aminoheterocyclic amine compounds, which have been shown to be promutagens (Yamazoe *et al.*, 1981; Gorrod, this volume). In these compounds the double-bonded nitrogen is part of an aromatic system. If α-hydrogens are available (*N*-alkyl substituted benzamidines) rapid rearrangement of the amine cation radical to the carbon centred radical could prevent oxygenation of nitrogen. Oxygenation of the α-carbon to an intermediate carbinolamine could result in the formation of *N*-dealkylated products (Ziegler, this volume; Guengerich *et al.*, 1982). Studies with *N*-methylbenzamidine are currently in progress to investigate this *N*-dealkylation.

Whether the amidoximes are responsible for the toxic effects, reported for most amidines (Briggaman, 1977) is also being studied. Preliminary experiments indicate that *N*-hydroxylation of amidines is not a deactivation reaction. It is possible that the toxicity of amidoximes is the result of the formation of unstable conjugates.

3. Enzymology:

Studies have been performed to elucidate the enzymology of the new type of *N*-hydroxylation. For these experiments the *N*-oxidation of the unsubstituted benzamidine to benzamidoxime has been chosen.

It seemed very likely that either the cytochrome P-450 system or the flavin-containing monooxygenase (Ziegler & Mitchell, 1972) is involved in the *N*-hydroxylation. Tests with purified flavin-containing monooxygenase indicated that benzamidine is not a good substrate for this enzyme (Clement, 1983).

Before studies of the enzymology could be continued, a quantitative analysis of benzamidoxime had to be developed. The best results were obtained by reversed phase high-performance liquid chromatography, after extraction of benzamidoxime with ether. The formation of benzamidoxime in the reaction mixture after different time intervals and in relation to the concentration of substrates could be followed using this h.p.l.c. system. The reaction was linear for 30 minutes and followed Michaelis–Menten kinetics (Clement & Zimmermann, in preparation). The concentration of benzamidoxime formed after incubation of benzamidine under various conditions and in the presence of inhibitors was also determined. The results are summarized in Table 2. It is obvious that the benzamidoxime formation required the presence of microsomal enzymes, molecular oxygen and NADPH.

The direct involvement of cytochrome P-450 in this *N*-hydroxylation is supported by the observations that inhibitors of cytochrome P-450, such as carbon monoxide, cyanide and SKF 525A, markedly decreased the rate of *N*-oxidation.

All these results were obtained with livers from non-induced rabbits. The inducing effect in rabbits by phenobarbital and 3-methylcholanthrene treatment is being studied at present. Studies with purified cytochrome P-450-isoenzymes are in progress.

Table 2

Characteristics of the microsomal *N*-hydroxylation of benzamidine I to
benzamidoxime II

Enzyme	Incubation mixture	nmol benzamidoxime/ min/mg protein
9000*g* supernatant (rabbit, not induced)	complete[a] pH 7.4	0.24[b]
	complete pH 8.6	0.20
	−NADPH	0.00
	−Mg^{2+}	0.14
	−O$_2$	0.00
	+5% O$_2$, 5% CO, 90% N$_2$	0.14
	+SKF 525A (5 mM)	0.00
	+KCN (25 μM)	0.12
Microsomes	complete	0.16[c]
100 000*g* supernatant	complete	0.00

Results are presented as the mean of three determinations. All experiments were performed with preparations from the same liver.
[a]Benzamidine HCl 10 μM in 1 ml, 1 ml liver preparation, 1 ml cofactor solution (containing NADPH, 2 mg; MgCl$_2$ 1.9mg) 3 ml phosphate buffer.
[b]The mean value of determinations with liver preparations from four different rabbits is 0.40.
[c]0.17 nmol benzamidoxime/min/nmol cytochrome P-450.

Preliminary experiments with other species indicate that well defined species differences in the *N*-hydroxylation of benzamidine occur, and that the rat appears to be defective in this respect (Clement & Zimmermann, in preparation).

In Table 3 the results on the *N*-oxidation of functional groups with strong basic properties including aliphatic amines are summarized. This table was constructed to develop a predictive concept for the metabolism of basic nitrogen-containing functional groups with regard to the two different mono-oxygenases. It is obvious that the involvement of the enzymes depends partly on the type of substitution at the nitrogen atoms. Primary alkylamines can be *N*-hydroxylated by the cytochrome P-450 system. Duncan & Cho (1980) used a reconstituted cytochrome P-450 system for the *N*-hydroxylation of phenter-mine. Usually P-450 is not involved in *N*-oxygenation of secondary and terti-ary alkylamines. Ziegler *et al.* (1973) have demonstrated that these com-pounds are oxidized by the purified amine oxidase, whereas primary alkylamines can only activate this enzyme. Our studies provided evidence for the *N*-oxidation of *N*-unsubstituted benzamidines by the cytochrome P-450 system. It is obvious that the cytochrome P-450 system preferentially oxidizes NH$_2$-groups. The results obtained with amidines are comparable to the *N*-oxidation of alkylamines. Both functional groups are *N*-hydroxylated by

Table 3
N-Oxygenation of functional groups with strong basic properties

Functional group	Cytochrome P-450 oxygenation	Flavin-containing monooxygenase	
		Catalytic site oxygenation	Regulatory site activation
Primary alkylamines	+[a]	−[b]	+[b]
Secondary alkylamines	−[c]	+[b]	+[b]
Tertiary alkylamines	−[c]	+[b]	+[b]
Benzamidines *N*-unsubstituted	+[d]	−[d]	?[e]
Benzamidines *N*-alkyl	−[d]	−[d]	?[e]
Guanidines	−[d]	−[d]	+[b]

[a]Duncan & Cho (1982).
[b]Ziegler *et al.* (1973).
[c]Not reported.
[d]This study.
[e]Study in progress.

cytochrome P-450 if a NH_2-group is present. This difference between amidines and guanidines can not be fully explained by convincing arguments at this stage of the study. It is possible that the electrons in guanidines are too delocalized to allow attack by electrophilic enzymes.

The pK_a-values do not appear to play an important part since the less basic *N,N″*-diphenylguanididine is not *N*-oxidized but the more basic amidines are. The state of hybridization does not appear to play an important part either, since primary alkylamines use more sp^3 orbitals and amidines more sp^2 orbitals (Häfelinger, 1975). Both are *N*-oxidized by the cytochrome P-450 system. It has been shown by Ziegler *et al.* (1973) that guanidines, like primary alkylamines, can activate the flavin-containing monooxygenase from hog liver. Whether the same applies to amidines has yet to be investigated.

This enzyme does not oxidize guanidines and amidines. In water solution the nucleophilicity of amidines and guanidines might be too small to allow attack by this enzyme. It is known that this enzyme can only oxidize compounds which are at least as nucleophilic as secondary amines (Ziegler & Poulsen, 1978). It has also been reported for other nitrogen- and sulphur-containing compounds that delocalization of electrons prevents interaction with the catalytic site (Poulsen, 1981). Therefore, in order to predict N-oxygenation of basic nitrogen-containing functional groups, the type of substitution at the nitrogen atoms, the preferred tautomeric forms, delocalization of electrons and nucleophilicity have to be considered.

ACKNOWLEDGEMENTS

This study was supported by the Deutsche Forschungsgemeinschaft.

REFERENCES

Albert, A., Goldacre, R. & Phillips, J. (1948), *J. chem. Soc.*, 2240.

Ashley, J. N., Barber, H. J., Ewins, A. J., Newbery, G. & Self, A. D. H. (1942), *J. chem. Soc.*, 103.

Briggaman, R. A. (1977), *Int. J. Dermatol.*, **16**, 155.

Clement, B. (1983), *Xenobiotica*, **13**, 467.

Clement, B. (1984), *Arch. Pharmazie.*, **317**, in press.

Coutts, R. T. & Beckett, A. H. (1977), *Drug Met. Rev.*, **6**, 51.

Damani, L. A. (1982), *Metabolic Basis of Detoxification* (eds. Jakoby, W. B. Bend, J. R. & Caldwell, J.) Academic Press, New York, p. 127 and references cited therein.

Dann, O. (1975), in *Development of Chemotherapeutic Agents for Parasitic Diseases* (ed. Mario, M.) North-Holland Publishing Company, Amsterdam, p. 101.

Dann, O. (1978), *Dt. Apothztg.*, **118**, 1674.

De Wolfe, R. H. (1975), *The Chemistry of Amidines and Imidates* (ed. Patai, S.) John Wiley & Sons, Chichester, p. 356.

Dignam, K. J., Hegarty, A. F. & Begley, M. J. (1980), *J. chem. Soc. Perkin II*, 704, and references cited therein.

Duncan, J. D. & Cho, A. K. (1982), *Mol. Pharmacol.*, **22**, 235.

Guengerich, F. P., MacDonald, P. L., Burka, L. T., Miller, R. E., Liebler, P. C., Zirvi, K., Frederick, C. B., Kadlubar, F. F. & Prough, R. A. (1982), in *Cytochrome P-450, Biochemistry, Biophysics and Environmental Implications* (eds. Hietanen, E., Laitinen, M. & Hanninen, O.) Elsevier Press, New York, p. 27, and references cited therein.

Häfelinger, G. (1975), in *The Chemistry of Amidines and Imidates* (ed. Patai, S.), John Wiley & Sons, Chichester, p. 1.

Hull, N. F. (1930), *J. Am. Chem. Soc.*, **52**, 5115.

Poulsen, L. L. (1981), *Rev. Biochem. Tox.*, **3**, 33.

Reiter, A. J. & Horner, W. H. (1979), *Arch. Biochem Biophys.*, **197**, 126.

Schlittler, E., Druey, J. & Marxer, A. (1962), *Fortschr. Arzneimittlefur-schung*, **4**, 341.

Schoepke, H. G., Brondyk, H. D., Wiemeler, L. H. & Schmidt, J. L. (1965), *Arch. int. Pharmacodyn.*, **153**, 185.

Weast, R. C. (1974), *Handbook of Chemistry and Physics*, 55th Edition, CRC Press, Cleveland, D 126.

Wenner, W. (1965), *J. med. Chem.*, **8**, 125.

Yamazoe, Y., Kenji, I., Kamataki, T. & Kato, R. (1981), *Drug Met. Disp.*, **9**, 292.

Ziegler, D. M. & Mitchell, C. H. (1972), *Arch. Biochem. Biophys.*, **150**, 116.

Ziegler, D. M., McKee, E. M. & Poulsen, L. L. (1973), *Drug Met. Disp.*, **1**, 314.

Ziegler, D. M. & Poulsen, L. L. (1978), in *Methods in Enzymology* (eds. Fleischer, S. & Packer, L.) Academic Press, New York, Vol. 52, p. 142.

Chapter 29

THE BIOLOGICAL
N-OXIDATION OF HYDRAZINES
AND AZO-COMPOUNDS

R. A. Prough, Department of Biochemistry, The University of Texas Health Science Centre, Dallas, Texas 75235, USA

1. The structure–activity relationships for hydrazines are described for the *N*-oxygenation reactions catalysed by cytochrome P-450, the flavin-containing monooxygenase, monoamine oxidase, and prostaglandin synthetase. All of the hydrazines tested inhibited prostaglandin synthetase activity.
2. Cytochrome P-450 and monoamine oxidase catalyse the formation of stable azo derivatives of 1,2-disubstituted hydrazines. Mono- and 1,1-disubstituted hydrazines, as well as hydrazides, inhibit these enzymes due to the formation of diazenyl-haem complexes or to the modification of the haem/flavin prosthetic groups.
3. The flavin-containing monooxygenase readily oxidizes 1,1-disubstituted hydrazines due to their high affinity for the enzyme.
4. The conversion of azo to azoxy derivatives appears to be exclusively catalysed by cytochrome P-450.

INTRODUCTION

During the 1970s, considerable effort was directed toward establishing the significance of the biological oxidation reactions of various hydrazine compounds. These compounds are of interest due to their wide use in industry and therapy, as well as their known carcinogenicity (Toth, 1975) and toxicity

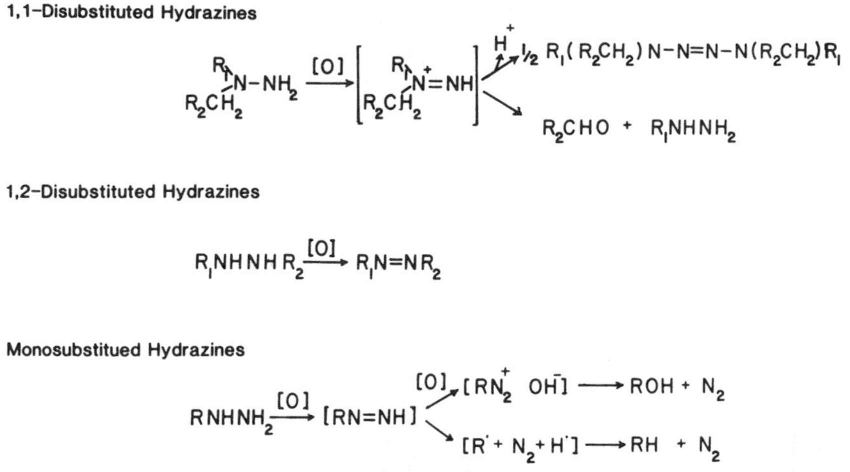

1,1–Disubstituted Hydrazines

$$\underset{R_2CH_2}{\overset{R_1}{\diagdown}}N-NH_2 \xrightarrow{[O]} \left[\underset{R_2CH_2}{\overset{R_1}{\diagdown}}\overset{+}{N}=NH \right] \overset{\overset{H^+}{\diagup}}{\underset{\diagdown}{}} \begin{array}{l} \tfrac{1}{2}\ R_1(R_2CH_2)\,N-N=N-N(R_2CH_2)R_1 \\[2mm] R_2CHO\ +\ R_1NHNH_2 \end{array}$$

1,2–Disubstituted Hydrazines

$$R_1NHNHR_2 \xrightarrow{[O]} R_1N=NR_2$$

Monosubstitued Hydrazines

$$RNHNH_2 \xrightarrow{[O]} [RN=NH] \overset{\overset{[O]}{\diagup}\ [R\overset{+}{N_2}\ O\bar{H}] \longrightarrow ROH\ +\ N_2}{\underset{\diagdown\ [R^{\cdot}\ +\ N_2 + H^{\cdot}] \longrightarrow RH\ +\ N_2}{}}$$

Fig. 1 – The expected products of the chemical oxidation of 1,1-di, 1,2-di, and mono-substituted hydrazines.

(Moloney & Prough, 1983). This review will attempt to encapsulate much of the information related to the enzymology of the biological *N*-oxidation of hydrazine and its derivatives.

Colvin (1969) and Juchau & Horita (1972) have addressed the possible role of biological oxidation reactions in hydrazine metabolism and toxicity. The occurrence of alkanes, arising from unstable monoalkyldiazenes formed during the microsomal metabolism of alkylhydrazines, was demonstrated by Prough *et al.* (1969, 1970). In addition, azo derivatives were shown to be formed from the biological oxidation of 1,2-disubstituted hydrazines (Druckrey, 1970; Fiala, 1977). The azo and diazene intermediates are the expected products formed during the chemical oxidation of hydrazines as seen in Fig. 1 (Smith, 1983). Mono- and 1,1-disubstituted hydrazines can form unstable diazenes which decompose to yield either hydrocarbons or hydrazones, respectively, while 1,2-disubstituted hydrazines are usually oxidized to stable azo derivatives.

THE ENZYMOLOGY OF THE *N*-OXIDATION OF HYDRAZINES

Microsomal Cytochrome P-450

Early reports of the interaction of hydrazine derivatives with microsomal cytochrome P-450 demonstrated that these compounds were potent inhibitors of this haem-dependent monooxygenase (Clark, 1967; Kato *et al.*, 1969). Reed's group observed that in the presence of rat liver microsomes, NADPH and oxygen, two routes of metabolism of monoalkylhydrazines existed: *N*-dealkylation and formation of the corresponding alkanes (Prough *et al.*, 1969; Wittkop *et al.*, 1969). However, these reactions were noted as

Table 1

Effect of inhibitors of cytochrome P-450 on the microsomal metabolism of 1,2-disubstituted hydrazines

Inhibitor	Metabolic rate[a] (nmol/min/mg)	
	1,2-Dimethylhydrazine	Procarbazine
None	6.9	9.4
$-$Oxygen	1.3	0.4
$-$NADPH	0.6	0.5
CO/O_2 (4:1 v/v)	2.5	4.0
Metyrapone (1 mM)	5.9	3.8
n-Octylamine (1 mM)	0.8	1.5
Anti-NADPH-cytochrome c (P-450) reductase globulin	1.1	2.3
Non-immune globulin	6.7	9.2

[a]The reaction mixtures contained liver microsomes from untreated rats.

proceeding at relatively low metabolic rates compared to the demethylation of other N-methyl drug substrates.

The metabolism of the 1,2-disubstituted hydrazine, 1,2-dimethylhydrazine (1,2-DMH), has been studied due to its unique organ specificity as a colon carcinogen. The azo metabolite of 1,2-DMH, azomethane, has been shown to be more potent as a carcinogen than the parent hydrazine (Druckrey, 1970). Fiala (1977) later demonstrated that *in vivo*, azomethane is rapidly formed from 1,2-DMH and is subsequently metabolized to azoxymethane at an appreciable rate. Further, the azoxy derivative is metabolized to CO_2 *in vivo*. Dunn *et al.* (1979) observed that the therapeutic 2-methylbenzylhydrazine, procarbazine or N-isopropyl-α-(2-methylhydrazino)-p-toluamide, was also converted to its azo and azoxy derivatives by liver microsomal fractions. This reaction was shown to be catalysed by cytochrome P-450 at a relatively high rate of 8–16 nmol product formed per min per mg protein.

In studies from our laboratory on the metabolism of 1,2-disubstituted hydrazines, we have concluded that cytochrome P-450 is the major hepatic enzyme that catalyses formation of the stable 1,2-disubstituted azo metabolites of these agents. In Table 1, the effects of various inhibitors on the microsomal metabolism of 1,2-dimethylhydrazine and procarbazine are shown. All of the agents which inhibit the activity of cytochrome P-450 caused decreases in the formation of the azo metabolites.

The effect of animal pretreatment regimen on the formation of azomethane and the azo derivative of procarbazine is shown in Table 2. Only phenobarbital was an effective inducing agent for these reactions. The observation that

Table 2

Effect of animal pretreatment on the microsomal metabolism of 1,2-
disubstituted hydrazines

Pretreatment	Metabolic rate[a] (nmol/min/mg)	
	1,2-Dimethylhydrazine	Procarbazine
None	6.8	9.1
Phenobarbital	13.7	16.5
5,6-Benzoflavone	6.1	9.0

[a]The reaction mixtures contained liver microsomes from untreated animals or animals pretreated
with 80 mg/kg phenobarbital (4 days) or 60 mg/kg 5,6-benzoflavone (4 days).

cytochrome P-450 catalyses the *N*-oxidation of 1,2-disubstituted hydrazines
at a significant rate indicates that these hydrazine derivatives are good substrates for this haem-containing microsomal monooxygenase. This result is in
part due to the fact that the azo products, azomethane or azo-procarbazine,
are stable intermediates which do not interfere with monooxygenase function.

It has been noted that the mono- and 1,1-disubstituted hydrazines are
metabolized by microsomal cytochrome P-450, albeit at much lower rates of
metabolism compared to the 1,2-disubstituted hydrazines. Reed's group
reported that liver microsomal fractions, in the presence of NADPH and O_2,
could catalyse an *N*-dealkylation reaction (Wittkop *et al.*, 1969) and a hydrazine oxidase reaction which converted the alkylhydrazine to its corresponding alkane (Prough *et al.*, 1969). Recent studies in this laboratory designed to
measure accurately the rates of aldehyde and alkane formation indicated that
these reactions proceed at a rate which is only 5–10% that of azo formation
from 1,2-DMH. Several of these reactions are described by Spearman *et al.*
(1985a) in a subsequent report in this volume.

The marked difference in the rates of metabolism between 1,2-disubstituted
and the mono- or 1,1-disubstituted hydrazines is most probably due to the
interaction of the reactive diazene or monoazo metabolites of 1,1-di and
mono-substituted hydrazines, respectively, with the haem group of cytochrome P-450. Hines & Prough (1980) observed that during the microsomal
metabolism of 1,1-disubstituted hydrazines, a unique spectral intermediate of
cytochrome P-450 was formed in the visible region of the spectra. Our subsequent work has demonstrated that this intermediate is abortive in nature and
terminates the normal function of the cyctochrome P-450 dependent monooxygenase. Based on the chemical literature, we have concluded that the oxidative product of 1,1-disubstituted hydrazines, a 1,1-disubstituted diazene, most
likely forms a haem–diazenyl metabolite complex. This complex is relatively

stable and prevents catalysis of normal monooxygenase reactions. Mansuy
et al. (1982) have demonstrated that stable haem–diazenyl complexes can be
formed with tetraphenylporphyrins and a 1,1-disubstituted diazene. Jonen
et al. (1982), Muakkassah *et al.* (1981a,b), and Moloney *et al.* (1984) have
shown that a number of monosubstituted hydrazines and hydrazides also form
inhibitory metabolic complexes with cytochrome P-450. Phenylhydrazine
metabolism results in formation of a unique absorbance maxima at 480 nm
and also results in considerable complexation of the cytochrome P-450 measured as the loss of CO-reactive cytochrome P-450 (Jonen *et al.*, 1982). Under
these conditions, considerable haem destruction occurs. Similar results were
noted for phenelzine, an antidepressant agent, except that a less distinct
absorbance maxima was noted (Muakkassah *et al.*, 1981a). Other hydrazines
and hydrazides also caused a loss of CO-reactive cytochrome P-450 without pronounced haem degradation (Muakkassah *et al.*, 1981b; Moloney *et al.*, 1984).
The only common feature observed after a study of the structure–activity relationship between hydrazine structure and inhibitory action was the correlation of the loss of CO-reactive cytochrome P-450 and the loss of absorbance
at 416–418 nm during metabolism (Moloney *et al.*, 1984). These results suggest that upon metabolism, the diazene or monoazo intermediate can complex to the haem iron and prevent formation of the Fe^{2+}-CO form of the
enzyme. Battioni *et al.* (1983) have provided evidence for the chemical
characteristics of the diazenyl–haem complexes of cytochrome P-450. These
complexes are unstable in the presence of oxygen and decompose to give the
normal decomposition products of diazenes and the free haemoprotein. The
hydrocarbon products described by Prough *et al.* (1969) are almost certainly
the result of formation of carbon-centred alkyl free radicals reported by
Augusto *et al.* (1981).

However, Lange & Mansuy (1981) and Ortiz de Montellano *et al.* (1981)
have provided evidence for a stable σ-iron–carbon bonded adduct formed
during metabolism of phenylhydrazine and have demonstrated that the
adduct can decompose either to give benzene or to phenylate the haem group
of cytochrome P-450. The existence of these inhibitory metabolite complexes
of mono- and 1,1-disubstituted hydrazine derivatives clearly explains why
these compounds are poor substrates for metabolism by cytochrome P-450.

Microsomal Flavin-Containing Monooxygenase
The flavin-containing monooxygenase described by Ziegler (1980) is capable
of *N*-oxygenating a number of *N*-alkyl compounds. Prough *et al.* (1981) have
described the metabolism of various hydrazines by this enzyme. For most
hydrazine derivatives, the pH-rate profile was identical with that seen for
amine substrates such as benzphetamine. The products of the reaction were
those expected for hydrazine oxidation products; namely, the hydrocarbon
product derived from a diazene intermediate or the aldehyde product derived
from a hydrazone in the case of methyl- or ethylhydrazine. 1,1-Disubstituted
hydrazines were the best substrates based on their low apparent K_m and high

Table 3
Effect of inhibitors on the microsomal metabolism of 1,1-dimethylhydrazine

Inhibitor	Metabolic rate[a] (nmol/min/mg)		
	Hamster	Hog	Rat
None	2.3	4.6	1.6
n-Octylamine (1 mM)	3.4	9.8	2.7
Metyrapone (1 mM)	2.0	4.4	1.4
Methimazole (1 mM)	0.7	1.5	0.4
Anti-NADPH-cytochrome *c* (P-450) reductase globulin	1.9	—[b]	1.3
Non-immune globulin	2.0	—[b]	1.4

[a]The liver microsomes were prepared from untreated animals.
[b]Not performed.

K_{cat} (V_{max}/K_m) values (the apparent K_m values were 100 μM or smaller). The V_{max} values were almost identical for all of the hydrazines studied and two known amine substrates, benzphetamine and *N,N*-dimethylaniline.

Because the 1,1-disubstituted hydrazines are inhibitors of cytochrome P-450 and substrates for the flavin-containing monooxygenase, we have established that the rates of metabolism of 1,1-dimethylhydrazine, in the presence of liver microsomes from hamster, hog, and rat, were not affected by inhibitors of cytochrome P-450 (Table 3). Yet, *n*-octylamine stimulated the activity as expected for the flavin-containing monooxygenase. In addition, animal pretreatment with sodium phenobarbital or 5,6-benzoflavone had little or no affect on the rate of microsomal demethylation of 1,1-dimethylhydrazine (Prough *et al.*, 1981). These results clearly indicate that in rodent liver microsomes, the principal enzyme involved in the metabolism of 1,1-disubstituted hydrazines is the flavin-containing monooxygenase due to the pronounced inhibition of cytochrome P-450 by the diazenyl metabolite–haem complex (Hines & Prough, 1980).

The 1,2-disubstituted hydrazines are poor substrates for the flavin-containing monooxygenase and the levels of this enzyme were low in young rats used for many of our studies (Prough *et al.*, 1981). Therefore, we could detect little or no involvement of the flavin-containing monooxygenase in the metabolism of the 1,2-disubstituted compounds. The 1,1-disubstituted compounds are potent inhibitors of both MAO and cytochrome P-450, further demonstrating the unique function of the flavin-containing monooxygenase in the disposition of the 1,1-disubstituted hydrazines.

Table 4

Subcellular fractionation of the *N*-oxidation of procarbazine by rat liver enzymes

Fraction	Metabolic rate[a] (nmol/min/mg)	
	No NADPH	NADPH
Homogenate	1.5	1.3
Mitochondrial	4.3	2.1
Microsomal	0.1	9.9
Cytosol	0.7	0.8

[a]The reaction mixtures contained an NADPH regenerating system and the various tissue fractions obtained from untreated rats (Prough *et al.*, 1981). NADPH was either added or omitted to measure the NADPH-dependent and independent activity. Nicotinamide and nicotinamide adenine nucleotides are inhibitors of MAO *in vitro* (Coomes & Prough, 1984).

Mitochondrial Monoamine Oxidase

The use of hydrazines as inhibitors of monoamine oxidase (MAO) has been established as a therapeutic regime for the treatment of depression (Zeller *et al.*, 1955). Several groups have suggested that the hydrazines are metabolized by the various amine oxidases, but at a much lower rate than that of the amine substrates (Hucko-Haas & Reed, 1970; Patek & Hellerman, 1974). These hydrazines appear to possess a high affinity for the enzyme and are only slowly metabolized by the flavoprotein oxidase.

In our survey of enzymes which are involved in the biological oxidation of 1,2-disubstituted hydrazine derivatives, we noted that up to 40% of the formation of stable 1,2-disubstituted azo derivatives in isolated hepatocytes was due to monoamine oxidase (Coomes & Prough, 1984). As seen in Table 4, the two major enzyme activities in the rat hepatocyte for *N*-oxidation of 1,2-dimethylhydrazine and the 1-methyl-2-benzylhydrazine, procarbazine, were apparently due to monoamine oxidase (NADPH-independent) and cytochrome P-450 (NADPH-dependent). Interestingly, there was little hydrazone formation from these two hydrazines, suggesting that formation of the stable azo intermediate is preferred (Coomes & Prough, 1984).

The monosubstituted hydrazines and hydrazides are also inhibitors of monoamine oxidase (Zeller *et al.*, 1955). In our studies with MAO, we have shown that alkylhydrazines are metabolized to yield the corresponding alkane. During this reaction, the MAO activity is drastically inhibited. Patek & Hellerman (1974) have demonstrated that during metabolism of phenylhydrazine by MAO, the flavin prosthetic group becomes phenylated.

These results suggest that the hydrazines are most likely converted to diazenes which may subsequently decompose to yield carbon-centred free radicals. These radicals may add covalently to the isoalloxazine ring of the flavoprotein.

Despite the inhibiting (suicide) action of hydrazines on MAO, one must realize that mitochondria are ubiquitously distributed in the mammal and certain organs may have sufficient MAO activity associated with their mitochondria to bioactivate the hydrazine derivatives. For example, lung possesses a reasonably high content of mitochondria which can catalyze the metabolism of hydrazines. In fact, our studies suggest that the MAO activity of lung mitochondria may be less susceptible to inhibition by hydrazines than is that of liver (Prough *et al.*, 1983). Therefore, MAO activity may be an important factor in the carcinogenicity and toxicity of hydrazines in extrahepatic tissue.

Prostaglandin H Synthetase
We have also evaluated the arachidonic acid-dependent cooxidation of hydrazine derivatives by this enzyme. The conversion of ethylhydrazine to ethane is stimulated in the presence of arachidonic acid and ram seminal vesicles (Prough *et al.*, 1983). This activity is also partially sensitive to *in vitro* addition of indomethacin. However, the reaction rates on a per mg of protein basis are very low compared to the activity of MAO or cytochrome P-450 in tissues such as rat lung. Although a small role of this enzyme in hydrazine metabolism cannot be precluded, it should be remembered that the hydrazines are potent inhibitors of this enzyme and cause a decrease in the arachidonate-stimulated oxidative burst of the prostaglandin H synthetase activity of seminal vesicle microsomes (Prough *et al.*, 1983).

CHARACTERIZATION OF THE ENZYMES INVOLVED IN HYDRAZINE OXIDATION

The preferred substrates and inhibitors of the various enzymes involved in hydrazine metabolism are listed in Table 5. The 1,2-disubstituted hydrazines are preferred substrates for monoamine oxidase and cytochrome P-450, whereas the 1,1-disubstituted hydrazines are preferred substrates for the flavin-containing monooxygenase. The roles of the various structural classes of the hydrazines as inhibitors are also shown.

N-OXYGENATION OF AZO COMPOUNDS

In our work, we have sought to characterize the enzymology of the formation of azoxy derivatives from the stable azo compounds (Dunn *et al.*, 1979; Wiebkin & Prough, 1980). As ^{14}C-labelled azo-procarbazine can be prepared, we have tested liver subcellular fractions for the presence of azo *N*-oxygenase activities. Only the microsomal fraction in the presence of NADPH and

Table 5
Enzymes catalysing hydrazine *N*-oxidation

Enzyme	Preferred substrates	Inhibitory action
Cytochrome P-450	1,2-Disubstituted hydrazines	Mono- and 1,1-di-substituted hydrazines, hydrazides
FAD-containing mono-oxygenase	1,1-Disubstituted hydrazines	Arylhydrazines ?
Monoamine oxidase	1,2-Disubstituted hydrazines	Mono- and 1,1-di-substituted hydrazines, hydrazides
Prostaglandin synthetase	Monosubstituted hydrazines	All

oxygen catalyses the formation of the two azoxy isomers of procarbazine from azo-procarbazine (Dunn *et al.*, 1979; Wiebkin & Prough, 1980). In addition, we have evaluated the possible role of the H_2O_2- and arachidonate-dependent peroxidases in this reaction; little or no activity in forming azoxy derivatives is found and the azo derivative of procarbazine seems refractory to metabolism by these enzymes. Recently, we have addressed the form specificity of cyto-chrome P-450 in the formation of azoxy metabolites. The major isozyme induced by polycyclic aromatic hydrocarbons and a minor isozyme induced by phenobarbital principally catalyse this reaction (Prough *et al.*, 1984). Studies are in progress to establish which enzymes metabolize and activate the azoxy derivatives.

ACKNOWLEDGEMENTS

Supported in part by grants from the American Cancer Society (BC336) and The Robert A. Welch Foundation (1–616). The Author wishes to thank his past and present collaborators for their aid in many of the studies presented in this publication.

REFERENCES

Augusto, O., Oritz de Montellano, P. R. & Quantanhilha, A. (1981), *Biochem. Biophys. Res. Commun.*, **101**, 1324.

Battioni, P., Mahy, J.-P., Delaforge, M. & Mansuy, D. (1983), *Eur. J. Biochem.*, **134**, 241.

Clark, B. (1967), *Biochem. Pharmacol.*, **16**, 2369.

Colvin, L. B. (1969), *J. Pharm. Sci.*, **58**, 1433.

Coomes, M. W. & Prough, R. A. (1984), *Drug Metab. Disp.*., **11**, 550.

Druckrey, H. (1970). in *Carcinoma of the Colon and Antecedent Epithelium*, (ed Burdette, W. J. & Thomas, C. C.) Springfield, p. 267.

Dunn, D. L., Lubet, R. A. & Prough, R. A. (1979), *Cancer Res.*, **39**, 4555.

Fiala, E. S. (1977), *Cancer*, **40**, 2436.

Fiala, E. S., Bobotas, G., Kulakis, C., Wattenberg, L. W. & Weisburger, J. H. (1977), *Biochem. Pharmacol.*, **26**, 1763.

Hines, R. N. & Prough, R. A. (1980), *J. Pharmacol. Exp. Therap.*, **214**, 80.

Hucko-Haas, J. & Reed, D. J. (1970), *Biochem. Biophys. Res. Commun.*, **39**, 396.

Jonen, H. G., Werringloer, J., Prough, R. A. & Estabrook, R. W. (1982), *J. Biol. Chem.*, **257**, 4404.

Juchau, M. R. & Horita, A. (1972), *Drug Metab. Rev.*, **1**, 71.

Kato, R., Takanaka, A. & Shoji, H. (1969), *Jap. J. Pharmacol.*, **19**, 315.

Lange, M. & Mansuy, D. (1981), *Tetrahedron Lett.*, **22**, 2561.

Mansuy, D., Battioni, P. & Mahy, J. P. (1982),*J. Am. Chem. Soc.*, **104**, 4487.

Moloney, S. J. & Prough, R. A. (1983), in *Reviews in Biochemical Toxicology*, Vol. 5 (eds. Hodgson, E., Bend, J. R. & Philpot, R. M.) Elsevier Biomedical, New York, p. 313.

Moloney, S. J., Snider, B. J. & Prough, R. A. (1984), *Xenobiotica*, 14, 803.

Maukkassah, S. F. & Wang, W. C. T. (1981a), *J. Pharmacol. Exp. Therap.*, **219**, 147.

Muakkassah, S. F., Bidlack, W. R. & Wang, W. C. T. (1981b), *Biochem. Pharmacol.*, **30**, 1651.

Ortiz de Montellano, P. R. & Kunze, K. L. (1981), *J. Am. Chem. Soc.*, **103**, 6534.

Patek, D. R. & Hellerman, L. (1974), *J. Biol. Chem.*, **249**, 2373.

Prough, R. A., Wittkop, J. A. & Reed, D. J. (1969), *Arch. Biochem. Biophys.*, **131**, 369.

Prough, R. A., Wittkop, J. A. & Reed, D. J. (1970), *Arch. Biochem. Biophys.*, **140**, 540.

Prough, R. A., Freeman, P. C. & Hines, R. N. (1981), *J. Biol. Chem.*, **256**, 4178.

Prough, R. A., Brown, M. I., Amrhein, C. A. & Marnett, L. J. (1983) in *Extrahepatic Drug Metabolism and Chemical Carcinogenesis* (eds. Rydstrom, J., Montelius, J. & Bengtsson, M.) Elsevier, Amsterdam, p. 489.

Prough, R. A., Brown, M. I., Dannan, G. A. & Guengerich, F. P. (1984), *Cancer Res.*, **44**, 543.

Smith, P. A. S. (1983), in *Derivatives of Hydrazine and Other Hydronitrogens Having N-N Bonds* Benjamin/Cummings, Reading MA., p. 142.

Spearman, M. E., Franck, T. G., Moloney, S. J. & Prough, R. A. (1985a), this volume.

Spearman, M. E., Moloney, S. J. & Prough, R. A. (1985b), *Mol. Pharmac.*, in press.

Toth, B. (1975), *Cancer Res.*, **35**, 3693.

Wiebkin, P. & Prough, R. A. (1980), *Cancer Res.*, **40**, 3524.

Wittkop, J. A., Prough, R. A. & Reed, D. J. (1969), *Arch. Biochem. Biophys.*, **134**, 308.

Zeller, E. A., Barsky, J. & Berman, E. R. (1955), *J. Biol. Chem.*, **214**, 267.

Ziegler, D. M. (1980), in *Enzymatic Basis for Detoxication*, Vol. 1 (ed. Jakoby, W. B.) Academic Press, New York, p. 201.

Chapter 30

METABOLIC *N*-HYDROXYLATION OF ACETOPHENONE IMINES

Maro Christou[†] and **J. W. Gorrod**, Department of Pharmacy, Chelsea College, University of London, Manresa Road, London SW3 6LX, UK

1. Previous work on the metabolism of 2,4,6-trimethylacetophenone imine is reviewed.
2. The chemical stability of acetophenone imines and the geometric isomerism of their potential *N*-hydroxylated metabolites, i.e. acetophenone oximes are discussed.
3. Metabolic *N*-hydroxylation of a series of chemically stable acetophenone imines to mixtures of *E* and *Z* isomeric oximes is described, both *in vitro* using liver microsomes from various species, and in the intact animal.
4. The major characteristics of the enzymes catalysing acetophenone imine *N*-hydroxylation are described and compared to those for enzymes catalysing primary aromatic amine and tertiary heteroaromatic amine *N*-oxidation.

INTRODUCTION

The biological oxidation of the nitrogen in an imine (R—C=NH), or an imino group (C=N), as part of a more complex structure, has not been

---
[†]Present address: Department of Pharmacology, Medical Sciences Centre, University of Wisconsin Medical School, Madison, WI, USA.
  Some of the data presented will appear in greater detail in other publications.

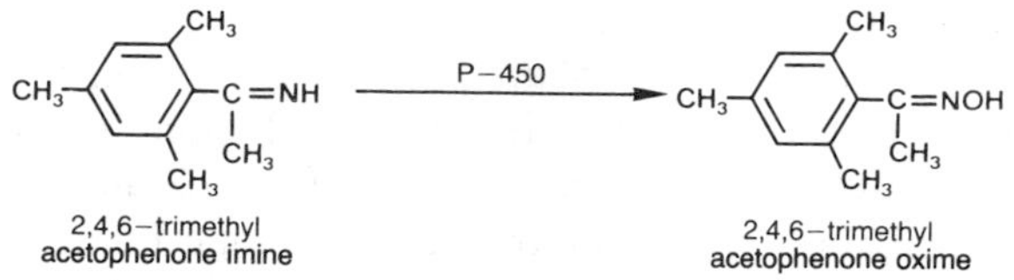

Fig. 1 – Metabolic *N*-hydroxylation of 2,4,6-trimethylacetophenone imine.

extensively investigated. The limited literature on the enzymic *N*-hydroxylation of imines is largely due to the inherent instability of the majority of these compounds (Moureau & Mignonac, 1913, 1920), rendering them unsuitable for use as substrates in metabolic studies.

The common occurrence of C=N in pharmacologically active xenobiotics (Gorrod & Christou, 1985a), and suggestions that imines are metabolic intermediates in the deamination of aliphatic amines such as amphetamine (Parli *et al.*, 1971a; Hucker *et al.*, 1971), focused interest on the metabolic transformation of imino nitrogen. In a study directed towards substantiating the role of imines as metabolic precursors of oximes, Parli *et al.* (1971b) were the first to demonstrate the metabolic *N*-hydroxylation of a stable imine (2,4,6-trimethylacetophenone imine) to the corresponding oxime by rat and rabbit liver microsomes (Fig. 1). This reaction was inhibited by DPEA, SKF 525A and carbon monoxide, but was enhanced following animal pretreatments with phenobarbital, but not 3-methylcholanthrene. Based on these findings and the intermediate basicity of 2,4,6-trimethylacetophenone imine ($pK_a = 7.0$), it was postulated that imine *N*-hydroxylation may be catalysed by enzyme systems similar to those mediating the *N*-hydroxylation of primary aromatic amines (Gorrod, 1982).

This paper reviews the findings of a more recent systematic study of the metabolic *N*-hydroxylation of a series of substituted acetophenone imines by various mammalian species both *in vitro* and *in vivo*, and the characterization of the enzyme systems involved.

EXPERIMENTAL

Acetophenone Imine Substrates

Unsubstituted and monosubstituted acetophenone imines are so readily hydrolysed to the corresponding ketones that hydrolytic parameters for *ortho*-, *meta*-, and *para*-methyl substituted acetophenone imines at 37 °C can only be measured in the presence of 10% water and 90% methanol (Christou, 1982). Under these conditions half-lives of 3, 0.5 and 1 minute respectively were obtained. Contrary to these acetophenone imines, aqueous solutions of *ortho*-, *meta*-, and *para*-methylsubstituted benzophenone imines exhibit half-lives of 160, 10 and 20 minutes respectively (Culbertson, 1951). Additional steric crowding in the vicinity of the carbimino group achieved by di-*ortho* substitution causes a much greater retardation of hydrolysis. As a result, aqueous solutions of di-*ortho*-substituted benzophenone imines show no measurable hydrolysis even when heated to 100 °C (Culbertson *et al.*, 1962).

Based on the above findings a series of acetophenone imines with at least two *ortho* substituents, namely, 2,6-dimethyl-, 2,6-dichloro-, 2,3,5,6-tetramethyl- and 2,3,4,5,6-pentamethylacetophenone imines were prepared using methods previously described for other ketimines (Pickard & Tolbert, 1961). Their spectral characteristics and physico-chemical properties have been described (Christou, 1982; Gorrod & Christou, 1985a,b). These substituted acetophenone imines showed no appreciable hydrolysis over several hours and were therefore chosen as model substrates for metabolic studies.

Geometric Isomerism of Acetophenone Oximes

The potential *N*-hydroxylated metabolites of acetophenone imines, i.e. oximes, were also prepared and characterized by elemental analyses and spectroscopic methods (Christou, 1982; Gorrod & Christou, 1985a). Aliphatic ketoximes exist in two geometric isomeric forms, *Z* and *E* (IUPAC, 1970). This isomerism was first established by n.m.r. methods based on the fact that the resonances of protons of carbon atoms adjacent to C=NOH appear in the spectrum twice (Lustig, 1961). It has also been reported that in the preparation of oximes from phenylketones of the type ArRC=O, the nature of the R group determines the orientation of the OH-group (Wittig, 1930). If R is a methyl group the ketoxime product is predominantly the least sterically hindered *E* (*anti*-phenyl) isomer.

Direct gas-chromatographic analysis of synthetic substituted acetophenone oximes using an OV 1 column and optimum conditions of separation first described by Gorrod & Christou (1982) enabled separation of two oxime chromatographic peaks. Combined g.c./m.s. analysis indicated that both peaks give identical mass spectra with a molecular ion (M^+) equivalent to the molecular weight of each oxime. The quantitatively predominant isomer has been assigned the less sterically hindered *E* configuration. With substituted acetophenone oximes the relative proportion of the more sterically hindered *Z* isomer is dependent on both the number and nature of the substituents on the phenyl group and decreases in the order 2,6-dimethyl- >2,4,6-trimethyl- >2,4,5,6-tetramethyl- >2,6-dichloro- >2,3,4,5,6-pentamethylacetophenone imine (Gorrod & Christou, 1985a).

RESULTS

In vitro *N*-Hydroxylation of Acetophenone Imines

Liver microsomes from various mammalian species catalysed the metabolic *N*-hydroxylation of a series of substituted acetophenone imines to mixtures of *E* and *Z* isomeric oximes, separated and identified by combined g.c./m.s. analysis as shown in Fig. 2 for *E* and *Z* isomeric 2,6-dimethylacetophenone oximes. Both chromatographic peaks exhibit mass spectral characteristics identical to those of the synthetic oxime. Direct g.c. methods that enable routine separation and quantitation of each isomeric form have been

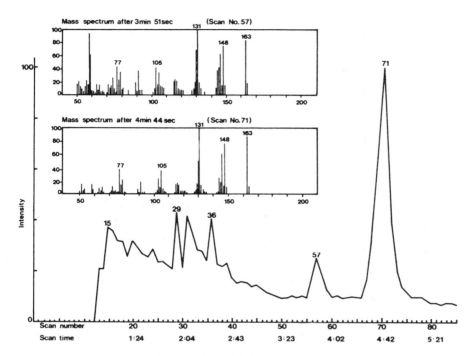

Fig. 2 – GC/MS separation and identification of two isomeric forms of 2,6-dimethylacetophenone oxime formed as *in vitro* metabolites of the corresponding imine by rat liver microsomes.

described in detail elsewhere (Gorrod & Christou, 1985a). The initial rates of formation of E, Z and total 2,6-dimethylacetophenone oxime by liver microsomes from various species are summarized in Table 1. The highest total acetophenone imine N-hydroxylase activity was exhibited by rabbit liver microsomes. Microsomes from hamster, rat and mouse livers exhibited activities equivalent to 50–60% of the activity in rabbit. Guinea-pig liver microsomes exhibited the lowest imine N-hydroxylase activities with all substituted acetophenone imine substrates examined (data not shown).

Comparison of the time courses for N-hydroxylation of various substituted acetophenone imines by rabbit liver microsomes (Fig. 3), indicates that an increase in the number of methyl substituents, and presumably increased lipophilicity, decreases both the initial rate and total extent of imine N-hydroxylation. Little difference was seen between the N-hydroxylation of the 2,6-dichloro analogue compared with the 2,6-dimethyl compound.

The E (*anti*-phenyl), oxime was the predominant isomer formed metabolically by liver microsomes of all species studied ($E/Z = 5$–15). A marked preference towards formation of the E isomeric form is consistent with the lower steric hindrance associated with this configuration. Despite the predominance of the E isomer the relative proportion of the more sterically hindered Z isomer is not constant and is in fact dependent on the species catalysing acetophenone imine N-hydroxylation (Table 1). It was observed

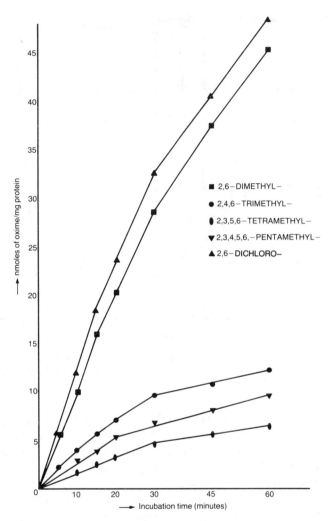

Fig. 3 – Effect of incubation time on the *in vitro N*-hydroxylation of substituted
acetophenone imines by rabbit liver microsomes.

that certain species such as mouse and guinea-pig exhibit proportionately
higher activities towards formation of the *Z* isomer compared with other
species which may be attributed to qualitative and quantitative differences
between hepatic microsomal acetophenone imine *N*-hydroxylases from different species.

In vivo N-Hydroxylation of Acetophenone Imines
Intraperitoneal administration of substituted acetophenone imines to various
species, results in urinary excretion of variable amounts of the
N-hydroxylated metabolites (Table 2). It is apparent that excretion of free

Table 1
The *in vitro* microsomal N-hydroxylation of 2,6-dimethylacetophenone imine by various mammalian species[a]

Species	2,6-Dimethylacetophenone oxime pmol/mg/min			E/Z Ratio[b]
	Z	E	Total	
Rabbit	100	1000	1100	10 ± 2
Hamster	50	680	730	15 ± 4
Rat	50	600	650	14 ± 4
Mouse	90	460	550	5 ± 1
Guinea-pig	40	280	320	7 ± 2
Ferret	30	260	290	9 ± 2

[a]Incubation was carried out for 10 minutes.
[b]E to Z ratio, calculated using the amounts of isomeric oximes formed after 10 minutes of incubation.

isomeric oximes in the urine is quantitatively important only in the rat where it represented 9% of the administered dose of acetophenone imine. It is of interest that the rat is a species allegedly poor in primary aliphatic amine N-hydroxylase activity as demonstrated by its inability to catalyse N-hydroxylation of chlorphentermine (Caldwell, 1978). Excretion of 0.5% or less of the administered imine as free oxime by all the other species examined, may indicate a more efficient conjugation of oximes to water soluble products in these species.

The composition of mixtures of E and Z isomeric oximes extracted from rat urine was different from that obtained using rat liver microsomes (Fig. 2). It is apparent that the intact rat excretes a considerably higher proportion of the more sterically hindered Z isomer compared to the E isomer in the urine (E/Z < 1). This difference between *in vivo* and *in vitro* observations may be explained by differences in the extent of conjugation of the two isomers. In the case of the Z isomeric form, steric crowding in the vicinity of the –OH group may prevent extensive phase II biotransformation and thereby result in the excretion of higher proportions of this isomer as free oxime. In contrast the –OH group in the E configuration is more accessible and may undergo more extensive conjugation and therefore not be excreted as free oxime.

The effects of treatment of urine from rats and rabbits with either β-glucuronidase or sulphatase, is summarized in Table 3. Neither of the two hydrolytic enzymes affected the amounts of free acetophenone oximes extractable from rat urine. Similarly β-glucuronidase had no effect on that extracted from rabbit urine. Hydrolysis with sulphatase, however, in this species resulted in a threefold increase in the total amount of the Z isomer

Table 2

Urinary excretion of E and Z isomeric forms of 2,6-dimethylacetophenone oxime following administration of 2,6-dimethylacetophenone imine (40 mg/kg i.p.) to various animal species

Species (male)	Number of animals	% Dose excreted as free 2,6-dimethylacetophenone oxime								
		0–24		24–48		48–100		0–100 hr		Total
		Z	E	Z	E	Z	E	Z	E	
Rat	2	1.5	1.3	2.6	0.6	1.1	0.2	5.2	2.1	7.3
Rabbit	1	0.06	0.001	ND		ND		0.06	0.001	0.06
Hamster	2	0.03	ND	ND		ND		0.03	ND	0.03
Mouse	2	0.60	ND	ND		ND		0.60	ND	0.60
Guinea-pig	1	0.02	ND	ND		ND		0.02	ND	0.02

ND = below detection limits

Table 3

The effect of β-glucuronidase and sulphatase on total amounts of extractable free isomeric oximes from the urine of rats and rabbits administered 2,6-dimethylacetophenone imine

Species and enzyme treatment	Free 2,6-dimethylacetophenone oxime nmol/ml urine		
	Z	E	Total
Rat			
Control	37	9	46
+ β-glucuronidase[a]	41	9	50
+ sulphatase[b]	45	9	54
Rabbit			
Control	3	< 0.05	3
+ β-glucuronidase	3	< 0.05	3
+ sulphatase	9	65	74

[a] 2500 units at 37 °C for 18 hours.
[b] 300 units at 37 °C for 24 hours.

and a 1000-fold increase in the total amount of the E isomer, resulting in the latter becoming the predominant isomer ($E/Z = 7$). These results indicate that the rabbit excretes a large proportion of the E isomeric acetophenone oxime as a sulphate conjugate which is not formed in the rat. It is possible that the rat excretes other types of conjugates resistant to both β-glucuronidase and sulphatase and that may account for the higher propor-

Table 4

Inhibition of the *in vitro* N-hydroxylation of 2,6-dimethyl- and 2,6-dichloroacetophenone imines by carbon monoxide

Gaseous phase	N-hydroxylation (control = 100%)					
CO:N_2:O_2 by volume	2,6-Dimethylacetophenone oxime			2,6-Dichloroacetophenone oxime		
	Z	E	Total	Z	E	Total
0:4:1	100	100	100	100	100	100
5:4:1	40	70	65	45	44	44
10:4:1	37	67	63	42	38	39

Table 5

The effects of metabolic inhibitors or activators on the *in vitro* N-hydroxylation of 2,6-dimethyl- and 2,6-dichloroacetophenone imines by rat liver microsomes

Compound	Concentration mM	N-hydroxylated metabolite					
		2,6-Dimethyl			2,6-Dichloro		
		syn(Z)	anti(E)	Total	syn(Z)	anti(E)	Total
Nil	—	100 (50)[a]	100 (680)	100 (730)	100 (140)	100 (1800)	100 (1940)
SKF 525A[b]	0.1	68	77	76	94	90	90
	1.0	ND	ND	53	ND	ND	76
DPEA[c]	0.1	62	54	54	80	79	79
	1.0	ND	ND	26	ND	ND	75
n-Octylamine	0.1	67	81	78	90	95	95
	1.0	ND	ND	44	ND	ND	75
1-Naphthylthiourea	1.0	50	140	130	46	146	133
Methimazole	0.1	40	78	73	50	90	95
Bromazepam	0.1	40	63	61	50	71	68
α-Naphthoflavone	0.1	108	113	112	108	104	104
Aniline	0.1	95	100	100	ND	ND	ND
	1.0	75	90	90	65	80	80
N-Benzylaniline	0.1	ND	ND	ND	28	58	56

Results are expressed as percentages of activity in controls ( )[a] amounts of oxime expressed as pmole/mg/min.
[b]2-Dimethylaminoethyl-2,2-diphenylvalerate.
[c]2,4-Dichloro-6-phenylphenoxyethylamine.
ND = Not determined.

tions of the Z isomer excreted as free oxime in urine. Alternatively, the rat may excrete an inhibitor of sulphatase in urine and prevent the release of the free oxime by sulphatase in our experiments.

Characterization of the Enzyme(s) Mediating Acetophenone Imine N-Hydroxylation

Results from selective inhibition studies (*in vitro*) and induction studies (both *in vitro* and *in vivo*), enabled determination of the major characteristics of the enzymes catalysing acetophenone imine N-hydroxylation.

Inhibition by carbon monoxide (Table 4) and concentration dependent inhibition by SKF 525A, DPEA and *n*-octylamine (Table 5) are consistent with the involvement of cytochromes P-450 in this type of N-hydroxylation. The MC-inducible cytochrome P-448 known to mediate the N-oxidation of aromatic amides to N-hydroxy metabolites (Lotlikar *et al.*, 1967), do not play a role in acetophenone imine N-hydroxylation as manifested by the lack of inhibition by α-naphthoflavone (Table 5) and the absence of an inducing effect following animal pretreatment with 3-methylcholanthrene (Table 6). In contrast, phenobarbital pretreatment resulted in two- to fourfold stimulation of N-hydroxylation of liver microsomes from various species (Table 6). Animal pretreatment with phenobarbital was also shown to stimulate both the rate of excretion and the total amounts of free E and Z isomeric oximes excreted in the urine of rats administered 2,6-dimethylacetophenone imine, as compared to a twofold decrease following MC-pretreatment (Table 7).

Table 6
The effects of animal pretreatment with 3-methylcholanthrene (MC) and phenobarbital (PB) on the *in vitro* N-hydroxylation of 2,6-dimethylacetophenone imine by various species

Species	Pretreatment	Cyt.P-450[a]	2,6-dimethylacetophenone oxime		
			Z isomer[a]	E isomer[a]	Total[a]
Rabbit	MC	80	80	80	80
	PB	200	180	180	180
Hamster	MC	130	105	125	125
	PB	210	220	190	200
Mouse	MC	125	125	125	125
	PB	200	280	370	360
Rat	MC	180	160	150	150
	PB	300	380	450	450
Ferret	MC	180	150	130	130
	PB	370	360	350	350

[a]Results expressed as percentages of activity found in microsomal incubates prepared from untreated animals.

Data presented here are consistent with the involvement of PB-inducible cytochrome P-450 isozymes in the metabolic *N*-hydroxylation of acetophenone imines. Involvement of the flavin-containing monooxygenase (Ziegler *et al.*, 1971), shown to catalyse the *N*-oxidation of secondary and tertiary aromatic amines (Gorrod & Gooderham, 1982; Gorrod & Patterson, 1983) is precluded.

The major characteristics of the PB-inducible cytochromes P-450 catalysing acetophenone imine *N*-hydroxylation are similar to those described for the enzymes catalysing *N*-oxidation of primary aromatic and tertiary heteroaromatic amines (Smith & Gorrod, 1978; Gorrod & Damani, 1979). However, activation of acetophenone imine *N*-hydroxylation by 1-naphthylthiourea and inhibition by bromazepam (Table 5), highlight two differences between acetophenone imine *N*-hydroxylases and the *N*-oxidases

Table 7

The effects of pretreatment with either 3-methylcholanthrene (MC) or phenobarbital (PB) on the urinary excretion of isomeric 2,6-dimethylacetophenone oximes in rats given 2,6-dimethylacetophenone imine

Pretreatment	% Dose excreted as free 2,6-dimethylacetophenone oxime					
	0–24 hours			0–100 hours		
	Z	*E*	Total	*Z*	*E*	Total
NONE	1.5	1.3	2.8	5.2	2.1	7.3
MC	0.8	0.6	1.4	3.0	1.5	4.5
PB	8.6	4.2	12.8	17.0	4.7	21.7

involved in the metabolism of primary aromatic and tertiary heteroaromatic amines. Furthermore the differential effects of some of the potential inhibitors, i.e. 1-naphthylthiourea, methimazole and bromazepam on the formation of the two oxime isomers (Table 5), suggests some heterogeneity in the cytochrome P-450 isozymes catalysing *N*-hydroxylation to *E* and *Z* isomeric oximes. Further studies utilizing purified forms of cytochrome P-450 in reconstituted systems may provide evidence in support of these suggestions and enable further understanding of the differences between imine and amine *N*-oxidases.

We conclude that acetophenone imine *N*-hydroxylation to mixtures of *E* and *Z* isomeric oximes is a general metabolic pathway both *in vitro* and *in vivo* and is catalysed by PB-inducible forms of cytochrome P-450.

ACKNOWLEDGEMENTS

This work was supported by a Chelsea College postgraduate studentship to M.C. The authors would like to thank the late Dr R. E. McMahon for kindly donating samples of authentic 2,4,6-trimethylacetophenone imine and oxime.

REFERENCES

Caldwell, J. (1978), in *Biological Oxidation of Nitrogen* (ed. Gorrod, J. W.) Elsevier/North Holland Biomedical Press, Amsterdam, p. 57.

Christou, M. (1982), PhD Thesis, University of London.

Culbertson, J. B. (1951), *J. Am. Chem. Soc.*, **73**, 4818.

Culbertson, J. B., Butterfield, D., Kolewe, D. & Shaw, R. (1962), *J. Org. Chem.*, **27**, 729.

Gorrod, J. W. (1982) in *Microsomes, Drug Oxidations and Drug Toxicity*, (eds. Sato, R. & Kato, R.) Japan Scientific Societies Press, Tokyo, p. 261.

Gorrod, J. W. & Christou, M. (1982) in *Microsomes, Drug Oxidations, and Drug Toxicity* (eds. Sato, R. & Kato, R.) Japan Scientific Societies Press, Tokyo, p. 563.

Gorrod, J. W. & Christou, M. (1985a), submitted to *Xenobiotica*.

Gorrod, J. W. & Christou, M. (1985b), submitted to *Xenobiotica*.

Gorrod, J. W. & Damani, L. A. (1979), *Xenobiotica*, **9**, 219.

Gorrod, J. W. & Gooderham, N. J. (1982), in *Microsomes, Drug Oxidations and Drug Toxicity* (eds. Sato, R. & Kato, R.) Japan Scientific Societies Press, Tokyo, p. 565.

Gorrod, J. W. & Patterson, L. H. (1983), *Xenobiotica*, **13**, 521.

Hucker, H. B., Michniewicz, B. M. & Rhodes, R. E. (1971), *Biochem. Pharmacol.*, **20**, 2123.

IUPAC (1970), '1968 Tentative rules, section E, Fundamental Stereochemistry' *J. Org. Chem.*, **35**, 2849.

Lotlikar, P. D., Enamoto, M., Miller, J. A. & Miller, E. C. (1967), *Proc. Soc. Expt. Biol. Med.*, **125**, 341.

Lustig, E. (1961), *J. Phys. Chem.*, **65**, 491.

Moureau, C. & Mignonac, G. (1913), *Compt. Rend.*, **156**, 1801.

Moureau, C. & Mignonac, G. (1920), *Ann. Chim.*, **14**, 322.

Parli, C. J., Wang, N. L. & McMahon, R. E. (1971a), *Biochem. Biophys. Res. Commun.*, **43**, 1204.

Parli, C. J., Wang, N. L. & McMahon, R. E. (1971b), *J. Biol. Chem.*, **246**, 6953.

Pickard, P. L. & Tolbert, T. L. (1961), *J. Org. Chem.*, **26**, 4886.

Smith, M. R. & Gorrod, J. W. (1978), in *Biological Oxidation of Nitrogen* (ed. J. W. Gorrod) Elsevier/North Holland, Amsterdam, p. 65.

Wittig, G. (1930), in *Stereochimie*, Akademische Verlags Gesellschaft, M.B.H., Leipzig, p. 182.

Ziegler, D. M., Poulsen, L. L. & McKee, E. M. (1971), *Xenobiotica*, **1**, 523.

Chapter 31

INFLUENCE OF CYTOSOLIC FACTORS ON HYDRAZINE AND HYDRAZIDE METABOLISM

M. E. Spearman, T. G. Franck, S. J. Moloney and **R. A. Prough**, Department of Biochemistry, University of Texas Health Science Centre, Dallas, Texas, 75235 USA

1. Alkyl hydrazines, hydrazides and hydrazones are metabolized by microsomal cytochrome P-450 to hydrocarbon products; addition of various thiols including reduced glutathione (GSH) stimulated hydrocarbon formation from all of the hydrazine derivatives tested two- to threefold.
2. Cytosolic enzymes, specifically the cytosolic glutathione S-transferases, in the presence of GSH were able to decrease hydrocarbon formation from iproniazid and isopropylhydrazine, but not from the other compounds tested. The mechanism of this alteration in product formation appeared to be due to the formation of an S-(2-propyl)GSH adduct from iproniazid.
3. GSH partially protected cytochrome P-450 from the loss of CO-reactive haem during the microsomal metabolism of iproniazid, ethylhydrazine, and the hydrazone derivative of procarbazine, but not from the other compounds tested. Cytosolic protein further protected from the loss of CO-reactive cytochrome P-450 found during the metabolism of iproniazid.
4. These data support the existence of azo, azo ester, or diazene intermediates of these hydrazine derivatives that may subsequently form complexes at the active site of the enzyme, decompose via a carbon-centred free radical to form hydrocarbons, or interact with the cytosolic glutathione S-transferases to form glutathione conjugates.

INTRODUCTION

Hydrazine and hydrazide derivatives are useful pharmacological and industrial chemicals whose uses range from the therapy of tuberculosis, mental depression, cancer, hypertension, and Parkinson's disease to uses as rocket propellants and synthetic chemical intermediates. However, all hydrazines and hydrazides tested have been reported to be carcinogenic (Toth, 1975), although most are toxic only in moderately large concentrations *in vitro* and *in vivo* (Nelson *et al.*, 1978; Wiebkin *et al.*, 1982).

The initial *N*-oxidation of the hydrazine portion of these molecules leads to formation of azo, diazene, azo ester, or hydrazone derivatives (Moloney & Prough, 1983a). Further metabolism or decomposition of these intermediates may lead to the production of carbon free radicals, as reported for iproniazid (Sinha, 1983) and ethylhydrazine (Augusto *et al.*, 1981). These free radicals may abstract a hydrogen atom from various donors to yield the corresponding hydrocarbons (Kosower & Kosower, 1976). Alternatively, they may bind covalently to cellular proteins or macromolecules. In addition, the *N*-oxidized intermediates may decompose to or be further metabolized to species capable of modifying microsomal cytochrome P-450 (Moloney *et al.*, 1984). Cytosolic factors, such as reduced glutathione (GSH), may alter the modification of proteins by these *N*-oxidized intermediates (Muakkassah *et al.*, 1981; Spearman *et al.*, 1984). The effect of GSH and other cytosolic constituents on hydrocarbon formation and cytochrome P-450 loss during hydrazine metabolism are considered in this report.

EXPERIMENTAL

The sources and purities of chemicals were as previously described (Dunn *et al.*, 1979; Moloney *et al.*, 1984; Spearman *et al.*, 1984). Azoprocarbazine (*N*-isopropyl-α-(2-methylazo)-*p*-toluamide) and hydrazoprocarbazine (*N*-isopropyl-α-(2-methylhydrazo)-*p*-toluamide) were prepared as described previously (Dunn *et al.*, 1979). *S*-Methylglutathione was purchased from Sigma Chemical Co. (St. Louis, MO). Sources and treatment of animals and preparation of tissue fractions were as previously described. All hydrazines were prepared daily in deoxygenated 100 mM-potassium phosphate buffer (pH 7.4) and the pH was adjusted with KOH as necessary. Ammonium sulphate precipitated cytosol was prepared by mixing the cytosol with ammonium sulphate (60 g/litre) for 2 hr, centrifuging the solution for 20 min at 12,000 **g**, and discarding the pellet. Subsequently, sufficient ammonium sulphate was dissolved in the supernatant to obtain a concentration of 440 g/litre and the sample was mixed and centrifuged as above. The pellet was then dissolved in a small volume of 100 mM-potassium phosphate buffer (pH 7.4) and was dialysed for 24 hr against this buffer. Cytosolic glutathione *S*-transferases were prepared according to Spearman & Liebman (1984).

Hydrocarbon production was quantitated by gas chromatographic analysis of the head space above the incubation mixtures according to Moloney &

Prough (1983b). Production of GSH conjugates was assayed by the method of Spearman *et al.* (1984) for *S*-(2-propyl)GSH conjugates or by a modification of this method (an isocratic mobile phase of 1% acetic acid in water) for *S*-methyl-GSH. Cytochrome P-450 spectra were recorded with an Aminco DW-2 spectrophotometer. A suspension of liver microsomes (1.5 mg/ml) from phenobarbital-treated rats was added in 0.1 M-Tris-HCl, 0.15 M-KCl, and 10 mM-MgCl$_2$ (pH 7.4) to the reference and sample cuvettes. NADPH (1 mM) was added to the sample cuvette and the solutions were allowed to incubate for 15 min. The various hydrazines were added to the sample cuvette in the presence or absence of GSH and cytosolic protein. Following this, sodium dithionite was added to the sample and reference cuvettes, the sample cuvette was bubbled with carbon monoxide, and the spectra were recorded. Subsequently, CO was bubbled into the reference cuvette and the spectra were recorded again to measure the amount of loss of CO-reactive hemoprotein.

RESULTS

GSH and Cytosol Effects on Microsomal Metabolism of Hydrazines
The results of studies on the effects of GSH and cytosolic protein on production of propane and propylene from iproniazid and isopropylhydrazine, ethane and ethylene from ethylhydrazine, and methane from methylhydrazine, azoprocarbazine, and hydrazoprocarbazine are summarized in Table 1. With all of the hydrazines and hydrazides tested, GSH stimulated hydrocarbon production about two- or threefold. These results are consistent with the intermediacy of putative carbon free radicals during hydrocarbon production. Cytosol, in the presence of 5 mM-GSH, depressed hydrazine metabolism by about 70% and 30% from iproniazid and isopropylhydrazine, respectively, compared with the GSH-stimulated rate but had no effect on hydrocarbon production from the other hydrazines. The concentration dependence of this cytosolic effect (half maximal inhibition occurred at 0.1–0.2 mM-GSH) and the dependence on GSH (and not other thiols) indicated that the cytosolic glutathione *S*-transferases (GST) might be responsible for this effect. Indeed, experiments with purified GST isozymes AA, A, B, C, and E indicated that isozymes A and B potently depressed hydrocarbon production from iproniazid whereas isozymes AA, C, and E were without significant effect (Spearman *et al.*, 1984). When concentrations of these isozymes estimated to exist in cytosol were reconstituted in liver microsomal reaction mixtures, the effects found approximated those found with the crude cytosolic fraction. Thus these GST isozymes appeared to be exclusively responsible for the effects of cytosol on altering hydrocarbon production from iproniazid during oxidative microsomal metabolism.

GSH Conjugate Formation from Hydrazines and Hydrazides
To study the potential mechanism of the effect of GST on iproniazid and isopropylhydrazine metabolism, *S*-(2-propyl)glutathione was synthesized as

Table 1
Hydrocarbon formation from microsomal metabolism

Hydrazine/hydrazide[a]	Metabolic activity (pmol/min/mg protein)		
	Control	Plus GSH[b]	Plus GSH + cytosol[b]
Iproniazid	455	1110 (240)	340 (75)
Isopropylhydrazine	410	920 (220)	650 (160)
Ethylhydrazine	90	150 (170)	150 (170)
Methylhydrazine	100	200 (200)	200 (200)
Azoprocarbazine	45	95 (210)	110 (240)
Hydrazoprocarbazine	185	580 (315)	510 (275)

[a]The concentration of hydrazines utilized were iproniazid and isopropylhydrazine 12 mM, ethylhydrazine and methylhydrazine 2 mM, azo- and hydrazoprocarbazine 0.5 mM. The incubation conditions and hydrocarbon quantitation methods utilized were as previously described (Moloney et al., 1984).
[b]The GSH concentration utilized was 5 mM and the cytosolic protein concentration was 1 mg/ml. The values in parentheses are the formation of hydrocarbon expressed as a percentage of control.

described by Spearman et al. (1984). Following the incubation of iproniazid with microsomal protein, NADPH, cytosol, and [^{3}H]GSH, a material could be isolated from the h.p.l.c. effluent that co-migrated with the synthetic S-(2-propyl)GSH and that also contained the radioactivity of [^{3}H]GSH. Upon mass spectral analysis, this biologically produced conjugate was found to be identical to the synthetic material. Similar studies performed with isopropylhydrazine failed to demonstrate significant formation of a GSH conjugate as suggested by the earlier results of Nelson et al. (1976). Similarly, studies with methylhydrazine, azoprocarbazine, or hydrazoprocarbazine did not result in formation of the corresponding S-methylglutathione conjugate (data not shown). Thus, conjugate formation with GSH occurred with an N-oxidized iproniazid metabolite, but not with metabolites of the other hydrazines.

GSH and Cytosol Effects on Cytochrome P-450 Loss
Typical difference spectra obtained during the microsomal metabolism of the benzaldehyde hydrazone derivative of procarbazine are shown in Fig. 1. In addition, the results of studies on the effects of GSH and cytosol on the loss of CO-reactive cytochrome P-450 mediated by other hydrazines and hydrazides are summarized in Table 2. Incubation of the hydrazone derivative of procarbazine with liver microsomes caused a marked loss of CO-reactive cytochrome P-450 when compared with the control incubation containing microsomes, NADPH, and buffer alone. The addition of 5 mM-GSH had a

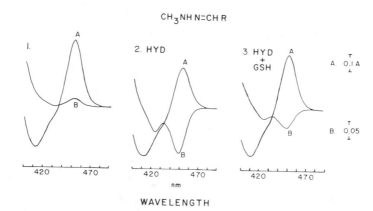

$CH_3NH N= CH R$

Fig. 1 – Loss of CO-reactive cytochrome P-450 during the microsomal metabolism of the hydrazone derivative of procarbazine with rat liver microsomes. In all three panels, the samples were preincubated as indicated in Experimental, both sample and reference cuvettes were reduced with sodium dithionite, and CO was bubbled into the sample cuvette (line A). Subsequently, CO was bubbled into the reference cuvette and the spectra again recorded (line B). In panel 1, the control incubation is shown in which both cuvettes contained microsomal protein and buffer and the sample cuvette contained 1 mM-NADPH alone. In panel 2, the sample cuvette also contained 0.5 mM-hydrazoprocarbazine. In panel 3, the sample cuvette contained 0.5 mM-hydrazoprocarbazine and 5 mM-GSH.

Table 2
Effect of GSH and cytosol on loss of CO-reactive cytochrome P-450

Hydrazine/hydrazide[a]	CO-reactive cytochrome P-450 lost (nmol of cytochrome P-450 lost/mg protein)		
	Control	Plus GSH[b]	Plus GSH + cytosol[b]
Iproniazid	0.45	0.27	0.02
Isopropylhydrazine	0.54	0.60	0.53
Ethylhydrazine	1.20	0.89	0.61
Methylhydrazine	0.97	1.00	0.47
Procarbazine	0.30	0.29	0.33
Azoprocarbazine	0.24	0.26	0.23
Hydrazoprocarbazine	1.10	0.65	0.64

[a]The concentrations of the hydrazines used were the same as in Table 1. Control incubation mixtures consisted of 1.5 mg/ml microsomal protein (cytochrome P-450 content 2.6 nmol P-450/mg), 1.0 mM-NADPH, and buffer in a 3-ml cuvette that was preincubated for 15 min at room temperature.
[b]The GSH concentration was 5 mM and the cytosol concentration was 1 mg/ml.

pronounced protective effect on the loss of cytochrome P-450. As summarized in Table 2, GSH also protected from a portion of the loss of CO-reactive cytochrome P-450 mediated by iproniazid and ethylhydrazine, but not by isopropylhydrazine, methylhydrazine, procarbazine, or azoprocarbazine. Cytosolic protein, in the presence of GSH, was able to further protect from the loss of CO-reactive cytochrome P-450 mediated by incubation of iproniazid, ethylhydrazine, or methylhydrazine with microsomes and NADPH. However, no effect of cytosol was found with reaction mixtures containing isopropylhydrazine, procarbazine, azoprocarbazine, or hydrazoprocarbazine.

DISCUSSION

Hydrazines and hydrazides have been shown to be N-oxidized to azo, diazene, azo ester, and hydrazone metabolites (Moloney & Prough, 1983a). Many of these are chemically unstable and may decompose spontaneously (or may be further metabolized) to products capable of covalent binding to DNA and protein or capable of covalent modification of cytochrome P-450. Thiols, such as GSH, may chemically react with species produced from these N-oxidations, such as free radicals. An inverse relationship between hydrocarbon formation and DNA or protein covalent binding has previously been noted with iproniazid and isopropylhydrazine by Nelson *et al.* (1976) and with procarbazine and azoprocrabazine by Moloney *et al.* (1985). Similar inverse relationships may exist with ethylhydrazine, methylhydrazine, and hydrazoprocarbazine. Although thiols stimulated hydrocarbon formation and can decrease protein covalent modification, they protected from cytochrome P-450 loss with only certain hydrazines. This may reflect the nature of the N-oxidized intermediate formed. For example, methylhydrazine N-oxidation has been proposed to form a diazene derivative capable of subsequently forming a stable iron porphyrin–diazenyl complex (Battioni *et al.*, 1983). Other monosubstituted hydrazines, such as ethylhydrazine, may undergo analogous reactions at the active site of the enzyme but may be relatively resistant to the protective effects of GSH on cytochrome P-450. Thus, the protective effect of thiols may be found only with N-oxidized intermediates that are sufficiently chemically stable to be released from the active site of the enzyme and that may be less effective in forming abortive complexes with cytochrome P-450.

   Several hydrazine and hydrazide metabolites also appeared to be substrates for certain cytosolic enzymes. With iproniazid, this effect appeared to be mediated by the cytosolic glutathione S-transferases and led to formation of an S-(2-propyl)GSH conjugate. However, with other hydrazines tested, little or no conjugate was formed although cytosol clearly exerted an effect on hydrocarbon formation or loss of CO-reactive cytochrome P-450 such as found with isopropylhydrazine, ethylhydrazine, and methylhydrazine.

In summary, the *N*-oxidation of hydrazines and hydrazides may produce reactive metabolites that are capable of DNA or protein covalent binding, hydrocarbon formation via free radicals, or complexation with cytochrome P-450. These cytochrome P-450 complexes may significantly direct enzyme function. GSH and the cytosolic GST may alter metabolic pathways away from production of these reactive species toward production of hydrocarbon products or GSH conjugates. Therefore, the toxicity of hydrazines and hydrazides might be expected to increase if GSH levels are depleted or if glutathione *S*-transferase enzymic activity is inhibited.

ACKNOWLEDGEMENT

Supported by American Cancer Society Grant BC-336 and Robert A. Welch Grant I-616. MES and SJM were R.A. Welch Fellows and TGF was an undergraduate Research Fellow of the Southwestern Graduate School of Biomedical Sciences.

REFERENCES

Augusto, O., Ortiz de Montellano, P. R. & Quintanilha, A. (1981), *Biochem. Biophys. Res. Commun.*, **101**, 1324.

Battioni, P., Mahy, J. P., Delaforge, M. & Mansuy, D. (1983), *Eur. J. Biochem.*, **134**, 241.

Dunn, D. L., Lubet, R. A. & Prough, R. A. (1979), *Cancer Res.*, **39**, 4555.

Kosower, N. S. & Kosower, E. M. (1976), in *Free Radicals in Biology*, Vol. 2 (ed. Pryor, W. A.) Academic Press, New York, p. 55.

Moloney, S. J. & Prough, R. A. (1983a), in *Reviews in Biochemical Toxicology*, Vol. 5 (eds. Hodgson, E., Bend, J. R. & Philpot, R. M.) Elsevier, New York, p. 313.

Moloney, S. J. & Prough, R. A. (1983b), *Arch. Biochem. Biophys.*, **221**, 577.

Moloney, S. J., Snider, B. J. & Prough, R. A. (1984), *Xenobiotica*, **14**, 803.

Moloney, S. J., Wiebkin, P., Cummings, S. W. & Prough, R. A. (1985), *Carcinogenesis*, in press.

Muakkassah, S. F., Bidlack, W. R. & Yang, W. C. T. (1981), *Biochem. Pharmacol.*, **30**, 1651.

Nelson, S. D., Mitchell, J. R., Timbrell, J. A., Snodgrass, W. R. & Corcoran, G. B. (1976), *Science*, **193**, 901.

Nelson, S. D., Mitchell, J. R., Snodgrass, W. R. & Timbrell, J. A. (1978), *J. Pharmacol. Exp. Therapeut.*, **206**, 574.

Sinha, B. (1983), *J. Biol. Chem.*, **258**, 796.

Spearman, M. E. & Liebman, K. C. (1984), *Drug Metab. Dispos.*, **12**, 661.

Spearman, M. E., Moloney, S. J. & Prough, R. A. (1984), *Mol. Pharmacol.*, **26**, 566.

Toth, B. (1975), *Cancer Res.*, **35**, 3693.

Wiebkin, P., Sieg, M. S., Nelson, R. E., Hines, R. N. & Prough, R. A. (1982), *Biochem. Pharmacol.*, **31**, 2921.

Chapter 32

TRIAZENE *N*-OXIDES

D. E. V. Wilman, Cancer Research Campaign Laboratory, Department of
Biochemical Pharmacology, Institute of Cancer Research, Clifton Avenue, Sutton,
Surrey, SM2 5PX, UK

1. 1-Aryl-3,3-dialkyltriazenes require oxidative metabolism to a 1-methyl-3-aryltriazene in order to demonstrate antitumour activity. No *N*-oxide metabolite has yet been identified.
2. The expected 1-aryl-3,3-dialkyltriazene-3-oxide is not available synthetically but 1-aryl-3,3-dialkyltriazene-1-oxides are.
3. 1-Alkyl-3-aryltriazene-1-oxides are readily synthesized and have wide ranging biological properties.
4. One of these, 3-(4-carbamoylphenyl)-1-methyltriazene-1-oxide is an effective antitumour agent, whereas its *O*-methylated analogue 1-(4-carbamoylphenyl)-3-methoxy-3-methyltriazene has little or no activity.

INTRODUCTION

Triazenes in general and DTIC (**I**, 5-(3,3-dimethyl-1-triazene)imidazole-4-carboxamide) in particular have generated considerable interest as cancer chemotherapeutic agents (Wilman & Farmer, 1985). DTIC is still regarded as the most effective single agent in the treatment of malignant melanoma, and is also useful in the treatment of Hodgkin's disease, neuroblastoma and rhabdomyosarcoma. Studies on the metabolism and mechanism of action of analogues of DTIC have led us to the conclusion that the structural requirements for aryldialkyltriazenes to be antitumour agents are as displayed in Fig. 1 (Connors *et al.*, 1976). These are a carrying group (R) at N^1, a methyl group at N^3 and a readily metabolized group (R') also at N^3. The carrying

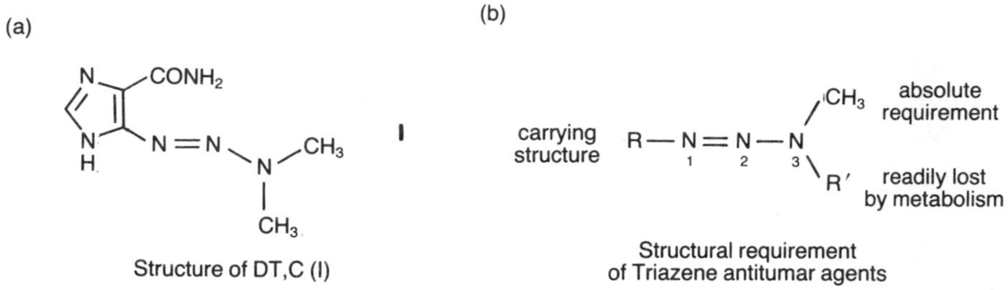

(a)   (b)

Structure of DT,C (I)

Structural requirement
of Triazene antitumar agents

Fig. 1 – Structure of DTIC (a) and (b) structural requirements of triazene antitumour agents.

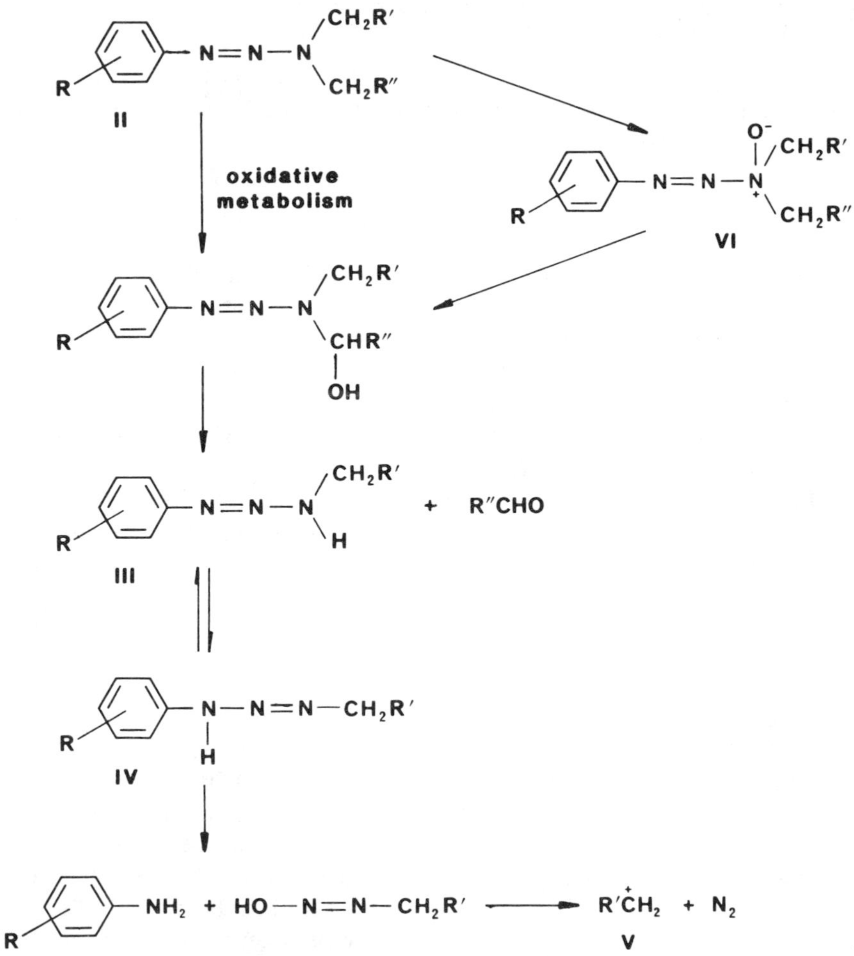

Fig. 2 – Metabolism of 1-aryl-3,3-dialkyltriazenes.

group may be a heterocycle as in the case of the imidazole group of DTIC, or an aryl group as in our own work. Di- and trialkyltriazenes have proved to be inactive, probably due to chemical instability.

In the case of 1-aryl-3,3-dialkyltriazenes (**II**), oxidative metabolism in the liver, of one of the alkyl groups at N^3, is followed by chemical loss of the resulting α-hydroxyalkyl group, as the appropriate alkyl aldehyde (Fig. 2). The monoalkyltriazene (**III**) so formed tautomerizes to the 3-aryl-1-alkyltriazene form (**IV**) which is inherently unstable to chemical hydrolysis. The rate of this hydrolysis is dependent on the pH of the medium and the electron donating properties of the aromatic ring.

Hydrolytic degradation of the monoalkyltriazene (**IV**) leads, via an intermediate diazo species, to the corresponding alkylcarbonium ion (**V**). All triazenes which can produce a monoalkylmetabolite demonstrate similar toxicity, mutagenicity and carcinogenicity but only those which are capable of forming a monomethyltriazene (**III**, R' = H) exhibit antitumour activity. Hence the basic structural requirement for a methyl group.

In considering the metabolism of the dialkyltriazenes it is interesting to speculate on the intermediacy or otherwise of an N-oxide species. The obvious position at which such oxidation might be expected to occur is N^3, in a similar fashion to that seen in the case of dialkylanilines and similar compounds. To date, no reports have appeared to suggest that such a metabolic intermediate (**VI**, Fig. 2) might exist. However, the present synthetic chemical evidence is not in favour of such an intermediate. Although the standard work on nitrogen chains (Smith, 1983) refers to the synthesis of such compounds from an aryldiazonium ion and N,N-dimethylhydroxylamine, the references quoted do not contain such a synthesis. In fact there appears to be no evidence for the existence of such compounds, and a second treatise (Benson, 1984) makes no mention of trisubstituted triazene-3-oxides. A recent structural investigation by ^{1}H- and ^{13}C-n.m.r. of a variety of synthetic triazene-N-oxides (Giumanini et al., 1983), failed to produce evidence for a triazene-3-oxide.

Despite the current lack of evidence for the existence of 1-aryl-3,3-dialkyltriazene-3-oxides, 1-aryl-3,3-dialkyltriazene-1-oxides (**VII**) are known and have been claimed to have anti-inflammatory properties (Miesel, 1976). Compounds of this type are readily prepared by the reaction of nitrosoben-

zene derivatives with the appropriate 1,1-disubstituted hydrazine (Miesel, 1976; Hoesch & Köppel, 1981). Such compounds are currently being investigated for antitumour activity, and their crystal structure determined to provide conclusive evidence for their structure, which currently relies on n.m.r. evidence (Giumanini *et al.*, 1983).

Although only one of the two expected trisubstituted triazene *N*-oxides is known and has been little investigated, very much more is known about the 1,3-disubstituted triazene *N*-oxides.

EXPERIMENTAL

While further investigating the range of possible triazene structural types which might possess antitumour activity, we synthesized two compounds containing a nitrogen–oxygen bond rather than all nitrogen–carbon bonds. These were 1-(4-carbamoylphenyl)-3-methoxy-3-methyltriazene (**VIII**, CB 10-363) and 1-(4-carbamoylphenyl)-3-methyltriazene (**IX**, CB 10-339) (Connors *et al.*, 1976). The synthesis of such compounds involves the coupling of a suitably substituted hydroxylamine with the appropriate aryldiazonium salt, in this case that derived from 4-aminobenzamide. This synthetic route parallels that for the aryldialkyltriazenes in which a dialkylamine replaces the hydroxylamine.

Although synthesized from *N*-methylhydroxylamine and commonly written in the *N*-hydroxy form, such compounds, as CB 10-339 appear on the basis of infrared (Mitsuhashi *et al.*, 1965) and n.m.r. spectra (Giumanini *et al.*, 1983) to exist in the tautomeric *N*-oxide form. In the specific instance of CB 10-339 this is 3-(4-carbamoylphenyl)-1-methyltriazene-1-oxide (**X**). This is structur-

ally reminiscent of the bonding arrangement in the alkylaryltriazenes (**III**). The methoxy derivative CB 10-363 of course is not an *N*-oxide, but a true trisubstituted triazene. X-Ray crystallography of these compounds is being undertaken to confirm their respective structures.

RESULTS AND DISCUSSION

A range of biological activities had previously been reported for 1,3-disubstituted triazene-1-oxides analogous to CB 10-339, but no information on antitumour testing was available. Apart from acaricidal, insecticidal and phytotoxic activities, a range of potential pharmaceutical uses have been reported. These include analgesic, anticonvulsant, antihypotensive, anti-inflammatory, antituberculotic, bronchodilatory, immunosuppressive and smooth muscle relaxant properties. However, none of these have been pursued, presumably as a result of the potential carcinogenicity of compounds of this type. In this context, it is interesting to note that 3-hydroxy-3-methyl-1-

Table 1

Activity of CB 10-339 and CB 10-363 towards the TLX5 lymphoma

Dose daily ×5 (mg/kg)	Increase in life span (%)	
	CB 10-339	CB 10-363
12.5		−2.9
25		−2.9
50	18.8	−4.8
100	18.8	−2.9
200	36.6	−6.7
400	54.4	−2.9
800	50.4	
1600	−20.8	

phenyltriazene, the parent of this series, and its 4-chlorophenyl and 2,4,6-trichlorophenyl analogues show no mutagenicity in the TA 1530 strain of *Salmonella typhimurium* alone or in the presence of the 9000 g fraction of mouse liver (Malaveille *et al.*, 1982).

Antitumour testing in two tumour systems has revealed a marked difference between the compounds CB 10-339 and CB 10-363. CB 10-339 is active against the TLX5 lymphoma (Table 1) and the AdjPC6/A plasmacytoma (Table 2). Both of these murine tumours are in general sensitive to alkylating agents such as melphalan and cyclophosphamide, and show no response to the antimetabolites methotrexate and 5-fluorouracil. CB 10-363, however, is inactive against the TLX5 lymphoma and shows much reduced activity towards the AdjPC6/A tumour. This distinct difference between the two compounds, particularly noticeable in the TLX5 tumour system, may be

Table 2

Activity of CB 10-339 and CB 10-363 against the AdjPC6/A plasmacytoma

	Dose daily ×5 (mg/kg)	
	CB 10-339	CB 10-363
LD_{50}	330	330
ID_{90}	15.5	86
TI	21.3	3.8

explained on the basis of the available routes of metabolism of the two compounds.

As was indicated earlier, in order for the triazenes to show antitumour activity, they must be capable of chemical or biological degradation to a monomethyltriazene. Metabolism of CB 10-363 is only possible by *N*-demethylation, resulting in the formation of an inactive *N*-methoxytriazene. It is therefore surprising that any activity is seen against the AdjPC6/A tumour. That which is evident may be due to a small amount of *O*-demethylation resulting in the formation of active CB 10-339.

As regards the possible metabolic profile of CB 10-339, consideration must be given to the chemical structure of the molecule before postulating a possible metabolic route. As the compound would appear to exist as the 1-oxide tautomer (**X**), it is easy to envisage the formation of a monomethyltriazene by reduction of the *N*-oxide moiety. No investigation of this possibility has been undertaken, thus proof of such a metabolic pathway is still awaited.

Another series of triazene *N*-oxides is available synthetically, these are the 3-aryl-1,3-dialkyltriazene-1-oxides, which in the case of 3-(1-carbamoyl-phenyl)11,3-dimethyltriazene-1-oxide (**XI**) is derived from CB 10-339 (**X**) by alkylation with methyl iodide in the presence of sodium hydride as a proton acceptor.

The obvious conclusion to be drawn from the preceding discussion is that although a wide variety of triazene *N*-oxides are known synthetically, some with interesting biological activity, little or nothing is known of their metabolism or of their involvement in the metabolism of triazenes in general.

REFERENCES

Benson, F. R. (1984), *The High Nitrogen Compounds*, John Wiley & Sons Inc., New York.

Connors, T. A., Goddard, P. M., Merai, K., Ross, W. C. J. & Wilman, D. E. V. (1976), *Biochem. Pharmacol.*, **25**, 241.

Giumanini, A. G., Lassiani, L., Nisi, C., Petrie, A. & Stanovnik, B. (1983), *Bull. Chem. Soc. Jpn*, **56**, 1887.

Hoesch, L. & Koppel, B. (1981), *Helv. Chim. Acta*, **64**, 864.

Lucas, V. S. & Huang, A. T. (1982), in *Clinical Management of Melanoma*, (ed. Seigler, H. F.) Martinus Nijhoff, The Hague, p. 381.

Malaveille, C., Brun, G., Kolar, G. & Bartsch, H. (1982), *Cancer Res.*, **42**, 1446.

Miesel, J. L. (1976), 3,3-Dialkyl-1-(substituted-phenyl)-triazene-1-oxides US Pat. 3,989,680.

Mitsuhashi, T., Osamura, Y. & Simamura, O (1965), *Tet. Letts*, **30**, 2593.

Smith, P. A. S. (1983), *Derivatives of Hydrazine and other Hydronitrogens having N–N Bonds*, Benjamin/Cummings Publishers Co., Reading, Mass., p. 152.

Wilman, D. E. V. & Farmer, P. B. (1985), in *Progressive Stages of Neoplastic Growth* (ed. Kaiser, H. E.) Pergamon Press, Oxford.

Chapter 33

BIOTRANSFORMATION OF SUBSTITUTED BENZOTRIAZOLES

H. Hoffmann, Institut für Pharmazeutische Chemie, Universität Frankfurt/Main, Georg-Voigt-Straße 14, D 6000 Frankfurt, FRG

1. Benzotriazole is used on a large scale in industry, and substituted derivatives have been recently investigated as potential drugs.
2. *N*-1-Alkylsubstituted benzotriazoles undergo extensive *N*-dealkylation *in vitro*, but very little aromatic ring hydroxylation is observed.
3. *N*-2-Alkylsubstituted benzotriazoles on the other hand have a very different metabolic pattern, being readily and extensively converted to dihydrodiols, presumably via epoxide intermediates.
4. No evidence for metabolic *N*-oxidation or triazole ring cleavage was obtained.

INTRODUCTION

Metabolic reactions do not usually occur on the skeleton of a molecule, but functional groups such as amino, alcohol, carboxyl and nitro groups are usually more reactive. Biotransformation of the ring skeleton, when it occurs at all, tends only to result in minor metabolites, which can nevertheless gain in significance depending on the pharmacokinetic and pharmacological character of a drug used over a long time. A systematic investigation of metabolic changes at the basic structure of a drug assumes the absence of other easily transformable groups.

Substituted benzotriazoles have been tested by various groups as potential drugs. For example, alizapirid has been introduced as an antiemetic and 2-propyl-4-*p*-tolylaminobenzotriazole was tested as an antiarrhythmic. Unsubstituted benzotriazole is used on a large scale in industry, where it is still not possible to avoid considerable skin and mucous membrane contact. Details of its human toxicity are not available, but experiments on animals have produced leukaemia and adipose liver dystrophy.

METABOLISM OF BENZOTRIAZOLE

When the metabolism of the unsubstituted benzotriazole is studied, the high metabolic stability of this compound with respect to oxidation is evident. The heterocyclic 5-membered ring does not alter at all; neither an *N*-oxidation nor a ring opening can be observed. The extent of hydroxylation at the carbocycle is small; when benzotriazole was incubated with 10,000 **g** supernant of rat liver homogenates or with microsomes, it did not exceed 5% of the initial substrate concentration. A linear correlation was found between the hydroxylation rate and the cytochrome P-450 concentration when experiments were repeated with preparations containing various amounts of cytochrome P-450. All the values given in Table 1 were obtained under the same conditions and with the same liver preparation and are therefore comparable.

Both 4- & 5-hydroxybenzotriazoles are formed as metabolites, of which the former is less polar, because of a hydrogen bridge existing between the OH-group and the nitrogen atom in position 3. Nevertheless, the stabilization of the 4-hydroxy compound does not lead to a significant formation of this metabolite. The ratio of the amounts of benzotriazole and 1-alkylbenzotriazoles hydroxylated in position 4 to those hydroxylated in position 5 is about 1:5 in all cases. The reason for this may be that the carbonium ion in position 4, i.e. with the hydroxy group in position 5, which is formed by the ring opening, has a lower energy than the carbonium ion in position 5, so that 5-hydroxybenzotriazoles can mainly be expected. An oxidation was observed only on the neighbouring carbon atoms 4 and 5, indicating, that an epoxide is probably formed as an intermediate.

METABOLISM OF 1-ALKYLBENZOTRIAZOLES

The slight oxidation of the carbocyclic ring is considerably diminished by an alkylation at the N-1 of the benzotriazole. Whilst 0.2% and 0.7% of the two phenols are formed from the 1-methyl compound, the percentage is usually much less with most of the other compounds tested (Table 1). The indication 'less than 0.1%' signifies that although these metabolites were qualitatively detected by t.l.c., h.p.l.c. and mass spectrometry, using reference substances especially prepared for this purpose, further figures have been dispensed with because of the small amounts of these metabolites.

Table 1

Biotransformation of 1-alkylbenzotriazoles:
Oxidation of the carbocyclic ring

Substrates	M e t a	b o	l i t e s		
$\begin{array}{c} R \\ \text{(benzotriazole)} \end{array}$	4-OH	5-OH	Benzotriazole		
			un-subst.	4-OH	5-OH
-H	0,7%	3,6%	-	-	-
-CH$_3$	0,2%	0,7%	6,5%	-	0,2%
-CH$_2$-CH$_3$	<0,1%	<0,1%	51,5%	0,25%	1,7%
-CH$_2$-CH$_2$-CH$_3$	<0,1%	<0,1%	47,1%	0,25%	1,7%
$\begin{array}{c} \text{CH}_3 \\ -\text{CH}-\text{CH}_2-\text{CH}_3 \end{array}$	<0,1%	<0,1%	39,1%	0,2%	1,3%
$\begin{array}{c} \text{CH}_3 \\ -\text{C}-\text{CH}_3 \\ \text{CH}_3 \end{array}$	0,2%	0,85%	-	-	-
$\begin{array}{c} \text{CH}_3 \\ -\text{CH}_2-\text{CH}-\text{CH}_3 \end{array}$	-	-	6,6%	-	0,2%
-CH$_2$-CH$_2$-CH$_2$-CH$_3$	<0,1%	<0,1%	32,4%	0,2%	1,2%

With regard to the amount of phenols formed, 1-tertiary butylbenzotriazole and 1-isobutylbenzotriazole are the exceptions. Comparable amounts of phenol were obtained both from the tertiary butyl compound as well as from 1-methylbenzotriazole, whilst the amount of the phenols which might be obtained from the 1-isobutylbenzotriazole is lower than the limit of detection.

A reason for the decrease of aromatic ring hydroxylation by substitution at N-1 could be the dominance of another faster reaction in the substituted compound. Column 4 of Table 1 shows that due to the oxidation at the carbon atom of the side chain, which is immediately linked with the benzo-triazole, N-dealkylation takes place. N-Dealkylation decreases as lipophilicity increases (the substrates are listed in order of increasing lipophilicity). Only when oxidative N-dealkylation is impeded, or is impossible, as with tertiary butylbenzotriazole, does aromatic ring hydroxylation take place in any appreciable amount. In addition, 4- and 5-hydroxybenzotriazole also results as

Table 2

Biotransformation of 1-alkylbenzotriazoles:
Oxidation of the sidechain

Substrates	Metabolites				
(benzotriazole core with R on N)	Benzo-triazole	$-C$	$-C$	$-C$	$-C$
$-CH_3$	$-H$ 6,5%	$-$			
$-CH_2-CH_3$	$-H$ 51,5%	$-$	$-$		
$-CH_2-CH_2-CH_3$	$-H$ 47,1%	$-$	OH 14,1%	OH 2,2%	
$\overset{CH_3}{-CH}-CH_2-CH_3$	$-H$ 39,1%	$\overset{CH_2OH}{-CH-}$ 1,1%	OH 18,2%	OH 0,9%	
$\underset{CH_3}{\overset{CH_3}{-C}}-CH_3$	$--$	$-$	$-$		
$-CH_2-\overset{CH_3}{CH}-CH_3$	$-H$ 6,6%	$-$	$\overset{CH_3}{\underset{OH}{-C-}}$ 47,8%	$-$	
$-CH_2-CH_2-CH_2-CH_3$	$-H$ 32,4%	$-$	OH 1,3%	OH 26,7%	OH 3,9%

secondary metabolites from the benzotriazole formed by N-dealkylation of the N-alkylbenzotriazoles.

Table 2 shows once again the amount of benzotriazole, which is formed by oxidative N-dealkylation. The third column shows further metabolites which have been formed by hydroxylation of the side chains. 1-Isobutyl-2-hydroxybenzotriazole is clearly an important metabolite. In general, oxidation largely occurs in the penultimate CH_2-group; methyl groups on the other hand are more difficult, or impossible, to oxidize. With the exception of the 1-methyl- and the tertiary butylbenzotriazole, 1-alkylbenzotriazoles are extensively metabolized (around 50–60% of substrate).

As a preliminary conclusion it may be stated that the formation of phenols by hydroxylation of benzotriazole occurs when no other easier biotransformation reactions are possible.

METABOLISM OF 2-ALKYLBENZOTRIAZOLES

The 2-alkylbenzotriazoles (Table 3) clearly show a different metabolic pattern. Although oxidation in the side chain corresponds qualitatively to the ratios obtained with the 1-alkylcompounds, i.e. it mainly occurs at C-1 (i.e. N-dealkylation) and at the penultimate carbon atom, the amounts formed are much smaller.

The differences between the 1- and 2-isomers are particularly obvious when oxidative N-dealkylation is examined. For example, 1-n-butylbenzotriazole yields 32.4%, but only 2.1% of benzotriazole is formed from the 2-isomer.

The reason for this change in the metabolic pattern is the ready formation of dihydrodiols with compounds substituted at N-2. About 20% of the substrate in each case is transformed into the corresponding dihydrodiol, and the yield reaches 33.6% with tertiary butylbenzotriazole, where the dihydrodiol is the only metabolite. The small amount afforded from 2-ethylbenzotriazole is attributed to a large extent to the extensive oxidative N-dealkylation of this compound.

Table 3

Biotransformation of 2-alkylbenzotriazoles

Substrates	Metabolites			
$\begin{smallmatrix} N \\ N-R \\ N \end{smallmatrix}$	Dihydrodiol $\begin{smallmatrix} N \\ N-R \\ N \end{smallmatrix}$ H OH	Benzo-triazole	-C - C - C - C		
$-CH_3$	–	-H 3,1%	–		
$-CH_2-CH_3$	4,5%	-H 26,4%	–	–	
$-CH_2-CH_2-CH_3$	17,2%	-H 9,3%	–	OH 6,2%	OH 1,1%
$\begin{smallmatrix} CH_3 \\ -CH-CH_2-CH_3 \end{smallmatrix}$	23,5%	-H 4,9%	CH_2OH -CH- 4,1%	OH 10,3%	OH 0,5%
$\begin{smallmatrix} CH_3 \\ -C-CH_3 \\ CH_3 \end{smallmatrix}$	33,6%	–	–	–	
$\begin{smallmatrix} CH_3 \\ -CH_2-CH-CH_3 \end{smallmatrix}$	20,5%	–	–	$\begin{smallmatrix} CH_3 \\ -C- \\ OH \end{smallmatrix}$ 38,6%	
$-CH_2-CH_2-CH_2-CH_3$	26,6%	-H 2,1%	–	OH 1,2%	OH 13,8% OH 3,9%

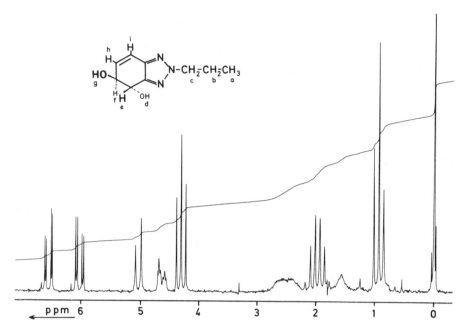

Fig. 1 – ¹H-n.m.r. Spectrum of 2-propyl-4,5-dihydrobenzotriazolediol.

On the basis of the experiments carried out so far it is still not possible to give a convincing explanation as to why dihydrodiol is not made from 2-methylbenzotriazole. With the exception of isobutylbenzotriazole, the percentage of the dihydrodiol formed increases with the lipophilicity of the substrate. This could indicate an increasing affinity of the substrate for the enzyme.

Various spectrometric methods were used to identify the dihydrodiols. Before i.r., mass and n.m.r. spectra were recorded, isolation and purification was undertaken by extracting the incubation mixtures using Extrelut ᴿ-columns and by t.l.c. and h.p.l.c.-separation. The h.p.l.c.-eluates were carefully brought to dryness in vacuum over a molecular sieve.

Figure 1 shows an n.m.r.-spectrum of the 2-propyl-4,5-dihydrobenzo-triazolediol, which provided the final proof of the structure, supplementary to the results of the u.v., i.r. and high resolution mass spectra (data not shown). The coupling constant of 9 Hz of the protons e and f is significant, because it confirms the assumed *trans* position. As far as we know, the formation of dihydrodiols from benzotriazoles and related compounds has not been described previously.

CONCLUDING REMARKS

After we had finished our investigations on the formation of dihydrodiols with 2-alkylbenzotriazoles, Tada *et al.* (1982) published a communication on the formation of dihydrodiol from quinoline in PCB- and MC-induced rats. The possibility of a connection between the formation of an epoxide and the proven carcinogenicity of quinoline was pointed out. A potential carcinogenicity problem should also be considered with 2-alkylbenzotriazoles. The large amount of dihydrodiol which is produced shows how easy it is to produce epoxides from these compounds.

The results clearly show that simple changes in a molecule can open up other ways for biotransformation, which may lead to undesirable metabolites. The demand to 'stop wasting drugs by metabolism' is double-edged. Structural changes made on molecules with the aim of suppressing a possible biotransformation reaction could be dangerous, when, as described, attempts to suppress side chain hydroxylation to an alcohol, leads to an increased formation of an epoxide.

REFERENCES

Tada, M., Takahashi, K. & Kawzoe, Y. (1982), *Chem. Pharm. Bull., Tokyo*, **30**, 3834.

Part 7

PROSTAGLANDIN H SYNTHETASE MEDIATED *N*-OXIDATIONS

Chapter 34

METABOLISM OF AROMATIC AMINES BY PROSTAGLANDIN H SYNTHETASE

T. E. Eling, J. A. Boyd, R. S. Ktrauss and **R. P. Mason**, National Institute of Environmental Health Sciences, Laboratory of Molecular Biophysics, P.O. Box 12233, Research Triangle Park, N.C. 27709, USA

1. Aromatic amines are excellent substrates for prostaglandin H synthetase (PHS), being oxidized to free radical metabolites.
2. Aminopyrine is demethylated by PHS. The mechanism is by oxidation to a radical cation followed by rearrangement to the iminium cation. The iminium cation is hydrolysed to formaldehyde.
3. 2-Aminofluorene (2-AF) is oxidized by PHS to 2-nitrofluorene, 2-azobis-fluorene, 2-aminodifluorenylamine and other products. The reaction proceeds via a free radical mechanism.
4. 2-AF is metabolized to electrophilic metabolites that covalently bind to DNA. The DNA-2-AF adducts were isolated, separated and found to be distinct from the DNA adducts derived from *N*-hydroxy-2-AF.
5. These unique DNA-2-AF adducts may serve as a useful index for assessing the role of PHS *in vivo*.

INTRODUCTION

Metabolism of chemicals to electrophilic metabolites is an important determinant in the development of chemically induced toxicity or neoplasia. We have investigated prostaglandin H synthetase (PHS) as a metabolic activating enzyme in extrahepatic tissue (Eling *et al.*, 1983; Marnett & Eling, 1983).

PHS metabolizes arachidonic acid to prostaglandin H_2, the pivotal intermediate from which all other prostaglandins arise. PHS catalyses two different enzymatic reactions: the cyclooxygenase converts arachidonic acid to PGG_2, while the hydroperoxidase reduces the hydroperoxide PGG_2 to the corresponding alcohol PGH_2.

Metabolism of chemicals is catalysed by the hydroperoxidase of PHS. Thus, metabolism is dependent on the formation of PGG_2, a hydroperoxide. However, other lipid hydroperoxides or H_2O_2 will also support catalysis (Eling et al., 1983; Marnett & Eling, 1983). To catalyse the reduction of the peroxides to alcohols, reducing cofactors are required by the enzyme. These cofactors donate single electrons to the peroxidase and in turn are converted to electron deficient metabolites which can be chemically reactive. Aromatic amines are excellent co-substrates for PHS peroxidase. Studies on the metabolism of aromatic amines by PHS are divided into two categories of chemicals: those compounds which undergo N-demethylation such as secondary and tertiary aromatic amines, and carcinogenic primary aromatic amines which undergo one electron oxidation.

N-DEALKYLATION OF AROMATIC AMINES BY PEROXIDASES—MECHANISMS

We have reported (Sivarajah et al., 1982) that a wide variety of monomethyl and dimethyl substituted anilines and aminopyrine are N-demethylated by arachidonic-acid-fortified ram seminal vesicle microsomes, a rich source of PHS. While the mechanism for N-demethylation of PHS appears to be a one-electron oxidation, it is different from the oxygen insertion mechanism proposed for P-450 catalysed N-demethylation. Several mechanisms for the N-dealkylation of aromatic amines by peroxidase have been proposed, using aminopyrine as model substrate. Griffin & Ting (1978), using horseradish peroxidase, proposed that aminopyrine undergoes a one-electron oxidation to a cation free radical which is then further oxidized to an iminium cation. Hydrolysis of the iminium cation yields the monomethyl amine and formaldehyde as depicted in Scheme 1.

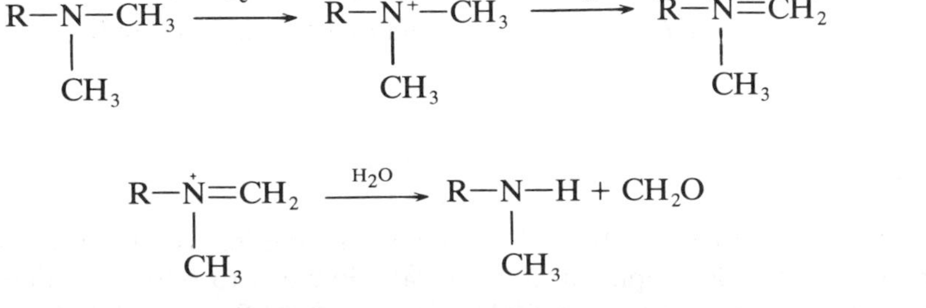

Scheme 1: Proposed mechanism for the horseradish peroxidase mediated N-demethylation of aminopyrine (Griffin & Ting, 1978)

We have proposed a different mechanism for aminopyrine demethylation by prostaglandin H synthetase in which aminopyrine is oxidized to a free radical cation, which then disproportionates to the iminium cation radical and aminopyrine. The iminium cation is subsequently hydrolysed to the mono-methyl amine and formaldehyde as depicted in Scheme 2.

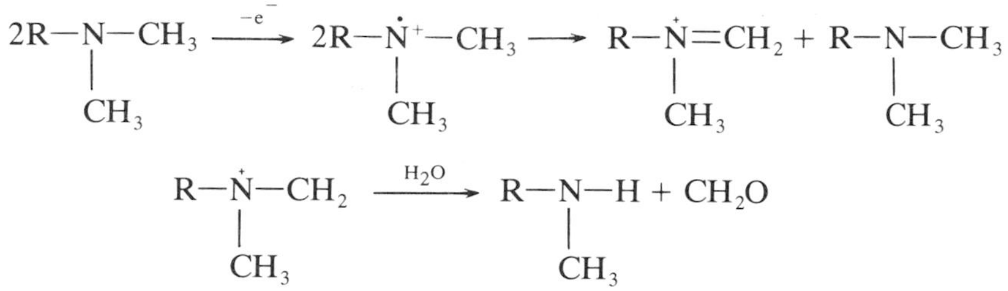

$$2R{-}\underset{\underset{CH_3}{|}}{N}{-}CH_3 \xrightarrow{-e^-} 2R{-}\underset{\underset{CH_3}{|}}{\overset{.}{N}{}^+}{-}CH_3 \longrightarrow R{-}\underset{\underset{CH_3}{|}}{\overset{.}{N}}{=}CH_2 + R{-}\underset{\underset{CH_3}{|}}{N}{-}CH_3$$

$$R{-}\underset{\underset{CH_3}{|}}{\overset{.}{N}}{-}CH_2 \xrightarrow{H_2O} R{-}\underset{\underset{CH_3}{|}}{N}{-}H + CH_2O$$

Scheme 2: Proposed mechanism for the prostaglandin H synthetase mediated *N*-demethylation of aminopyrine.

Our proposal is based on the finding that the steady-state concentration of the radical is linearly related to the square root of the enzyme concentration (Lasker *et al.*, 1981) and is consistent with a bimolecular decay. We have measured (Eling *et al.*, 1985) the decay of the aminopyrine free radical directly, taking advantage of the very rapid aminopyrine oxidation by prostaglandin H synthetase and inactivation of the enzyme by 15-hydroperoxy-arachidonic acid. The aminopyrine free radical decay obeyed second order kinetics, indicating a bimolecular decay, consistent with the disproportation proposal.

Glutathione rapidly reduced the aminopyrine free radical to aminopyrine. During this process, glutathione was oxidized to the thiyl radical. Previously, Moldeus and coworkers (Moldeus *et al.*, 1983) have shown that glutathione is oxidized to glutathione disulphide by either the aminopyrine or dimethylaniline cation free radical. This reaction was presumed to proceed via a thiyl radical (GS˙) (see equation 1 below).

$$A\overset{..}{P} + GSH \rightarrow AP + GS^{.}$$

$$2GS^{.} \rightarrow GSSG \tag{1}$$

We have spin trapped the GS˙ radical produced by the oxidation of GSH by the aminopyrine free radical. Our data therefore support the hypothesis of Moldeus and coworkers (Moldeus *et al.*, 1983) and suggest that amine cation free radicals generated by intact cells which contain GSH will result in the formation of GS˙. These results support the proposed mechanism in which aminopyrine is oxidized by PHS hydroperoxidase to the aminopyrine free radical, which then disproportionates to the iminium cation. The iminium cation is further hydrolysed to the demethylated amine and formaldehyde. Glutathione reduces the aminopyrine radical to aminopyrine with the con-comitant oxidation of GSH to its thiyl radical (Fig. 1).

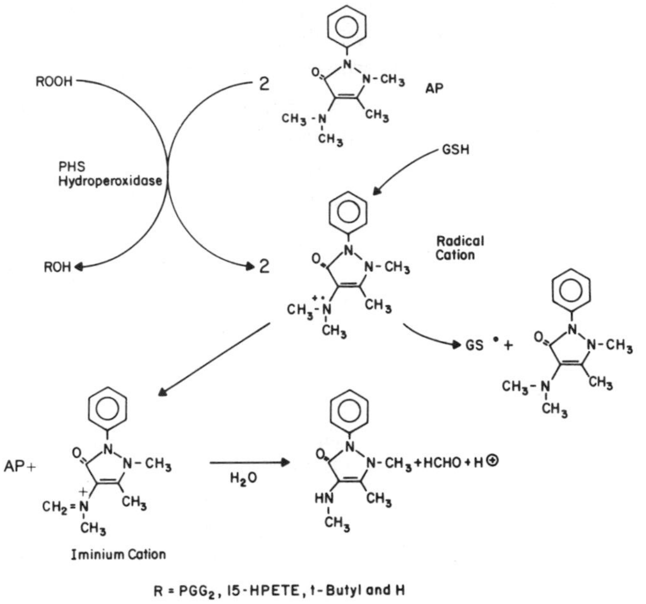

Fig. 1 – Aminopyrine N-demethylation by prostaglandin H synthetase. AP is oxidized to a free radical cation (2), which then disproportionates to the iminium cation radical and AP. The radical cation may be reduced to AP by GSH.

PROSTAGLANDIN H SYNTHETASE METABOLISM OF BENZIDINE AND 2-AMINOFLUORENE

Investigations on the metabolism of primary aromatic amines by PHS have centred on two compounds: benzidine and 2-aminofluorene. Both of these compounds are potent carcinogens and benzidine induces carcinomas of the urinary bladder in humans. PHS may serve as an enzyme for the metabolic activation of these compounds to their ultimate carcinogenic forms in target tissues.

Josephy and coworkers (Josephy et al., 1983; Josephy et al., 1982a; Josephy et al., 1982b) have extensively investigated the metabolism of benzidine and its structural analogue 3,5,3′,5′-tetramethylbenzidine by horseradish peroxidase and PHS peroxidase. Both enzymes metabolize benzidine to a radical cation which is in equilibrium with a blue-coloured charge transfer complex of benzidine itself and its two electron oxidation product, a di-imine dication (Fig. 2). Azobenzidine was the only stable end product of peroxidation of benzidine identified, while the metabolites of 3,5,3′,5′-tetramethylbenzidine have not been identified. Benzidine oxidized by PHS in the presence of nucleic acids in vitro, yields reactive intermediate(s) that bind covalently to the nucleic acid with very high efficiency (Morton et al., 1983) and are mutagenic in bacterial test systems (Robertson et al., 1983).

2-Aminofluorene is also metabolized by PHS (Boyd et al., 1983; Boyd & Eling, 1984). The major isolatable metabolites were 2,2′-azobisfluorene,

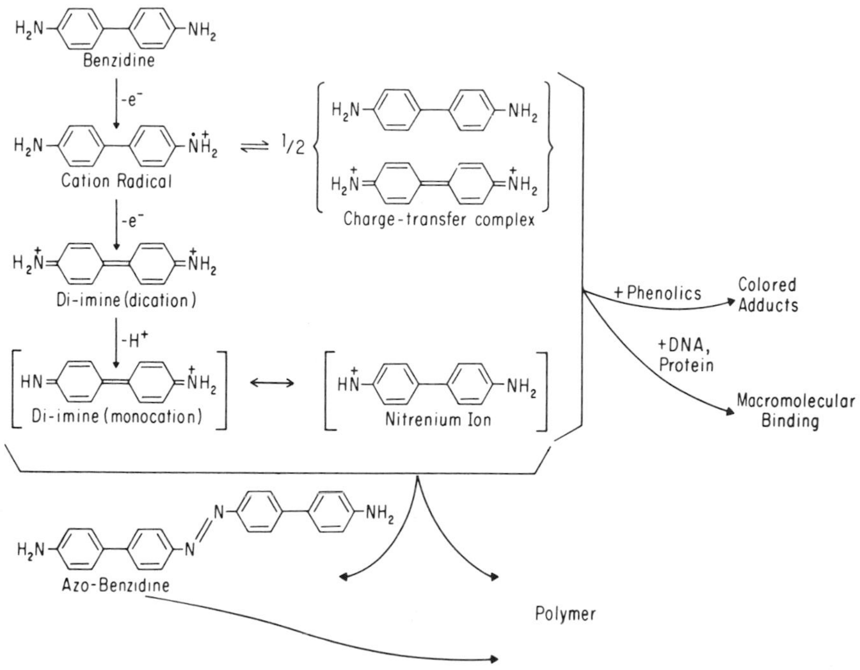

Fig. 2 – Metabolism of benzidine by horseradish peroxidase and prostaglandin H synthetase peroxidase. Azobenzidine was the only stable end-product of peroxidation of benzidine identified.

2-nitrofluorene, 2-aminodifluorenylamine, polymeric material and material covalently bound to microsomal protein. The reaction was dependent on archidonic acid, was inhibited by the addition of indomethacin to the incubation system, and was very rapid. Hydrogen peroxide also supported metabolism. N-Hydroxy-2-aminofluorene and 2-nitrosofluorene were extremely rapidly oxidized by PHS to 2-nitrofluorene (Fig. 3). Horseradish peroxidase also oxidized 2-aminofluorene to 2-nitrofluorene and 2,2′-azobisfluorene, but chloroperoxidase oxidized 2-aminofluorene primarily to 2-nitrosofluorene. This result indicates that PHS-dependent oxidation of 2-aminofluorene may proceed through a free radical mechanism similar to that of horseradish peroxidase. Metabolism of 2-aminoflurorene by PHS also results in formation of intermediates which are mutagenic and bind to nucleic acids (Morton *et al.*, 1983). Formation of N-hydroxy-2-aminofluorene could not be detected during PHS-dependent oxidation of 2-aminofluorene. The reactive intermediates responsible for the binding to nucleic acids could be free radicals or the 2-aminofluorene nitrenium ion. Experiments have demonstrated that N-(deoxyguanosin-8-yl)-2-aminofluorene (the sole DNA adduct formed by the 2-aminofluorene nitrenium ion) represents only a small fraction of the 2-aminofluorene/DNA adducts formed by PHS and that several unique but unidentified DNA adducts were also formed (Krauss & Eling, 1985). This

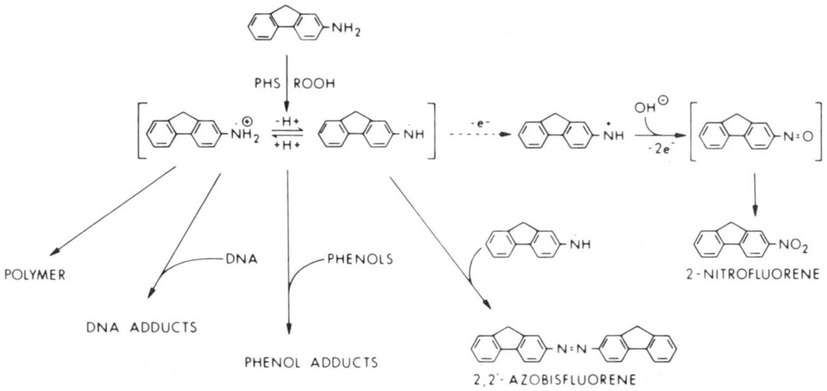

Fig. 3 – Metabolism of 2-aminofluorene by prostaglandin H synthetase peroxidase.

suggests that the nitrenium ion is not an important intermediate in the metabolism of 2-aminofluorene by PHS. Further studies aimed at clarifying this point are under way

In summary, aromatic amines are excellent substrates for PHS and undergo metabolism by a one electron oxidation mechanism. The stable metabolites isolated depend on the chemical nature of the electron deficient intermediates. PHS may represent an important enzyme system for converting carcinogenic aromatic amines to their ultimate carcinogen metabolites in extrahepatic tissues.

ACKNOWLEDGEMENTS

The authors wish to acknowledge Dr K. Sivarajah and Dr P. D. Josephy for their contribution to this investigation. We also wish to thank Ms Peggy Ellis for preparing this manuscript.

REFERENCES

Boyd, J. A., Harvan, D. & Eling, T. E. (1983), *J. Biol. Chem.*, **258**, 8246.
Boyd, J. A. & Eling, T. E. (1984), *J. Biol. Chem.*, **259**, 13885.
Eling, T. E., Boyd, J. A., Reed, G. A., Mason, R. P. & Sivarajah, K. (1983), *Drug Metab. Rev.*, **14**, 1203.
Eling, T. E., Mason, R. P. & Sivarajah, K. (1985) *J. Biol. Chem.*, **260**, 1601.
Griffin, B. W. & Ting, P. L. (1978), *FEBS Lett.*, **74**, 139.
Josephy, P. D., Eling, T. E. & Mason, R. P. (1982a), *J. Biol. Chem.*, **257**, 3669.
Josephy, P. E., Mason, R. P. & Eling, T. E. (1982b), *Cancer Res.*, **42**, 2567.
Josephy, P. D., Eling, T. E. & Mason, R. P. (1983), *J. Biol. Chem.*, **258**, 5561.

Krauss, R. S. & Eling, T. E. (1985), *Cancer Res.*, **45**, 1680.

Lasker, J. M., Sivarajah, K., Mason, R. P., Kalyanaraman, B., Abou-Donia, M. B. & Eling, T. E. (1981), *J. Biol. Chem.*, **256**, 7764.

Marnett, L. J. & Eling, T. E. (1983), in *Review in Biochemical Toxicology* Vol. V (eds. Hodgson, E., Bend, J. R. & Philpot, R. M.) Elsevier, New York, p. 135.

Moldeus, P., O'Brien, P. J., Thor, H., Berggren, M. & Orrenius, S. (1983), *FEBS Letters*, **16**, 411.

Morton, K. C., King, C. M., Vaught, J. B., Wang, C. Y., Lee, M.-S. & Marnett, L. J. (1983), *Biochem. Biophys. Res. Commun.*, **111**, 96.

Robertson, I., Sivarajah, K., Eling, T. E. & Zieger, E. (1983), *Cancer Res.*, **43**, 476.

Sivarajah, K., Lasker, J. & Eling, T. (1982), *Mol. Pharmacol.*, **21**, 133.

Chapter 35

N-OXIDATION OF 4-CHLOROANILINE BY THE MICROSOMAL PROSTAGLANDIN SYNTHETASE SYSTEM: REDOX CYCLING OF A NITROXIDE INTERMEDIATE

P. Hlavica and **Ines Golly**, Institut für Pharmakologie und Toxikologie der Universität, Nussbaumstrasse 26, D-8000 München 2, FRG

1. 4-Chloroaniline undergoes N-oxidation in ram seminal vesicle microsomal preparations fortified with arachidonic acid to yield N-(4-chlorophenyl)-hydroxylamine and 1-chloro-4-nitrosobenzene. Hydrogen peroxide also supports metabolism of the amine substrate to the same organic solvent-extractable products, suggesting that the hydroperoxidase activity of prostaglandin endoperoxide synthetase is responsible for the co-oxidation.
2. Analysis of the reaction mixtures by e.s.r. spectrometry reveals the formation of a radical intermediate bearing the characteristics of a strongly immobilized nitroxide.
3. Arylamine-stimulated O_2^- release can be observed when the arachidonate-containing incubation media are supplemented with NADPH. Redox cycling of the nitroxide/hydroxylamine couple is presumed to represent the major source of O_2^-, but additional mechanisms, such as redox changes of nitro anion radicals resulting from potential further metabolism of 1-chloro-4-nitrosobenzene, cannot be excluded.
4. The concerted action of carrier-bound nitroxides and O_2^- in initiating damage of cellular macromolecules is discussed.

INTRODUCTION

The metabolic N-oxidation of primary aromatic amines has been recognized as resulting in the formation of more toxic and/or carcinogenic products (Weisburger & Weisburger, 1973). The most extensively studied enzyme catalysing bioactivation of this type of nitrogenous compound is the cyto-chrome P-450-dependent microsomal monooxygenase system, located predominantly in the liver.

Recently, interest has focused on microsomal prostaglandin endoperoxide synthetase (EC 1.14.99.1), which may provide a complementary system for amine activation (Kadlubar *et al.*, 1982; Boyd *et al.*, 1983; Josephy *et al.*, 1983) in many extrahepatic tissues exhibiting low levels of monooxygenase(s).

Using 4-chloroaniline (4-CA) as a model substrate, we report on the occurrence, during prostaglandin endoperoxide synthetase-mediated N-oxidation of the amine, of a strongly immobilized nitroxide species. Further, NADPH-dependent O_2^- production during co-oxidation of the arylamine will be described.

EXPERIMENTAL

Ram seminal vesicle microsomal fractions were prepared by the method of Boyd *et al.* (1983). The preparations were devoid of detectable amounts of cytochrome P-450, but contained significant NADPH-cytochrome *c* reductase (EC 1.6.2.4) activity.

Partial succinoylation of cytochrome *c* was performed as indicated by Finkelstein *et al.* (1981).

N-Oxidation of 4-CA was performed in reaction mixtures containing 0.1 M-phosphate buffer, pH 7.8, 0.1 mM-arachidonic acid (or 0.1 mM-H_2O_2), 1 mM-amine substrate and ram seminal vesicle microsomal protein as indicated. Incubations were carried out for 1 min at 37 °C.

Substrate-induced, NADPH-sustained formation of O_2^- was measured by recording the increase in absorbance at 550 nm in assay mixtures containing ram seminal vesicle microsomal protein (0.5 mg/ml), 0.1 mM-arachidonic acid, 0.1 mM-NADPH and 20 μM-succinoylated ferricytochrome *c* in 50 mM-phosphate buffer, pH 7.7. Reactions were initiated by the addition of 1 mM-4-CA to the sample cell. Alternatively, arachidonate was omitted, and reactions were started by the addition of either 1-chloro-4-nitrosobenzene or 1-chloro-4-nitrobenzene to the experimental cuvette.

Nitroxide free radicals were generated in a system containing ram seminal vesicle microsomal protein (3.3 mg/ml), 1.3 mM-arachidonic acid and 3.3 mM-4-CA in 0.1 M-phosphate buffer, pH 7.8. The microsomal pellet was transferred to a quartz capillary tube, and e.s.r. spectra were recorded at 25 °C with a Bruker ER 200 SRD spectrometer equipped with a TM cavity.

Nitrosoarene was quantified by the method of Herr & Kiese (1959).

RESULTS AND DISCUSSION

As shown in Fig. 1, addition of 4-CA to aerobic ram seminal vesicle microsomal suspensions fortified with arachidonic acid results in *N*-oxidation of
the arylamine; the process is severely blocked by the presence of indomethacin, suggesting the reaction to be catalysed by the prostaglandin endoperoxide
synthetase system. Indeed, H_2O_2 used instead of arachidonic acid also sus-
tains amine oxidation (Fig. 1). Microsomal cytochrome P-450 can be ruled
out as accounting for the observed peroxidase activity, since the microsomal
preparations used were devoid of measurable amounts of the pigment.
Accordingly, no NADPH/O_2-dependent *N*-oxidation of 4-CA was observed
(Fig. 1).

Analysis of the reaction mixtures with and without prior $K_3Fe(CN)_6$
treatment reveals that 19% of the material trapped by the colorimetric reagent originates from *N*-(4-chlorophenyl)hydroxylamine, the remainder being
attributable to the corresponding nitroso derivative. The low steady-state
level of the hydroxylamine suggests rapid conversion of the compound to the

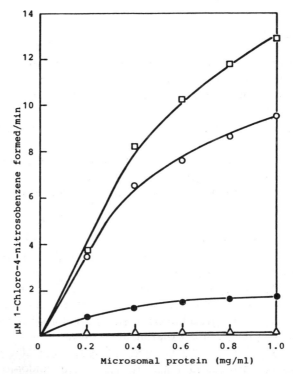

Fig. 1 – *N*-Oxidation of 4-chloroaniline by ram seminal vesicle microsomal
fractions. Assay mixtures composed as described in the Experimental section were
incubated in the absence (○) or presence (●) of 0.1 mM-indomethacin. In some
experiments, either 0.1 mM-H_2O_2 (□) or NADPH/O_2 (△) were used in place of
arachidonic acid.

nitroso analogue. This process gives rise to the formation of a radical intermediate characterized by the e.s.r. spectrum depicted in Fig. 2(c). The signal
($g = 2.0050$) is typical of a strongly immobilzed nitroxide free radical (Floyd
et al., 1978). The enzymic nature of nitroxide formation is indicated by the
inability of boiled microsomal fraction to sustain radical production (Fig. 2d).

The concurrence, during co-oxidation of 4-CA, of hydroxyamine and nitroxide intermediates as well as the presence, in ram seminal vesicle microsomal fractions, of significant amounts of NADPH-cytochrome c reductase
prompted us to study the possibility of amine-induced O_2^- release in aerobic
microsomal preparations fortified with arachidonic acid and NADPH; partially succinoylated ferricytochrome c was used as an indicating scavenger for
O_2^-. As shown in Fig. 3(a), addition to the reaction mixtures of 4-CA elicits
reduction of the cytochrome c derivative, and this process is markedly
blocked by the presence of superoxide dismutase; boiling of the latter enzyme

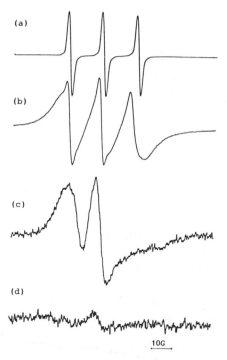

Fig. 2 – Typical 9.35 GHz electron spin resonance spectrum arising from oxidation
of 4-chloroaniline by prostaglandin endoperoxide synthetase. (a) Spectrum of free
2,2,6,6-tetramethylpiperidine-1-oxyl(TMPO) used as an external standard. (b)
Spectrum of partially membrane-bound TMPO. (c) Metabolic spectrum generated
upon the incorporation of 4-CA into ram seminal vesicle microsomal suspensions
fortified with arachidonic acid; for details see the Experimental section. (d)
Identical to (c), except that boiled microsomal fraction was used. The instrumental
conditions were as follows: magnetic field, 3348 G; scan range, 100 G; modulation
amplitude, 2.5 G; nominal microwave power, 20 mW; receiver gain, 0.32–160×10^4;
scan time, 100 s.

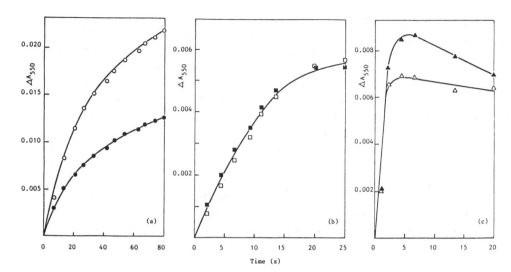

Fig. 3 – Amine-induced reduction of succinoylated ferricytochrome *c* in ram seminal vesicle microsomal preparations in the absence ($\bigcirc$, $\square$, $\triangle$) and presence ($\bullet$, $\blacksquare$, $\blacktriangle$) of superoxide dismutase (100 µg/ml). Microsomal suspensions supplemented with arachidonic acid, NADPH and modified ferricytochrome *c* as specified in the Experimental section were divided between two cuvettes. Amine substrate was added to the sample cell and the absorbance change at 550 nm monitored at 25 °C (a). In some experiments, either arachidonic acid (b) or NADPH (c) were omitted.

abolishes its inhibitory action (data not shown). These findings provide evidence for the involvement of $O_2^{-\cdot}$ (Kuthan *et al.*, 1982). The observed release of $O_2^{-\cdot}$ is believed to be closely related to oxidative transformation of the amine substrate by the prostaglandin endoperoxide synthetase system. This view is supported by the inability of 4-CA to induce the formation of superoxide dismutase-sensitive material when arachidonic acid is omitted from the incubation media (Fig. 3b). Further, $O_2^{-\cdot}$ production is completely dependent on the presence of reducing equivalents (Fig. 3c).

Ultimate proof of the ability of *N*-oxidized derivatives of 4-CA to sustain NADPH-dependent $O_2^{-\cdot}$ formation in ram seminal vesicle microsomal fractions comes from experiments with 1-chloro-4-nitrosobenzene. Results, not shown here, indicate that, in the presence of electron donor, catalytic amounts of the nitroso compound undergo enzyme-controlled reduction, probably via the nitroxide and hydroxyamine intermediates, to the parent amine. This metabolic event is associated with the release of $O_2^{-\cdot}$, as evidenced by the marked inhibition of superoxide dismutase of the reduction of modified ferricytochrome *c* (Fig. 4a). It has to be kept in mind that metabolically formed nitrosoarene might be further oxidized in the arachidonate-containing reaction mixtures to yield the nitro analogue. Nitroaromatics, in their turn, can undergo redox cycling to produce $O_2^{-\cdot}$ (Perez-Reyes *et al.*, 1980). Indeed, addition of 1-chloro-4-nitrobenzene to aerobic microsomal suspensions supplemented with NADPH results in $O_2^{-\cdot}$ release (Fig. 4b).

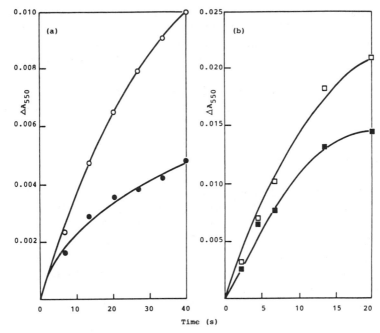

Fig. 4 – Nitroso- and nitroarenc-induced reduction of succinoylated ferricytochrome c in ram seminal vesicle microsomal preparations in the absence (○, □) and presence (●, ■) of superoxide dismutase (100 μg/ml). Microsomal suspensions supplemented with NADPH and modified ferricytochrome c as specified in the Experimental section were divided between two cuvettes. 1-Chloro-4-nitrosobenzene (a: final concentration 10 μM) or 1-chloro-4-nitrobenzene (b: final concentration 1 mM) was added to the sample cell and the absorbance change at 550 nm monitored at 25 °C.

However, it has to be recognized that nitro anion radicals, possibly arising from reduction of the nitroarene, exhibit considerably higher reactivity than $O_2^{-\cdot}$ towards ferricytochrome c, and this might simulate lower efficiency of superoxide dismutase.

In summary, our results establish evidence for the NADPH-dependent formation, during prostaglandin endoperoxide synthetase-mediated co-oxidation of 4-CA, of superoxide anion. Although redox cycling of the nitroxide/hydroxyamine couple arising from N-oxidation of the arylamine is believed to represent the major source of $O_2^{-\cdot}$ (Stier *et al.*, 1980), additional mechanisms can be envisaged. These include redox changes of nitro anion radicals potentially formed in the assay media or oxidation of metabolically generated aminophenols to the corresponding semiquinoneimine radicals. The latter species might oxidize NADPH to the pyridinium nucleotide free radical, which reacts with oxygen to produce $O_2^{-\cdot}$ (Mason & Chignell, 1982).

Although 4-CA itself belongs to the group of non-carcinogenic aromatic amines, it can be regarded as a model to assess general principles governing the bioactivation of amine compounds. With respect to this, the occurrence, in our metabolic system, of a carrier-bound primary aromatic nitroxide is of special interest, since this type of adduct, characterized by remarkable longevity, has been proposed to combine with DNA in living cells to cause distortion of the macromolecule (Stier *et al.*, 1982). Additional damage of critial cellular constituents may be afforded by the simultaneous action of OH⁻ generated from O_2^- via the iron-catalysed Haber–Weiss mechanism. Further work is needed to elucidate the biological importance of such processes in arylamine-induced cytotoxicity.

REFERENCES

Boyd, J. A., Harvan, D. J. & Eling, T. E. (1983), *J. Biol. Chem.*, **258**, 8246.

Finkelstein, E., Rosen, G. M., Patton, S. E., Cohen, M. S. & Rauckman, E. J. (1981), *Biochem. Biophys. Res. Commun.*, **102**, 1008.

Floyd, R. A., Soong, L. M., Stuart, M. A. & Reigh, D. L. (1978), *Arch. Biochem. Biophys.*, **185**, 450.

Herr, F. & Kiese, M. (1959), *Naunyn-Schmiedeberg's Arch. Exp. Pathol. Pharmakol.*, **235**, 351.

Josephy, P. D., Eling, T. E. & Mason, R. P. (1983), *J. Biol. Chem.*, **258**, 5561.

Kadlubar, F. F., Frederick, C. B., Weis, C. C. & Zenser, T. V. (1982), *Biochem. Biophys. Res. Commun.*, **108**, 253.

Kuthan, H., Ullrich, V. & Estabrook, R. W. (1982), *Biochem. J.*, **203**, 551.

Mason, R. P. & Chignell, C. F. (1982), *Pharmacol. Rev.*, **33**, 189.

Perez-Reyes, E., Kalyanaraman, B. & Mason, R. P. (1980), *Mol. Pharmacol.*, **17**, 239.

Stier, A., Clauss, R., Lücke, A. & Reitz, I. (1980), *Xenobiotica*, **10**, 661.

Stier, A., Clauss, R., Lücke, A. & Reitz, I. (1982), in *Free Radicals, Lipid Peroxidation and Cancer* (eds. McBrien, D. C. H. & Slater, T. F.) Academic Press, London, p. 329.

Weisburger, J. H. & Weisburger, E. K. (1973), *Pharmacol. Rev.*, **25**, 1.

Chapter 36

INVOLVEMENT OF PROSTAGLANDIN H SYNTHETASE IN AROMATIC AMINE-INDUCED TRANSITIONAL CELL CARCINOMA OF URINARY BLADDER

T. V. Zenser, **R. W. Wise** and **B. B. Davis**, Geriatric Research, Education and Clinical Centre, St. Louis University School of Medicine, St. Louis, Missouri 63125, USA

1. Dog bladder microsomal prostaglandin H synthetase (PHS) activated several carcinogenic aromatic amines to compounds which covalently bound to macromolecules.
2. FANFT-induced rat bladder cancer was significantly reduced by co-feeding aspirin.
3. These *in vivo* and *in vitro* studies indicate that PHS activation may be the initiating step in certain chemical-induced bladder cancers.
4. A model was developed illustrating PHS activation and proposed sites of prevention.

INTRODUCTION

Metabolism of aromatic amines by prostaglandin H synthetase (PHS) has been proposed as an additional pathway for carcinogen activation (Zenser *et al.*, 1979, 1980). The aromatic amines benzidine, 2-naphthylamine, and 2-aminofluorene are metabolized to mutagenic products by PHS (Robertson *et al.*, 1983). Mono- and diacetylated benzidine are poor substrates for PHS activation (Morton *et al.*, 1983), which is consistent with *N*-acetylation being a detoxification pathway for aromatic amine bladder carcinogenesis (Lower *et al.*, 1979). Target tissue metabolism is a salient feature of the proposed activation of aromatic amines by PHS. A model illustrating this hypothesis is depicted in Fig. 1. According to this model, benzidine is activated by the prostaglandin hydroperoxidase component of PHS. Activated benzidine binds covalently to critical cellular macromolecules. This binding is thought to be involved in the initiation of the carcinogenic process (Miller, 1970). There are four general sites (numbered 1, 2, 3, and 4 in Fig. 1) for potential inhibition of benzidine activation and subsequent covalent binding to critical cellular macromolecules.

The purpose of this review is to describe studies in which PHS activation of aromatic amines was assessed in different tissues from a species (dog) susceptible to bladder cancer (Wise *et al.*, 1984), and studies in which the existence of the sites illustrated in Fig. 1 were demonstrated (Zenser *et al.*, 1983). The application of these results to prevention of aromatic amine-induced bladder cancer is discussed.

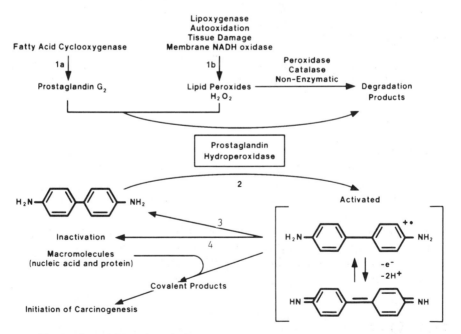

Fig. 1 – A model based on the hypothesis that the hydroperoxidase component of prostaglandin H synthetase is involved in the initiation of bladder cancer induced by benzidine and other chemical carcinogens. The numbered sites are described in the text (Zenser *et al.*, 1983).

ACTIVATION OF AROMATIC AMINES BY MICROSOMES PREPARED FROM DIFFERENT DOG TISSUES

The conversion of benzidine to aqueous-soluble products by dog microsomes was assessed (Wise *et al.*, 1984). Following addition of arachidonic acid, a significant increase in aqueous-soluble benzidine derivatives was observed with microsomes prepared from bladder transitional epithelium, inner medulla, and outer medulla. Arachidonic acid-initiated benzidine metabolism was not observed with hepatic or renal cortical microsomes. Furthermore, aqueous-soluble benzidine metabolites were not detected following addition of NADPH to liver, kidney, or bladder microsomes. The latter is consistent with previous studies demonstrating a lack of mixed-function oxidase metabolism of benzidine (Zenser *et al.*, 1979). The relative rate of tissue arachidonic acid-initiated metabolism of benzidine correlated with the relative rate of [^{14}C]PGE$_2$ synthesis from [^{14}C]arachidonic acid. The unsaturated fatty acid 11,14,17-eicosatrienoic acid could not substitute for arachidonic acid. Indomethacin inhibited benzidine metabolism. These results are consistent with benzidine metabolism by bladder transitional epithelial PHS. *o*-Dianisidine, 4-aminobiphenyl and 2-naphthylamine were also metabolized by dog bladder microsomal PHS.

To determine the relative contribution of mixed-function oxidases and PHS to aromatic amine activation, 2-naphthylamine metabolism by dog liver and bladder was assessed. In the liver, NADPH but not arachidonic acid initiated binding. Binding to protein and tRNA was 13.5 ± 2.2 and 0.04 ± 0.01 pmol/mg protein/min, respectively. NADPH-dependent binding was inhibited 90 to 100% by addition of 2-[(2,4-dichloro-6-phenyl)phenoxy]ethylamine (DPEA). In the bladder, arachidonic acid- but not NADPH-dependent metabolism initiated binding. Protein and tRNA binding was 120 ± 2.2 and 0.21 ± 0.01, respectively, and was completely inhibited by addition of indomethacin.

These data demonstrate that the relative amounts of PHS and mixed-function oxidase enzyme activities were different in dog liver and bladder. Hepatic mixed-function oxidase catalysed the macromolecular binding of 2-naphthylamine. Bladder mixed-function oxidase-catalysed binding of 2-naphthylamine was not detected. In contrast, PHS-catalysed binding of this carcinogenic aromatic amine was demonstrated with bladder transitional epithelium but not with liver. This is consistent with the observed lack of hepatic PHS metabolism of arachidonic acid. The rate of bladder PHS-catalysed 2-naphthylamine binding to macromolecules was nearly tenfold greater than hepatic mixed-function oxidase-catalysed binding. These results provide the first demonstration that bladder transitional epithelium from a species (dog) susceptible to aromatic amine-induced bladder cancer contains PHS and is capable of activating a variety of aromatic amine precarcinogens by a PHS-dependent pathway. Therefore, activation by bladder transitional epithelial PHS may be involved in the genesis of primary aromatic amine-induced bladder cancer (Wise *et al.*, 1984).

ASSESSMENT OF SPECIFIC SITES AT WHICH PROSTAGLANDIN H SYNTHETASE-CATALYSED ACTIVATION OF AROMATIC AMINES CAN BE PREVENTED

The effects of different test agents on benzidine binding were evaluated using either arachidonic acid or H_2O_2 as co-substrate (Zenser et al., 1983). Propyl-thiouracil, methimazole, glutathione and ascorbate inhibited benzidine binding whether H_2O_2 or arachidonic acid was used as the co-substrate. However, indomethacin only inhibited arachidonic acid and not H_2O_2-initiated binding. Subsequent studies demonstrated that only indomethacin inhibited the metabolism of [^{3}H]arachidonic acid. Glutathione inhibition of benzidine binding resulted in a corresponding increase in the aqueous TCA-soluble fraction.

In additional studies, the effects of certain agents of PHS-catalysed benzidine cation radical formation were assessed. Changes in the relative strength of the electron spin resonance signal were monitored with time after addition of enzyme, using H_2O_2 as co-substrate. Initial inclusion of indomethacin in the incubation mixture did not alter radical formation. Inclusion of 1 mM-KCN in the initial incubation prevented radical formation. Addition of KCN after 30 seconds stopped radical production but did not appear to alter the decay of the generated radical. By contrast, addition of ascorbate after the late addition of KCN caused a dose-dependent increase in benzidine radical decay. Ascorbate (0.5 mM) caused an immediate loss of blue colour and a return to the original colourless solution. Although an early addition of methimazole inhibited radical production, a late addition of methimazole did not alter the decay of the benzidine cation radical. These results indicate that ascorbate inhibits by reducing the radical, whereas KCN and methimazole inhibit radical production only.

It has been postulated that during enzymatic oxidation of benzidine a charge-transfer complex between the diamine and diimine is formed which is in equilibrium with a free radical cation (Josephy et al., 1982). The possible existence of this complex precluded unambiguous identification of the initial product of the enzymatic oxidation of benzidine at that time as formation of either species would give rise to the other. To examine this mechanism further, we determined the course of free radical and charge-transfer complex formation and decay during autoreduction of synthetic benzidinediimine in aqueous solutions. Benzidinediimine was shown to be reduced to the same free radical cation which was produced during peroxidatic activation of benzidine (Wise et al., in press). This radical was identified as the free radical cation of benzidine (Wise et al., 1983b). The time course of radical production and decay was not coincidental with charge-transfer complex formation and decay (Wise et al., in press). At pH 4.0, the free radical was decreasing at 5 minutes while the absorbance at 600 nm (approximate absorption maximum for the change-transfer complex) was increasing. Liquid chromatographic analysis of the solution after 20 minutes showed that only benzidine was present. The addition of ascorbate, but not aspirin or indomethacin, greatly facilitated the conversion of the diimine to diamine. Synthetic

[^{14}C]benzidinediimine binding to DNA was inhibited 90 to 100% with glutathione and ascorbate, respectively. While these results did not identify the product of benzidine peroxidation which binds DNA, they demonstrated that charge-transfer complex formation does not necessarily precede nor coincide with free radical generation during oxidation.

The diverse manner in which the test agents affected benzidine metabolism suggests that they are acting at several distinct sites (numbered 1, 2, 3 and 4) as illustrated in the model (Fig. 1). According to this model, prevention of benzidine binding to macromolecules can occur by (1) inhibiting the generation of peroxide co-substrate; (2) inhibiting hydroperoxidase-catalysed activation of carcinogen; (3) reduction of oxidized intermediate(s) back to the parent (procarcinogenic) compound; and (4) conjugation of activated intermediate(s).

Specific inhibitors for each one of these sites were demonstrated. Aspirin and other non-steroidal anti-inflammatory agents inhibit arachidonic acid metabolism and are acting at site 1a in Fig. 1. While phenidone, a lipoxygenase inhibitor (Blackwell & Flower, 1978), would act at site 1b. Antiperoxidase drugs, like propylthiouracil and methimazole (Engler *et al.*, 1982; Zelman *et al.*, 1984), act at site 2, regardless of the pathway (site 1a or 1b) used to generate the peroxide substrate. Site 2 inhibitors can function as competitive inhibitors of activation. Ascorbic acid rapidly reduces products of benzidine oxidation—quinonediimine and free radical—back to benzidine (site 3). Glutathione forms a conjugate with an activated benzidine intermediate(s) preventing binding to protein and nucleic acid (site 4). Sites 1 through 4 are amenable to therapeutic manipulation.

INVOLVEMENT OF PHS IN THE INITIATION OF CHEMICAL CARCINOGENESIS

Because of the lack of available animal models to assess aromatic amine-induced bladder cancer, the 5-nitrofuran N-[4-(5-nitro-2-furyl)2-thiazolyl]-formamide (FANFT) has been used to assess the involvement of PHS in bladder cancer. PHS activates both FANFT and its deformylated analogue ANFT. ANFT is thought to be the more proximate carcinogen in FANFT-induced bladder cancer. The only oxidative metabolism of ANFT demonstrated to date is PHS catalysed (Mattammal *et al.*, 1981). Mixed-function oxidase metabolism of ANFT has not been demonstrated. Horseradish peroxidase, lactoperoxidase and chloroperoxidase do not metabolize ANFT (Wise *et al.*, 1983a). ANFT oxidation appears, therefore, to be specific for the prostaglandin hydroperoxidase portion of PHS.

According to the model depicted in Fig. 1, if PHS is involved in the initiation of FANFT-induced bladder cancer, co-administration of aspirin (site 1a inhibitor) with FANFT should reduce the incidence of bladder cancer. Co-administration of 0.5% aspirin with 0.2% FANFT reduced the number of early morphological bladder lesions (Cohen *et al.*, 1981) and reduced the incidence of bladder cancer from 87% to 37% in rats (Murasaki *et al.*, 1984).

No bladder cancer was observed in control or aspirin-fed rats. Aspirin treatment was shown to reduce bladder prostaglandin E_2 synthesis significantly during the 12 weeks that FANFT was administered (Cohen *et al.*, 1981). These results are consistent with PHS involvement in genesis of FANFT-induced bladder cancer.

Our model suggests additional experiments to prove this hypothesis. The lack of complete inhibition of FANFT-induced bladder cancer may be due to peroxides being provided by site 1b. Aspirin is not an effective inhibitor of site 1b. Site 2 inhibitors may be more effective than aspirin in that they inhibit regardless of the source of peroxide. Site 2 inhibitors also have the advantage of not necessarily reducing prostaglandin synthesis as occurs with aspirin. Because prostaglandins have a variety of important physiological functions including promotion, this is an important quality of site 2 inhibitors. A glutathione conjugate of PHS activated ANFT has recently been identified in intact tissue (John Rice, unpublished data). This thioether conjugate may be a suitable biological marker for PHS activation *in vivo*. Such a marker might be useful in further assessing sites of prevention and choosing suitable inhibitors for more extensive tumorigenesis studies.

ACKNOWLEDGEMENTS

This research was supported by the Veterans Administration and the National Cancer Institute grant CA-28015. The authors wish to thank Miss Sharon Smith for secretarial assistance.

REFERENCES

Blackwell, G. J. & Flower, R. J. (1978), *Prostaglandins*, **16**, 417.
Cohen, S. M., Zenser, T. V., Murasaki, G., Fukushima, S., Mattammal, M. B., Rapp, N. S. & Davis, B. B. (1981), *Cancer Res.*, **41**, 3355.
Engler, H., Taurog, A. & Nakashima, T. (1982), *Biochem. Pharmacol.*, **31**, 3801.
Josephy, P. D., Eling, T. & Mason, R. P. (1982), *J. Biol. Chem.*, **257**, 3669.
Lower, G. M., Jr., Nilsson, T., Nelson, C. E., Wolf, H., Gamsky, T. E. & Bryan, G. T. (1979), *Environ. Health Perspect.*, **29**, 71.
Mattammal, M. B., Zenser, T. V. & Davis, B. B. (1981), *Cancer Res.*, **41**, 4961.
Miller, J. A. (1970), *Cancer Res.*, **30**, 559.
Morton, K. C., King, C. M., Vaught, J. B., Wang, C. Y., Lee, M.-S. & Marnett, L. J. (1983), *Biochem. Biophys. Res. Commun.*, **111**, 96.
Murasaki, G., Zenser, T. V., Davis, B. B. & Cohen, S. M. (1984), *Carcinogensis (Lond.)*, **5**, 53.
Robertson, I. G. C., Sivarajah, K., Eling, T. E. & Zeiger, E. (1983), *Cancer Res.*, **43**, 476.

Wise, R. W., Zenser, T. V. & Davis, B. B. (1983a), *Cancer Res.*, **43**, 1518.

Wise, R. W., Zenser, T. V. & Davis, B. B. (1983b), *Carcinogenesis (Lond.)*, **4**, 285.

Wise, R. W., Zenser, T. V., Kadlubar, F. F. & Davis, B. B. (1984), *Cancer Res.*, **44**, 1893.

Wise, R. W., Zenser, T. V. & Davis, B. B., *Carcinogensis*, *in press*.

Zelman, S. J., Rapp, N. S., Zenser, T. V., Mattammal, M. B. & Davis, B. B. (1984), *J. Lab. Clin. Med.*, **104**, 185.

Zenser, T. V., Mattammal, M. B., Armbrecht, H. J. & Davies, B. B. (1980), *Cancer Res.*, **40**, 2839.

Zenser, T. V., Mattammal, M. B. & Davis, B. B. (1979), *J. Pharmacol. Exp. Ther.*, **211**, 460.

Zenser, T. V., Mattammal, M. B., Wise, R. W., Rice, J. R. & Davis, B. B. (1983), *J. Pharmacol. Exp. Ther.*, **227**, 545.

Part 8

PEROXIDASE-H$_2$O$_2$-CATALYSED N-OXIDATIONS

Chapter 37

THE PEROXIDASE CATALYSED N-OXIDATION AND ACTIVATION OF N,N',N,N'-TETRAMETHYLBENZIDINE

P. J. O'Brien[a] and **B. Gregory**[b], Departments of Biochemistry[a] and Chemistry[b], Memorial University of Newfoundland, St. John's, Newfoundland, Canada A1B 3X9

1. N,N', N,N'-$(CH_3)_4$-Benzidine (TMB) is N-demethylated by a peroxidase–H_2O_2 system and results in some DNA binding and adduct formation with phenol derivatives.
2. High resolution mass spectrometry was used to show that N,N', N-$(CH_3)_3$-benzidine, N,N-$(CH_3)_2$-benzidine, N,N'-$(CH_3)_2$-benzidine and N-CH_3-benzidine were formed.
3. Peroxidase–H_2O_2 catalysed DNA binding or phenol adduct formation with N,N-$(CH_3)_2$-benzidine and N-CH_3-benzidine.
4. N-Demethylation is proposed to occur by hydrolysis of the iminium cation or diimine of TMB. The N,N', N-$(CH_3)_3$-benzidine formed then commences a new cycle of N-demethylation.

INTRODUCTION

Peroxidases have been implicated in the one-electron oxidative activation of diethylstilbestrol-induced uterine cancer (Metzler & McLachlan, 1978), *trans*-4-aminostilbene-induced Zymbal gland tumours (Osborne *et al.*, 1980), acetylaminofluorene-induced mammary tumours (Malejka-Giganti & Ritter, this volume), and benzene-induced leukaemia (Subrahmanyan & O'Brien, in press). The peroxidase activity of prostaglandin H synthetase have been implicated in phenacetin induced kidney necrosis (Andersson *et al.*, 1984) or bladder cancer induced by benzidine and nitrofuran derivatives (Zenser *et al.*, 1979; Wise *et al.*, 1984). A comparison of arylamine activation catalysed by peroxidase with that catalysed by a cytochrome P-450 dependent mixed function oxidase has been reviewed (O'Brien, 1984).

One approach to decreasing the carcinogenicity or toxicity of various chemicals is to substitute various groups on the molecule to prevent the peroxidase catalysed activation or decrease the reactivity of the products with DNA.

Benzidine *in vivo* is readily *N*-acetylated and whilst the *N*-acetyl derivative is readily *N*-hydroxylated by a mixed-function oxidase *in vivo* (Martin *et al.*, 1983), *N*-acetylarylamines are poorly oxidized by peroxidases (Marshall & O'Brien, 1984). However, they are readily oxidized by peroxidases in the presence of microsomes as a result of *N*-deacetylase activity (O'Brien, in press). It is significant that the rat Zymbal gland, an important target tissue for carcinogenic arylamines, is very high in deacetylase activity (Irving, 1979).

Benzidine congener-based azo dyes are widely used in textiles, the leather industry, paper printing, spray paints, etc. Intestinal bacteria readily reduce these dyes *in vivo* resulting in the release of carcinogenic benzidine derivatives (Cerniglia *et al.*, 1982) and DNA-bound metabolites in the liver (Martin *et al.*, 1983). These dyes are therefore a potential hazard. These benzidine derivatives also readily bind to DNA during a peroxidase catalysed oxidation. However, 3,3',5,5'-tetramethylbenzidine was rapidly oxidized but did not bind (Tsuruta *et al.*, 1984). The latter derivative is also non-carcinogenic in short-term tests (Garner *et al.*, 1975). Because of this, azo dyes have recently been prepared from this derivative with the hope of making non-carcinogenic azo dyes available (Josephy & Werrasoonija, 1984).

The amine nitrogen of the benzidine is the most active group on benzidine, so that methyl substitution of this nitrogen would be expected to offer some protection. In the present study, the activation of *N*,*N'*,*N*,*N'*-tetramethylbenzidine (TMB) by a peroxidase catalysed oxidation is investigated.

DNA BINDING STUDIES

Reaction mixtures (3 ml 0.1 M-phosphate buffer, pH 6.5) containing 0.1 mM-TMB, 0.2 mM-H$_2$O$_2$, 20 μg peroxidase and calf thymus DNA (3 mg) were incubated at 20 °C for different times. DNA was isolated as previously described after stopping the reaction with ascorbate (0.2 mM) and extracting with ethyl acetate (Subrahmanyan & O'Brien, in press). DNA binding was estimated from the absorbance maxima at 290 nm with an equal amount of unbound DNA in the reference cell. Little DNA binding occurred in the first 3 minutes but binding increased and reached a maximum at 30 minutes. The DNA was grey in colour. A similar level of DNA binding occurred if the DNA was added at 3 minutes and incubated for 30 minutes. However, no binding occurred if the DNA was added at 30 minutes and the reaction mixture incubated for a further 30 minutes. It was concluded that a product formed from TMB N-demethylation binds to DNA and that the polymeric end product does not bind to DNA. It was estimated that not more than 15% of the TMB was bound to DNA in 30 minutes. By contrast, up to 97% of benzidine binds to DNA in three minutes (Tsuruta *et al.*, 1984).

SPECTRAL ANALYSIS OF THE REACTION MIXTURE

The time sequence for the formation of various intermediates is shown in Fig. 1 for a reaction mixture containing 3 ml 0.2 M-sodium acetate pH 5, 0.1 mM-TMB, 0.2 mM-H_2O_2 and 2 μg peroxidase incubated at 20 °C. A yellow diimine with an absorbance maxima at 462 nm rapidly formed. In Fig. 1, the diimine was followed at 500 nm so as to bring the absorbance on scale. After 4 minutes, the mixture started to turn green with an absorbance maxima at 670 nm attributed to charge-transfer complex formation. At this time, TMB diimine concentration decreased and various benzidines (assayed at 290 nm) formed by TMB N-demethylation appeared. If the reaction mixture was monitored at 670 nm, a charge-transfer complex was also formed in the first 15 seconds as a result of TMB cation radical formation (cf. 3,3′,5,5′-$(CH_3)_4$-benzidine, Josephy *et al.*, 1982). Formaldehyde formation increased linearly with time and was nearly complete by 10 minutes.

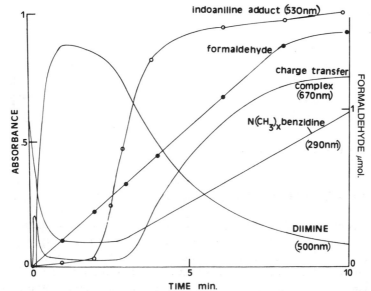

Reaction   conditions: 0.2 M   sodium   acetate,   pH 5,   100 μM [4]N-TMB, 200 μM H_2O_2, 2 μgm HRP, 3 ml, 20°C.

Fig. 1 – Time sequence of TMB oxidation—spectral analysis and formaldehyde determination.

PHENOLIC ADDUCTS

The phenolic antioxidants 3,5-di-*t*-butyl-4OH-benzoic acid or butylated hydroxyanisole were added at different times and the purple phenolic adducts were then extracted and measured at 530 nm. It was found that little phenolic adduct was formed unless the phenol was added after 2 minutes reaction time. Analysis by mass spectrometry of the former phenolic adducts separated by t.l.c. showed a major product mass number of 416 (*M*-2) which corresponds to an '*N,N*-(CH$_3$)$_2$ indoaniline derivative' structure as shown in Fig. 2. Another minor product with a mass number of 402 (*M*-2) was also identified and corresponds to a '*N*-CH$_3$-indoaniline derivative'. Clearly phenolic adducts are formed from *N,N*-(CH$_3$)$_2$-benzidine and *N*-CH$_3$-benzidine produced by TMB *N*-demethylation.

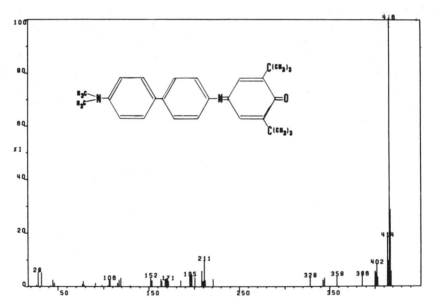

Fig. 2 – Mass spectrum of 3,5-di-*t*-butyl-4OH-benzoic acid adduct with TMB oxidation product.

ANALYSIS OF TMB *N*-DEMETHYLATION PRODUCTS

The concentrated ethyl acetate extracts from the reaction mixture at various reaction times were separated by t.l.c. silica gel chromatography using ethyl acetate: chloroform (2:8), or by h.p.l.c. on a Waters h.p.l.c. system using a C$_{18}$ μBondapak column with a 50–100% methanol gradient (No. 6). The various t.l.c. bands were analysed by high resolution mass spectrometry and u.v. spectroscopy. The R_f for these bands as separated by t.l.c. and the retention times for the peaks separated by h.p.l.c. are shown in Table 1. Exact mass measurements were carried out on the products using a VG Micromas

Table 1
Mass spectral analysis of TMB products

Benzidine derivative	M^+ calculated	M^+ observed	R_f^a	R_t^b
$N,N',N,N'\text{-}(CH_3)_4\text{-}$	240.1626	240.1602	0.54	17.5'
$N,N',N\text{-}(CH_3)_3\text{-}$	226.1470	226.1461	0.44	15'
$N,N'\text{-}(CH_3)_2\text{-}$	212.1313	212.1290	0.26	13.5'
$N,N\text{-}(CH_3)_2\text{-}$	212.1313	212.1306	0.17	13'
$N\text{-}CH_3\text{-}$	198.1157	198.1181	0.14	10'

^{a}Silica gel t.l.c.: solvent ethyl acetate: chloroform (2:8).
^{b}H.p.l.c. C_{18} μBONDAPAK column/50–100% MeOH.

7070MS mass spectrometer at 10,000 resolving power, by peak matching from the appropriate m/e of perfluorokerosene. The determined structure of the various products are shown (Table 1) and it can be seen that there is an excellent agreement between the calculated and observed mass numbers.

In Table 2, it can be seen that the amount of products formed depends on the time of incubation and the number of H_2O_2 equivalents. Thus, at an early time, $N\text{-}(CH_3)_3$-benzidine is the major product, whereas $N,N'\text{-}(CH_3)_2$-benzidine, $N,N\text{-}(CH_3)_2$-benzidine and $N\text{-}CH_3$-benzidine are formed at a later time. With two H_2O_2 equivalents, polymer is the final product.

Table 2
TMB products formed under various reaction conditions

Benzidine derivative	% PRODUCT		
	0.5 mM-H_2O_2		1.0 mM-H_2O_2
	10'	30'	30'
$N,N',N,N'\text{-}(CH_3)_4\text{-}$	40	35	11
$N,N',N\text{-}(CH_3)_3\text{-}$	32	6	7
$N,N'\text{-}(CH_3)_2\text{-}$	11	23	14
$N,N\text{-}(CH_3)_2\text{-}$	12	25	16
$N\text{-}CH_3\text{-}$	5	11	11
Polymer	—	—	41

TMB (0.5 mM), H_2O_2 (0.5 or 1.0 mM), HRP (20 μg) (Type V), 0.2 M-acetate buffer, pH 6.5 were incubated at 20 °C. The reaction was stopped with ascorbate (1.0 mM) and the mixture was extracted with ethyl acetate.

CONCLUSIONS

The mechanism for the activation of TMB to a DNA reactive species by a peroxidase–H$_2$O$_2$ system involves *N*-demethylation. The products *N*,*N*-(CH$_3$)$_2$-benzidine and *N*-CH$_3$-benzidine, after further oxidation by peroxidase–H$_2$O$_2$, can bind to DNA as well as form phenol adducts. A mechanism for TMB *N*-demethylation is presented in Fig. 3. The first formed cation radical, in equilibrium with a charge-transfer complex, can be further oxidized to a diimine or can disproportionate to an iminium cation. Hydrolysis of the latter leads to loss of a methyl group as formaldehyde and the *N*,*N'*,*N*-(CH$_3$)$_3$-benzidine formed commences a new cycle of *N*-demethylation. The diimine may also undergo hydrolysis leading to formaldehyde formation.

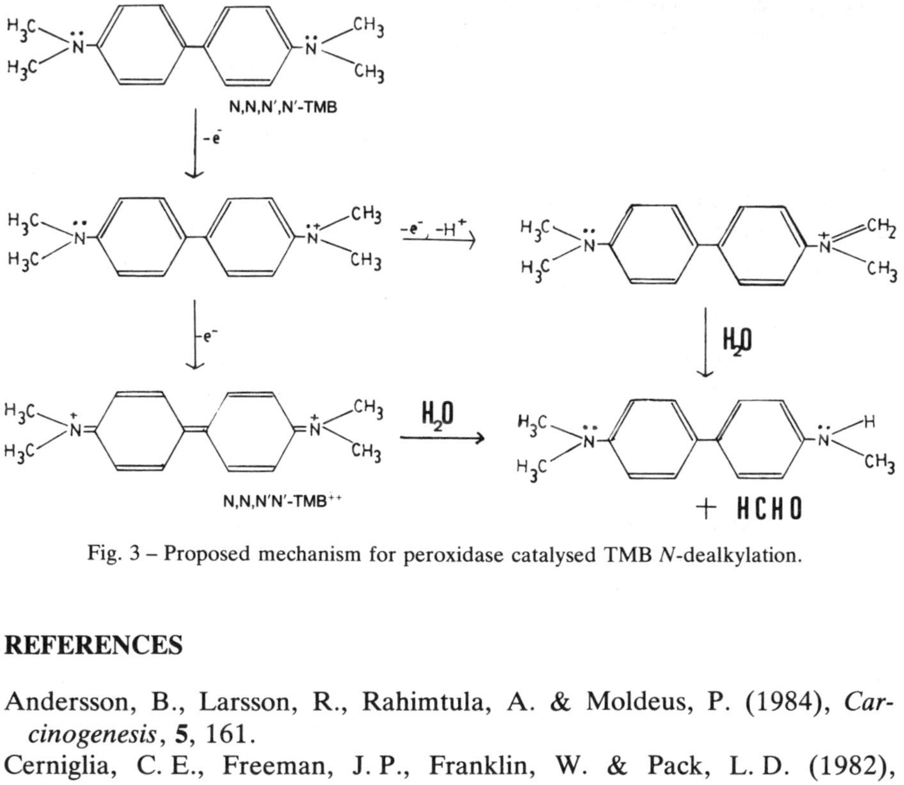

Fig. 3 – Proposed mechanism for peroxidase catalysed TMB *N*-dealkylation.

REFERENCES

Andersson, B., Larsson, R., Rahimtula, A. & Moldeus, P. (1984), *Carcinogenesis*, **5**, 161.

Cerniglia, C. E., Freeman, J. P., Franklin, W. & Pack, L. D. (1982), *Biochem. Biophys. Res. Comm.*, **107**, 1224.

Garner, R. C., Walpole, A. L. & Rose, F. L. (1975), *Cancer Lett.*, **1**, 39.

Irving, C. C. (1979). in *Carcinogenesis: Identification and Mechanism of Action* (eds. Griffen, A. C. & Shaw, C. R.) Raven Press, New York, p. 211.

Josephy, P. D., Eling, T. E. & Mason, R. P. (1982), *J. Biol. Chem.*, **257**, 3669.

Josephy, P. D. & Weersooriya, M. (1984), *Chem-biol. Interact.*, **49**, 375.

Malejka-Giganti, D. & Ritter, C. L., *this volume*.

Marshall, W. & O'Brien, P. J. (1984), in *Icosanoids and Cancer* (ed. Thaler, H.) Raven Press, New York.

Martin, C. N., Beland, F. A., Kennelly, J. C. & Kadlubar, F. F. (1983), *Environ. Health Perspect.*, **49**, 101.

Metzler, M. & McLachlan, J. A. (1978), *Biochem. Biophys. Res. Comm.*, **85**, 874.

O'Brien, P. J. (1984), in *Free Radicals in Biology*, Vol. VI (ed. Pryor, W.) Academic Press, New York, p. 289.

O'Brien, P. J., *J. Amer. Oil Chem. Soc.*, in press.

Osborne, J. C., Metzler, M. & Neumann, H. G. (1980), *Cancer Lett.*, **8**, 221.

Subrahmanyan, V. G. & O'Brien, P. J., *Xenobiotica*, in press.

Tsuruta, Y., Josephy, P. D., Rahimtula, A. D. & O'Brien, P. J. (1984), Proc. Amer. Assoc. Cancer Res., Abstr. 350.

Wise, R. W., Zenser, T. V. & Davis, B. B. (1984), *Cancer Res.*, **44**, 1893.

Zenser, T. V., Mattammal, M. B. & Davis, B. B. (1979), *J. Pharmacol. Therap.*, **211**, 460.

Chapter 38

PEROXIDASE *VERSUS* CYTOCHROME P-450 CATALYSED *N*-DEMETHYLATION OF *N,N*-DIMETHYLANILINE

P. J. O'Brien, **S. Forbes** and **D. Slaughter**, Department of Biochemistry, Memorial University of Newfoundland, St. John's, Newfoundland, Canada, A1B 3X9

1. N,N',N,N'-$(CH_3)_4$-benzidine (TMB) can be an intermediate in the N-demethylation of N,N-$(CH_3)_2$-aniline (DMA) by a peroxidase-H_2O_2 system, but not with a cytochrome P-450-cumene hydroperoxide system. At very low peroxidase concentrations, V_o TMB is proportional to the square of the peroxidase concentration, but is first order at higher enzyme concentrations. These results indicate that the dimerizing species is activated by peroxidase binding.

2. The N-demethylation of TMB and DMA by peroxidase-H_2O_2 is readily inhibited by a wide range of hydrogen donors in contrast to that with cytochrome P-450–cumene hydroperoxide. Clearly the latter system involves a different mechanism from that with peroxidase–H_2O_2.

INTRODUCTION

The cytochrome P-450 catalysed N-dealkylation of N,N-dimethylaniline (DMA) supported by cumene hydroperoxide involves a stoichiometric conversion to HCHO and N-methylaniline (Nordblom & Coon, 1974). Recently, a horseradish peroxidase (B–C) isozyme at a low concentration (0.15 μg/ml) was found to catalyse a similar stoichiometric conversion with H_2O_2 (Kedderis & Hollenberg, 1983). In the present investigation, using more sensitive techniques and purified horseradish peroxidase (Sigma Type V, A403/A280 = 3.2), different results were obtained.

EXPERIMENTAL

The products of DMA peroxidation were determined as follows: The reaction mixtures containing 3 ml 0.4 M-phosphate buffer pH 5.5, 3 μmole H_2O_2, 3 μmole DMA and peroxidase (0.01–20 μg) was terminated with dithionite (0.1 ml, 0.1 M) and extracted with 2 ml ethyl acetate. Silica gel t.l.c. of the extract was developed with hexane:ethyl acetate (88:12). It was found that N,N',N,N'-$(CH_3)_4$-benzidine (TMB) (R_f 0.27) was the major product after a 15-second reaction time with 20 μg peroxidase (HRP) and that all of the DMA was oxidized. E.p.r. spectral evidence confirms this (Griffin *et al.*, 1981). Other minor bands with a lower R_f were identified as N-dealkylation products of TMB (O'Brien & Gregory, 1985). However, with 0.01 μg HRP incubated for 1 hour, although N-methylaniline (R_f 0.32) was a major product, only 0.5% of the DMA (R_f 0.54) had been oxidized and extensive HRP inactivation occurred. N-Demethylation was monitored by formaldehyde production (Nash, 1953). Reaction mixtures containing 3 μmole H_2O_2 and 20 μg HRP incubated for 5 minutes produced 1.43 μmole HCHO from 3 μmole DMA whereas 2.78 μmole HCHO was formed from 1.5 μmole TMB. The above suggests that HCHO formation from DMA may be due to TMB formation.

A new highly sensitive method for determining the kinetics of TMB formation was developed. An ethyl acetate extract of the reaction mixture was dried under vacuum to remove DMA and N-methylaniline, redissolved in ethyl acetate and monitored for fluorescence (λ_{ex} = 316 nm, λ_{em} = 388 nm). A 20 nM solution of TMB gave a fluorescence 10% above background. Using this method, it was shown that TMB was also formed by DMA peroxidation catalysed by 0.01 μg HRP even though no TMB could be detected by t.l.c. or h.p.l.c.

RESULTS AND DISCUSSION

A graph of initial enzyme velocity with respect to TMB formation (V_o TMB) from DMA versus HRP concentration is convex at low concentrations of HRP but approaches linearity at higher enzyme concentrations (Fig. 1). The rate of TMB diimine formation (at 462 nm) from TMB is linear with respect to HRP concentration but a convex curve was obtained for TMB diimine formation from DMA. At very low HRP concentrations, there is a second order relationship between V_o TMB and enzyme concentration and V_o TMB is proportional to (HRP concentration)^2 (Fig. 1). At higher enzyme concent-

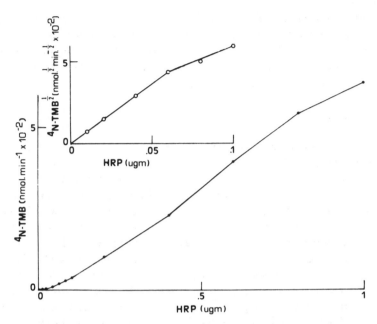

Fig. 1 – TMB formation (V_o/TMB) from DMA at low concentrations of HRP (inset) and at higher enzyme concentrations. Reaction mixtures contained 3 nmole DMA and 3 nmole H_2O_2.

rations, V_o TMB becomes first order. The time curve for TMB production is initially linear, even at low HRP concentrations. The initial rate of formaldehyde production (V_o HCHO) is first order with respect to HRP concentration (Kedderis & Hollenberg, 1983).

The following mechanism could explain this:

$$2 \text{ DMA} + 2 \text{ HRP} \underset{k_2}{\overset{k_1}{\rightleftharpoons}} 2 \text{ DMA} : \text{HRP} \xrightarrow{k_3} \text{TMB} + 2 \text{ HRP}$$

where (HRP) is the free enzyme concentration and (DMA.HRP) is the enzyme-substrate complex such that total enzyme $(\text{HRP}_t) = (\text{HRP}) + (\text{DMA.HRP})$. At very low (HRP_t) virtually all of the enzyme will be in the form of DMA.HRP so that

$$V_o \text{ TMB} = k_3 (\text{HRP}_t)^2,$$

giving the observed second order relationship.

Since DMA.HRP is maintained at a steady state concentration during the period of linearity between TMB production and time,

$$k_1(\text{HRP})(\text{DMA}) = k_2(\text{DMA.HRP}) + k_3(\text{DMA.HRP})^2.$$

At high enzyme concentrations, $(\text{HRP}) \rightarrow (\text{HRP}_t)$ and $k_2(\text{DMA.HRP})$ becomes negligible with respect to $k_3(\text{DMA.HRP})^2$ so that V_o TMB versus (HRP_t) becomes first order with enzyme concentration, consistent with the observed pattern (Fig. 1).

Table 1

Effect of inhibitors of N-demethylation

| INHIBITOR | DMA + ROOH/P-450 | % rate of formaldehyde formation | |
		DMA + H_2O_2/HRP	TMB + H_2O_2/HRP
None	100[a]	100[b]	100[c]
GSH (0.1 mM)	95	11	5
NADH (0.1 mM)	87	9	8
Ascorbate (0.1 mM)	97	21	24
Butylhydroxyanisole (0.1 mM)	78	7	2
2,3,6-$(CH_3)_3$-phenol (0.1 mM)	82	16	11
2,6-$(CH_3)_2$-phenol (0.1 mM)	92	13	15
SKF-525A (1 mM)	2	83	87

[a]88 nmol min^{-1} nmol P-450^{-1} [b]502 nmol min^{-1} nmol HRP^{-1} [c]756 nmol min^{-1} nmol HRP^{-1}

Reaction conditions: -[a]Sodium potassium phosphate buffer (0.4 M), pH 6.0, N,N-dimethylaniline (3.0 mM), rat liver microsomes (1 mg protein/ml), cumene hydroperoxide (0.2 mM) were incubated at 37 °C for 60 secs. [b,c]As a, but HRP (0.5 μg), DMA or TMB (0.1 mM), H_2O_2 (0.1 mM).

The two most relevant features of the proposed mechanism are (1) the DMA $p-p$ dimerization to TMB involves an interaction between two 'activated' substrate molecules and (2) that the dimerising species are enzyme bound. Models that allow dimerization to occur between one 'activated' and one unaltered substrate molecule or models that allow the dimerizing species to exist unbound to enzyme, predict kinetic relationships that are at variance with the observed data.

As shown in Table 1, the hydrogen donors glutathione, NADH, ascorbate and phenolic antioxidants markedly inhibit DMA N-demethylation by HRP but not by cytochrome P-450. A similar inhibition was seen with HRP catalysed TMB demethylation. In the absence of DMA or TMB, little donor oxidation occurred so that the donors were not acting as competitive inhibitors. The drug SKF 525A markedly inhibited the cytochrome P-450 reaction only.

Peroxidase and cytochrome P-450 therefore catalyse N-demethylation by different mechanisms. The peroxidase mechanism may involve dimerization of DMA cation radicals resulting in TMB formation. Subsequently, the TMB N-demethylates via disproportionation of its cation radical to an iminium ion. The hydrogen donors presumably inhibit by reducing the DMA and TMB cation radicals. At very low HRP concentration, some DMA N-demethylation to N-CH_3-aniline occurs. It has been proposed that cytochrome P-450 differs from HRP in that it catalyses an α-carbon deprotonation of DMA to an unstable carbinolamine (Kedderis & Hollenberg, 1983). However, their suggestion that a cation radical is the first formed intermediate is hard to reconcile with the lack of effect of hydrogen donors. These differences were also found for aminopyrine N-demethylation (Moldeus *et al.*, 1983) and the steady state free radical concentration was linearly related to the square root of peroxidase concentration suggesting that the N-demethylation involves the disproportionation of two radicals (Lasker *et al.*, 1981; O'Brien, 1984).

REFERENCES

Griffin, B. W., Davis, D. K. & Bruno, G. V. (1981). *Bioorg. Chem.*, **10**, 342.

Kedderis, G. L. & Hollenberg, P. F. (1983), *J. Biol. Chem.*, **258**, 8129.

Lasker, J. M., Sivarajah, K., Mason, R. P., Kalyanaraman, B., AbouDonia, M. B. & Eling, T. E. (1981), *J. Biol. Chem.*, **256**, 7764.

Moldeus, P., O'Brien, P. J., Thor, H., Berggren, M. & Orrenius, S. (1983), *FEBS Lett.*, **162**, 411.

Muira, G. T., Walsh, J. S., Kedderis, G. L. & Hollenberg, P. F. (1983), *J. Biol. Chem.*, **258**, 1445.

Nash, T. (1953), *Biochem. J.*, **55**, 416.

Nordblom, G. D., White, R. E. and Coon, M. J. (1974), *Arch. Biochem. Biophys.*, **175**, 524.

O'Brien, P. J. (1984), in *Free Radicals in Biology* (ed. Pryor, W.), Vol. VI,
    Academic Press, New York, p. 289.
O Brien, P. J. & Gregory, B., this volume.

Chapter 39

ON THE CHEMISTRY OF THE PEROXIDATIVE OXIDATION OF p-PHENETIDINE

T. Lindquist, **S.-E. Hillver**, **B. Lindeke**, **P. Moldéus**,[a] **D. Ross**[a] and **R. Larsson**[a],
Department of Organic Pharmaceutical Chemistry, Biomedical Centre, University of Uppsala, Box 574, S-751 23 Uppsala Sweden, and [a]Department of Forensic Medicine, Karolinska Institute, S-104 01 Stockholm, Sweden

1. The oxidation of p-phenetidine was investigated with prostaglandin synthetase (PGS), horseradish peroxidase (HRP) and a chemical system consisting of hydrogen peroxide–sodium tungstate.
2. Oxidation products identified were: 4-ethoxynitrosobenzene (**1**), 4-ethoxynitrobenzene (**2**), 4,4'-diethoxyazobenzene (**3**), 4-(4-ethoxyphenylimino)-2,5-cyclohexadien-1-one (**4**), 4,4'-diethoxyazoxybenzene (**5**), 4-ethoxy-4'-nitrosodiphenylamine (**6**) and 3,6-bis(4-ethoxyphenylimino)-4-ethoxy-1,4-cyclohexadienylamine (**7**).
3. Products **3**, **5** and **7** were isolated from the enzymic as well as the chemical system, whilst **1**, **2** and **6** have so far been recovered from the chemical and **4** from the enzymic system only.
4. The oxidations catalysed by PGS and HRP differed from that catalysed by tungstate. Few N-oxygenated compounds were recovered from the former systems while such compounds were abundant in the latter.
5. The products formed in the reactions catalysed by PGS and HRP are in agreement with those seen in anodic oxidation of anilines while it appears as if oxidation mechanisms catalysed by tungstate and monooxygenases are similar.

INTRODUCTION

p-Phenetidine, a primary metabolite of phenacetin, has been shown to be
metabolized to genotoxic as well as protein-binding products in peroxidase-
type reactions catalysed by enzymes such as prostaglandin synthetase (PGS)
and horseradish peroxidase (HRP) (Andersson et al., 1982; Andersson et al.,
1983; Andersson et al., 1984). Similar to peroxidase-catalysed oxidations of
certain other aromatic amines, e.g. benzidine, phenylenediamine and
aminophenol (O'Brien, 1984) the formation of substrate-free radicals seems
to be implied (Ross et al., 1985), and they could be responsible for the
genotoxic properties (Larsson et al., 1984). In addition, free radical oxidation of aromatic amines is bound to give rise to a myriad of more or less
reactive coupling products amongst which are potential protein binding
species (Larsson et al., 1984).

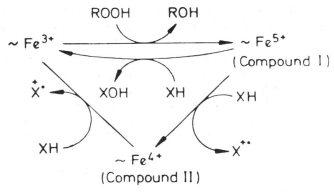

Fig. 1 – Proposed mechanisms of peroxidase-catalysed oxidations.

One requirement for a peroxidase to generate substrate-free radicals is that
the oxidized enzyme (known as compound I, Fig. 1) undergoes reduction by
two consecutive univalent steps, each of which is capable of generating the
radical (Metzler, 1977). The stoichiometry of the oxidation of p-phenetidine
is consistent with such a mechanism (Ross et al., 1985). Moreover, from the
pattern of oxidation products formed (Ross et al., 1985; Larsson et al.,
1984) it can be seen that the peroxidase-catalysed oxidation of p-phenetidine
in essence mimics its anodic oxidation (Nelson, 1974), the latter by definition
consisting of a stepwise removal of electrons.

The present contribution summarizes some recent findings with regard to
the chemistry of the peroxidative oxidation of p-phenetidine with PGS, HRP
and a chemical system containing hydrogen peroxide–sodium tungstate.

EXPERIMENTAL

Detailed experimental procedures are described in Ross et al. (1985). Microsomes from ram seminal vesicles (RSV) were isolated as described by Egan

et al. (1976). Incubations with RSV microsomes or HRP were performed in 0.1 M-phosphate buffer pH 8.0, supplemented with 1.0 mM-EDTA. Two-phrase oxidation with Na_2WO_4 in H_2O–CH_2Cl_2 was performed essentially as described by Levin & Skarogobatova (1976). The oxidation products were extracted into organic solvents, concentrated and separated on t.l.c.-silica plates generally using a mobile phase of chloroform–methanol 19:1. Product elucidation of the separated products retrieved from the t.l.c.-plates was done by mass spectrometric and nuclear magnetic resonance spectrometric analyses.

RESULTS

Oxidation products so far identified in the actual oxidation systems are summarized in Fig. 2. They are 4-ethoxynitrosobenzene (**1**), 4-ethoxynitrobenzene (**2**), 4,4'-diethoxyazobenzene (**3**), 4-(4-ethoxyphenylimino)-2,5-cyclohexadien-1-one (**4**), 4,4'-diethoxyazoxybenzene (**5**), 4-ethoxy-4'-nitrosodiphenylamine (**6**) and 3,6-bis(4-ethoxyphenylimino)-4-ethoxy-1,4-cyclohexadienylamine (**7**). Of these **3**, **5**, and **7** were isolated from the enzymic as well as the chemical system, whilst **1**, **2**, and **6** have so far been recovered from the chemical and **4** from the enzymic systems only.

Compounds **4** and **7**, which are orange and red respectively, are both decolorized on reduction by $S_2O_3^{2-}$, NADPH or reaction with GSH (Ross *et al.*, 1985; Larsson *et al.*, 1984). Experiments performed with ^{14}C-ring labelled

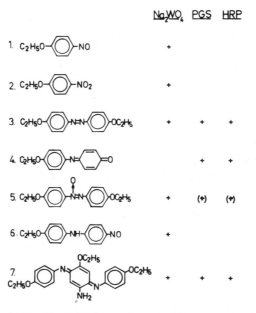

Fig. 2 – Structures identified after Na_2WO_4, PGS and HRP catalysed peroxidation of *p*-phenetidine.

p-phenetidine showed that, of the compounds identified in the enzymic systems, only **4** is covalently bound to protein (Larsson *et al.*, 1984). Compound **6**, formed in the tungstate–H_2O_2 system, oxidized GSH but did not seem to form conjugates with thiols.

In the enzymic systems the formation of **3**, **4** and **7** gradually increased with time, compound **4** being the most important, accounting for more than 50% of recovered radioactivity (Ross *et al.*, 1985). Also, incubation of [ethoxy-^{14}C]-*p*-phenetidine yielded **4** concomitant with the formation of one equivalent of [^{14}C]ethanol (Larsson *et al.*, 1984).

DISCUSSION

The products isolated in the various oxidation systems discussed in this study indicate that nitrogen oxygenation as well as nitrogen oxidation and coupling reactions occur (cf. Lindeke, this volume). However, the oxidations catalysed by PGS and HRP differ from that catalysed by tungstate inasmuch as none or few *N*-oxygenated products were recovered from the former systems while such compounds were abundant in the latter. Although the detailed mechanism of neither peroxidase (Metzler, 1977) nor tungstate-catalysed peroxidation (Ogata *et al.*, 1960) is as yet fully understood, the former is considered to involve two univalent reduction steps, (cf. Fig. 1) while the latter probably involves a direct two-electron reduction in the rate-limiting step:

$$R_3N: + T\text{–}OOH \longrightarrow R_3\overset{+}{N}\text{–}OH + T\text{–}O^- \qquad (1)$$

Thus, as an 'oxenoid' form of oxygen is transferred, the latter comes to resemble the monooxygenase reaction catalysed by cytochrome P-450.

Contrary to the monooxygenase type systems', the products formed in the reaction catalysed by PGS and HRP are in full agreement with those seen in anodic oxidation of anilines (Nelson, 1974; Bacon & Adams, 1968). Such oxidations are characterized by the initial formation of radicals which then undergo coupling reactions, directly or after delocalization of the unpaired electron away from the amine nitrogen. With *para*-substituted anilines the major products formed in aqueous solution tend to be dimers resulting from head to tail coupling (Fig. 3, and Fig. 2, compounds **4** and **6**). In these reactions coupling of two aniline radicals is considered to be favoured, but radical

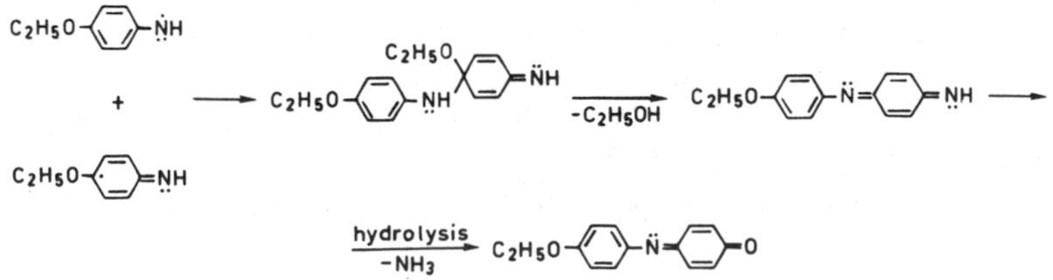

Fig. 3 – Proposed mechanism for the formation of 4-(4-ethoxyphenylimino)-2,5-cyclohexadien-1-one (**4**) from two *p*-phenetidine radicals.

attack on the parent amine cannot be excluded. Also, in the presence of base anodic oxidation in non-aqueous media preferentially results in head-to-head coupling (cf. Fig. 2, compounds **3** and **5**) (Wawzonek & McIntyre, 1967).

From, the results just discussed it is clear that oxidation mechanisms catalysed by tungstate and monooxygenases are similar, but these oxidations are different to those of anodic or peroxidative oxidation. Thus, metabolites formed from monooxygenases (i.e. cytochrome P-450) would differ from those generated by peroxidases (i.e. PGS), resulting in different toxicological consequences from either P-450 or peroxidative metabolism of p-phenetidine *in vivo*.

ACKNOWLEDGEMENTS

The work discussed in this contribution was supported by the Swedish Medical Research Council (grants Nos. 5645-05A and 5918-02A).

REFERENCES

Andersson, B., Nordenskjöld, M., Rahimtula, A. & Moldéus, P. (1982), *Mol. Pharmacol*, **22**, 479.

Andersson, B., Larsson, R., Rahimtula, A. & Moldéus, P. (1983), *Biochem. Pharmacol.*, **32**, 1045.

Andersson, B., Larsson, R., Rahimtula, A. & Moldéus, P. (1984), *Carcinogenesis*, **5**, 161.

Bacon, J. & Adams, R. N. (1968), *J. Am. Chem. Soc.*, **90**, 6596.

Egan, R. W., Paxton, J. & Kuehl, Jr., F. A. (1976), *J. Biol. Chem.*, **251**, 7329.

Larsson, R., Ross, D., Nordenskjöld, M., Lindeke, B., Olsson, L.-I. & Moldéus, P. (1984), *Chem.-Biol. Interact.*, **52**, 1.

Levin, A. & Skorabogatova, M. S. (1976), *Bull. Acad. Sci. USSR, Chem. Soc.*, **25**, 465.

Metzler, D. E. (1977), *Biochemistry, The Chemical Reactions of Living Cells.* Academic Press, New York, p. 571.

Nelson, R. F. (1974), in *Techniques of Chemistry*, Vol. V, Part I, (ed. Weinberg, N. L.) John Wiley & Sons, New York, p. 535.

O'Brien, P. J. (1984), in *Free Radicals in Biology*, Vol. VI, (ed. Pryor, W. A.) Academic Press, New York, p. 289.

Ogata, Y., Tomizawa, K. & Maeda, H. (1960), *Bull. Chem. Soc. Japan*, **53**, 285.

Ross, D., Larsson, R., Andersson, B., Nilsson, U., Lindquist, T., Lindeke, B. & Moldéus, P. (1985), *Biochem. Pharmacol.*, **34**, 343.

Wawzonek, S. & McIntyre, T. W. (1967), *J. Electrochem. Soc.*, **114**, 1025.

Chapter 40

PEROXIDASE/H$_2$O$_2$-CATALYSED OXIDATIONS OF N-SUBSTITUTED ARYL COMPOUNDS

Danuta Malejka-Giganti and **Clare L. Ritter**, Department of Laboratory Medicine and Pathology, University of Minnesota, and Veterans Administration Medical Centre, 54th Street and 48th Avenue South, Minneapolis, MN 55417, USA

1. Extracts of 105,000 g pellets of rat uterine (UT) and mammary gland (MG) homogenates in 0.01 M-Tris-HCl buffer, pH 7.2, with 0.5 M-CaCl$_2$ contained a haemprotein spectrally similar to that of lactoperoxidase and utilized guaiacol, 17β-estradiol and N-arylamines as hydrogen donors. The peroxidative activities of UT extracts were up to 20 times greater than those of MG extracts.

2. Oxidations of N-hydroxy-N-2-fluorenylacetamide (N-OH-2-FAA) (via nitroxyl free radical) to 2-nitrosofluorene (2-NOF) and N-acetoxy-2-FAA by horseradish peroxidase/ or lactoperoxidase/ or tissue extracts/H$_2$O$_2$ were markedly decreased in the buffer above suggesting chelation of N-OH-2-FAA by Ca^{2+}.

3. Extracts of UT or MG in 0.1 M-sodium phosphate buffer, pH 7.4, with 0.02% cetyltrimethylammonium bromide (Cetab) oxidized N-OH-2-FAA, but 2-NOF was the chief or the only product of the reaction, respectively.

4. The data show that peroxidative activities of UT and MG participate in oxidation of N-arylamines and N-OH-2-FAA and hence, may be of significance in tumorigenesis.

ONE ELECTRON OXIDATION OF N-FLUORENYLACETOHYDROXAMIC ACIDS—A POSSIBLE MECHANISM IN TUMORIGENESIS

Numerous aromatic amines, amides and hydroxamic acids are carcinogenic for the rat upon systemic administration (Clayson & Garner, 1976). Since the sites of tumour induction are distant from the site of application of the compounds, it has been postulated that tumorigenesis requires their metabolic activation (Miller, 1970). We found, however, that N-fluorenylacetohydroxamic acids, which were carcinogenic for the mammary gland when given systemically to the rat (Gutmann et al., 1970), were also carcinogenic upon direct application to the mammary gland (Malejka-Giganti et al., 1977). This contrasted with the marginal local tumorigenicity of their corresponding N-arylamides. Our results were recently confirmed by Allaben et al. (1982) who showed local mammary tumorigenicity of a series of N-2-substituted fluorenylhydroxamic acids. Since the hydroxamic acids are not reactive in vitro with cellular macromolecules presumed to be of critical importance in initiation of tumorigenesis, their metabolic conversion to reactive species, such as electrophiles, capable of covalent interaction with nucleic acids is necessary (Miller & Miller, 1976).

One sequence proposed for activation of the carcinogenic N-fluorenylacetohydroxamic acids is one-electron oxidation to nitroxyl free radicals followed by dismutation to nitrosofluorenes and the acetate esters (Bartsch et al., 1972). The nitroso compounds could interact with the $-SH$ groups on proteins to yield fluorenamine adducts (Barry et al., 1969) and/or with double bonds of lipids in an Alder-ene reaction to form a hydroxylamine derivative which undergoes oxidation to yield a nitroxyl free radical–lipid adduct (Sullivan, 1966; Floyd et al., 1978). The carcinogen–lipid adduct may cause changes in membrane structures. The acetate esters could react with DNA or other macromolecules (e.g. ribonuclease), and thus alter the transcription process. It is possible that both carcinogen–lipid and carcinogen–DNA interactions play a role in mammary tumorigenesis.

One electron oxidation of N-hydroxy-N-2-fluorenylacetamide (N-OH-2-FAA) and other hydroxamic acids has been shown to occur with various peroxidase/peroxide systems, and is particularly efficient with horseradish peroxidase (HRP)/H$_2$O$_2$ (Bartsch & Hecker, 1971; Floyd et al., 1976). More recently, a nitroxyl free radical was generated from N-OH-2-FAA by hepatic microsomes of the rat in the presence of peroxide, and its formation was ascribed to peroxidative activity of cytochrome P-450 or P-420 (Stier et al., 1980; Reigh & Floyd, 1981). Since we found only minute amounts of cytochrome P-450 in the rat mammary gland (Ritter & Malejka-Giganti, 1982), we felt that other haemproteins, such as cytochrome c, which is present in this tissue in relatively large amounts, may catalyse oxidation of the hydroxamic acids. Indeed, we detected electron spin resonance (e.s.r.) signals of nitroxyl free radicals of both isomeric hydroxamic acids, N-OH-2-FAA and N-OH-3-FAA, in the systems containing cytochrome c/H$_2$O$_2$ in 0.1 M-sodium acetate, pH 6.0, or in 0.05 M-phosphate buffer, pH 7.5, after incubation with Triton X-100 (Ritter *et al.*, 1983). The spectral characteristics of these signals were indistinguishable from those generated by HRP/H_2O_2, but the kinetics of their formation and decay were different. The dismutation products, 2- and 3-nitrosofluorene (2- and 3-NOF), reacted with membrane lipid, such as lecithin, to yield signals of the immobilized nitroxyl free radicals. The other dismutation products were the acetate esters, N-AcO-2-FAA and N-AcO-3-FAA. In addition, relatively large amounts of 2-FAA and 3-FAA were formed which suggested reduction of the nitroxyl free radicals to the amides by the reduced cytochrome *c*.

PEROXIDATIVE ACTIVITIES OF RAT UTERUS AND MAMMARY GLAND

To show whether one electron oxidation of N-OH-2-FAA occurs with rat mammary gland *in vitro*, it was necessary to obtain tissue preparations with peroxidative activity. Since the preparation of active peroxidase extracts from rat uterus had been described (Lyttle & DeSombre, 1977), we used this method with mammary gland and compared the characteristics and peroxidative activities of extracts from both tissues. Mammary glands and uteri were obtained from 70-day old Sprague–Dawley rats. The 105,000 **g** pellets of whole homogenates were washed with 1 mM-Tris-HCl buffer, pH 7.2, to remove haemoglobin (Matsubara *et al.*, 1974), and then extracted with 10 mM-Tris-HCl buffer containing 0.5 M-$CaCl_2$. The extracts contained a haemprotein with spectral characteristics resembling those of lactoperoxidase (LP) (Fig. 1). Uterine extracts contained twice as much protein as did mammary gland extracts per rat (Table 1). The mammary gland and uterine extracts utilized guaiacol (GU), *p*-phenylenediamine (PDA), 3,5,3',5'-tetramethylbenzidine (TMBD), and 17β-estradiol (ES) as hydrogen donors. The peroxidative activities of uterine extracts were about 20 times greater than those of mammary gland extracts with GU and ES, and about 12 times greater with the amines. These results indicated that $CaCl_2$ extracts of mammary gland contain haemprotein and have peroxidative activity, although much less than uterine extracts and probably with differing hydrogen donor specificities.

To examine peroxidative activities of mammary gland and uterine extracts with N-OH-2-FAA as a hydrogen donor, we used two h.p.l.c. systems for separation and quantitation of the oxidation products of N-OH-2-FAA. In system 1 (Wong *et al.*, 1982), we separated 2-FAA from N-AcO-2-FAA and 2-NOF; we monitored the first two compounds at 280 nm and 2-NOF at 360 nm. In system 2 (Ryzewski & Malejka-Giganti, 1982), we determined the unreacted substrate, N-OH-2-FAA, and confirmed the presence of 2-FAA and N-AcO-2-FAA. With the use of a diode array LC detector, we identified spectrally the three reaction products. As shown in our earlier work (Ritter *et al.*, 1983) and also here, HRP/H_2O_2 was particularly efficient in oxidizing N-OH-2-FAA in phosphate buffer, pH 7.4 (Table 2). Equimo-

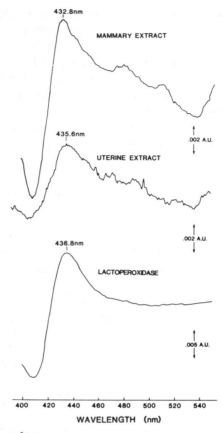

Fig. 1 – Fe^{3+}-CN vs Fe^{3+} Difference spectra of haemproteins. CaCl$_2$ extracts of rat mammary gland and uterus or bovine milk lactoperoxidase (Sigma) were treated with 1 mM-KCN and difference spectra recorded.

Table 1
Peroxidative activities of uterine (UT) and mammary gland (MG) extracts

Tissue	Protein of the extract mg/rat	Hydrogen donor:*			
		GU	PDA	TMBD	ES
UT	4.07**	0.324	0.194	0.058	0.021
MG	1.85	0.015	0.016	0.0046	0.0010
UT/MG	2.2	22	12	13	21

*GU, guaiacol; PDA, p-phenylenediamine; TMBD, 3,5,3′,5′-tetramethylbenzidine; ES, 17β-estradiol. Values are units/mg protein. Unit is the amount of extract giving Δ OD of 60/min (GU, Wagai & Hosoya, 1982) or μmole of product/min (PDA, De Wolf *et al.*, 1978; TMBD, Josephy *et al.*, 1982) or 100% of radioactivity bound to albumin (ES, Jellinck & Perry, 1967).
**All values are the means from three experiments.

lar amounts of N-AcO-2-FAA and 2-NOF were formed as well as small amounts of 2-FAA. However, when we used Tris-HCl buffer with 0.5 M-CaCl$_2$, i.e. the extraction medium for tissue peroxidative activity, the amounts of N-AcO-2-FAA and 2-NOF decreased markedly, by 83 and 73%, respectively, indicating that in the medium containing CaCl$_2$ the peroxidative utilization of N-OH-2-FAA by HRP/H$_2$O$_2$ is reduced. Addition of EDTA to the medium restored utilization by HRP/H$_2$O$_2$ completely based on the recovery of 2-NOF. However, the amounts of N-AcO-2-FAA were no longer equimolar to those of 2-NOF. LP was active toward N-OH-2-FAA with a 1000-times lower concentration of H$_2$O$_2$ than HRP. Again, the highest activity was measured in phosphate buffer. In buffer containing CaCl$_2$, 60% less of N-AcO-2-FAA and 2-NOF were formed. In contrast to HRP/H$_2$O$_2$ system, addition of EDTA did not restore product formation and appeared to inactivate LP. These data suggested that CaCl$_2$ chelated N-OH-2-FAA making it unavailable for oxidation. Chelation of CaCl$_2$ by EDTA prevented this with HRP, but not with LP. When relatively large amounts of uterine extract, in terms of guaiacol units per reaction was used, peroxidative activity toward N-OH-2-FAA could be shown in the buffer containing CaCl$_2$ (Table 3). Thus, small amounts of N-AcO-2-FAA and 2-NOF were found in addition to the relatively large amounts of 2-FAA. However, based on the added substrate, the recovery was only 75% in this system. These data indicated that the use of CaCl$_2$ for solubilization of tissue peroxidase(s) did not create favorable conditions for oxidation of the hydroxamic acid. Since Wagai & Hosoya (1982) used a cationic detergent, cetyltrimethylammonium bromide (Cetab), for solubilization of rat uterine peroxidase, we prepared mammary gland and uterine extracts in phosphate buffer containing this detergent. The 4% Cetab used for extraction was reduced by dialysis to 0.02% in the final extract. With these extracts, we showed peroxidative activity toward N-OH-2-FAA with 30 to 50 times fewer guaiacol units in the reaction mixture (Table 3). Optimal H$_2$O$_2$ for each extract was used. Uterine extract yielded both N-AcO-2-FAA and 2-NOF which suggested nitroxyl free radical of N-OH-2-FAA as their precursor. However, the ratio of the two dismutation products was 1:12. With mammary gland extracts, we found only 2-NOF and no N-AcO-2-FAA. We have no explanation for the excessive amounts of 2-NOF, but we are continuing the investigation. The reactivity of the ester with enzyme protein and/or decomposition of the compound in the buffer did not seem to account for this since we recovered nearly 100% of the unchanged N-AcO-2-FAA from this system.

The evidence that the formation of 2-NOF was accompanied in the uterine extract by N-AcO-2-FAA, was H$_2$O$_2$-dependent and was completely inhibited by KCN (data not shown) strongly suggested that 2-NOF arose, at least in part, from peroxidase-mediated oxidation of N-OH-2-FAA via a nitroxyl free radical mechanism. This activation may explain neoplastic changes in rat uterus resulting from i.p. administration of N-OH-2-FAA (W. T. Allaben, personal communication). Oxidation of N-OH-2-FAA by the rat mammary gland requires further study especially since Reigh et al. (1978) reported

Table 2
Peroxidase/H_2O_2-catalysed oxidation of N-OH-2-FAA

Enzyme system[*]	Guaiacol units per reaction	Buffer[**]	Products (nmoles)				Substrate (nmoles)	Total recovery (%)
			2-FAA	N-AcO-2-FAA	2-NOF			
HRP/H_2O_2 [0.9 mM]	0.008	A	0.99	6.17	6.48		54.8	109
		B	0.75	1.07	1.94		56.5	95
		C	0.69	2.48	7.17		57.2	107
LP/H_2O_2 [0.9 μM]	0.28	A	1.58	7.75	4.29		47.4	97
		B	1.34	3.13	1.71		57.1	100
		C	0	0	0		61.7	98

[*]HRP, horseradish peroxidase; LP, lactoperoxidase
[**]A: 0.1 M-sodium phosphate, pH 7.4
B: 0.01 M-Tris-HCl, 0.5 M-CaCl$_2$, pH 7.2
C: B + 0.01 M-EDTA

Table 3

Oxidation of N-OH-2-FAA by uterine (UT) and mammary gland (MG) extracts

Enzyme system*	Guaiacol units per reaction	Buffer**	Products (nmoles)				Substrate (nmoles)	Total recovery (%)
			2-FAA	N-AcO-2-FAA	2-NOF			
UT/H$_2$O$_2$	0.21	B	3.00	1.18	2.22		41.0	75
UT		B	0.26	0	0		45.0	72
UT/H$_2$O$_2$	0.007	D	1.58	1.31	15.5		38.2	90
UT		D	trace	0	0		56.5	90
MG/H$_2$O$_2$	0.004	D	0.59	0	3.50		53.5	91
MG		D	trace	0	0		60.9	97

*UT/H$_2$O$_2$ [0.9 mM], MG/H$_2$O$_2$ [0.3 mM].
**B: 0.01 M-Tris-HCl, 0.5 M-CaCl$_2$, pH 7.2.
  D: 0.1 M-sodium phosphate, pH 7.4, containing UT (0.015 ml) or MG (0.45 ml) extract in 0.02% Cetab per 1 ml incubation.

formation of nitroxyl free radical from N-OH-2-FAA and peroxides by mammary cell sonicates. Peroxidative activities of uterine and mammary extracts shown in this study toward PDA and TMBD, may be of importance in tumorigenesis by aromatic amines.

ACKNOWLEDGEMENTS

This work was supported by the US Veterans Administration and by PHS Grant CA-28000 awarded by the National Cancer Institute, DHHS.

REFERENCES

Allaben, W. T., Weeks, C. E., Weis, C. C., Burger, G. T. & King, C. M. (1982), *Carcinogenesis*, **3**, 233.

Barry, E. J., Malejka-Giganti, D. & Gutmann, H. R. (1969), *Chem.-Biol. Interact.*, **1**, 139.

Bartsch, H. & Hecker, E. (1971), *Biochem. Biophys. Acta*, **237**, 567.

Bartsch, H., Miller, J. A. & Miller, E. C. (1972), *Biochim. Biophys. Acta*, **273**, 40.

Clayson, D. B. & Garner, R. C. (1976), in *Chemical Carcinogens*, ACS Monograph 173, (ed. Searle, C. E.) American Chemical Society, Washington, DC, p. 366.

De Wolf, M. J. S., Lagrou, A. R., Hilderson, H. J. J., Van Dessel, G. A. F. & Dierick, W. S. H. (1978), *Biochem. J.*, **174**, 939.

Floyd, R. A., Soong, L. M. & Culver, P. L. (1976), *Cancer Res.*, **36**, 1510.

Floyd, R. A., Soong, L. M., Stuart, M. A. & Reigh, D. L. (1978). *Arch. Biochem. Biophys.*, **185**, 450.

Gutmann, H. R., Leaf, D. S., Yost, Y., Rydell, R. E. & Chen, C. C. (1970), *Cancer Res.*, **30**, 1485.

Jellinck, P. H. & Perry, G. (1967), *Biochim. Biophys. Acta*, **137**, 367.

Josephy, P. D., Eling, T. & Mason, R. P. (1982), *J. Biol. Chem.*, **257**, 3669.

Lyttle, C. R. & DeSombre, E. R. (1977), *Proc. Natl. Acad. Sci. USA*, **74**, 3162.

Malejka-Giganti, D., Rydell, R. E. & Gutmann, H. R. (1977), *Cancer Res.*, **37**, 111.

Matsubara, T., Prough, R. A., Burke, M. D. & Estabrook, R. W. (1974), *Cancer Res.*, **34**, 2196.

Miller, J. A. (1970), *Cancer Res.*, **30**, 559.

Miller, E. C. & Miller, J. A. (1976), in *Chemical Carcinogens*, ACS Monograph 173, (ed. Searle, C. E.) American Chemical Society, Washington, DC, p. 737.

Reigh, D. L. & Floyd, R. A. (1981), *Cancer Biochem. Biophys.*, **5**, 213.

Reigh, D. L., Stuart, M. & Floyd, R. A. (1978), *Experientia*, **34**, 107.

Ritter, C. L. & Malejka-Giganti, D. (1982), *Biochem. Pharmacol.*, **31**, 239.

Ritter, C. L., Malejka-Giganti, D. & Polnaszek, C. F. (1983), *Chem.-Biol. Interactions*, **46**, 317.

Ryzewski, C. N. & Malejka-Giganti, D. (1982), *J. Chromatogr.*, **237**, 447.

Stier, A., Clauss, R., Lücke, A. & Reitz, I. (1980), *Xenobiotica*, **10**, 661.

Sullivan, A. B. (1966), *J. Org. Chem.*, **31**, 2811.

Wagai, N. & Hosoya, T. (1982), *J. Biochem.*, **91**, 1931.

Wong, P. K., Hampton, M. J. & Floyd, R. A. (1982), in *Prostaglandins and Cancer: First International Conference*, (A. R. Liss, Inc.) New York, p. 167.

Part 9

INTERACTON OF *N*-OXIDIZED COMPOUNDS WITH CELLULAR CONSTITUENTS–TOXICOLOGICAL IMPLICATIONS

Chapter 41

REACTION OF *N*-OXIDATION PRODUCTS OF AROMATIC AMINES WITH NUCLEIC ACIDS

E. Kriek, **J. G. Westra** and **M. Welling**, The Netherlands Cancer Institute, Antoni van Leeuwenhoek Huis, 21 Plesmanlaan, 1066 CX, Amsterdam, The Netherlands.

1. Aromatic amines represent a group of compounds with well-known toxic, carcinogenic and mutagenic properties. The carcinogenic action of these compounds is thought to be initiated by modification of DNA after metabolic activation.
2. *N*-Oxidation, although quantitatively a minor reaction, is a necessary first step in the covalent binding, carcinogenicity and mutagenicity of aromatic amines and amides. The *N*-hydroxy derivatives formed can be further activated by a variety of secondary activation steps, including the formation of arylnitrenium-carbonium ions, *O*-acetylation and one-electron or radical oxidation.
3. These activations lead to the formation of reactive intermediates, capable of reacting with nucleophilic groups in nucleic acids and proteins, with the formation of covalent bonds. Favoured sites of reaction in DNA are the C-8 position of guanine and the exocyclic amino groups of the purines. Specificity of these reactions with guanine has been found *in vitro* as well as *in vivo* for the majority of aromatic amines.
4. In recent years highly sensitive and specific antibodies have become available for the detection and quantification of carcinogen-DNA adducts by immunological procedures. These methods have been applied successfully for the analysis of DNA modified with *N*-acetyl-2-aminofluorene and 2-aminofluorene.

INTRODUCTION

The group of chemicals known as *N*-substituted aryl compounds, which includes aromatic amines, amides and nitroaromatics, were among the first chemical compounds to be identified as carcinogens in animals and humans. Biochemical and biological studies of aromatic amines, together with other classes of carcinogens, have provided a sound basis for experimental cancer research as a scientific discipline, and have contributed enormously to our understanding of the mechanism of action of these chemicals in biological systems. Two recent international conferences have been devoted especially to *N*-substituted aryl compounds (Thorgeirsson *et al.*, 1981; Kadlubar & Beland, 1983). The importance of aromatic amines and related compounds as a general environmental problem has been discussed recently in an IARC Monograph (Egan *et al.*, 1981).

METABOLISM

Studies on the metabolism and carcinogenicity were initiated in Britain and the USA during the late 1940s and early 1950s. In Britain Boyland's group investigated 2-naphthylamine (NA) and 4-aminobiphenyl (ABP); these studies were directed primarily to the understanding of the induction of cancer of the urinary bladder by these agents. In the USA the Weisburgers and the Millers studied in detail the urinary excretion of the carcinogen *N*-acetyl-2-aminofluorene (AAF), after its administration to rats. The historical aspects of such studies on *N*-aryl carcinogens and their metabolism have been reviewed recently by the Millers (Miller & Miller, 1983) and will not be discussed here. The first evidence for *N*-hydroxylation as an activation pathway was obtained in Miller's laboratory in 1960, when a glucuronide conjugate of *N*-hydroxy-AAF was isolated and characterized. Subsequent studies with a variety of aromatic amines and amides have led to the generality that *N*-oxidation, although a relatively minor reaction, is the first stage in their metabolic activation to electrophiles. In this reaction arylamides are converted to arylhydroxamic acids, and arylamines to the corresponding arylhydroxylamines.

The *N*-hydroxy compounds may be further activated by a variety of enzymatic reactions, including *O*-esterification (acetate and sulphate), *N*,*O*-transacetylation and one-electron or radical activation. The mechanisms which generate ionic intermediates have been extensively studied during the past decades, but the radical activation has been neglected for many years. Recently, evidence has been obtained in several laboratories, that arylamines may also be activated through peroxidation, catalysed by prostaglandin hydroperoxide synthetase (PHS), an enzyme which is widely distributed in extrahepatic tissues. This type of activation is probably associated with the formation of radicals. The various activation reactions have been discussed elsewhere in this volume (see articles by Lindeke; Frederick *et al.*, Lotlikar,

Eling, and O'Brien & Gregory). For this review, three typical activation reactions of N-hydroxy compounds, which lead to the formation of covalent bonds with nucleic acids, have been selected. The reaction products formed *in vitro* in mutagenesis test systems and *in vivo* will be compared in those cases where they have been identified. Finally, recently developed methods for the detection and quantification of carcinogen-DNA adducts will be briefly discussed.

ACTIVATION MECHANISMS AND FORMATION OF ARYLAMINE-DNA ADDUCTS

Activation reactions leading to the formation of covalent bonds with nucleic acids can include, 1. Acid-catalysed formation of carbonium–nitrenium ions from arylhydroxylamines; 2. Formation of carbonium–nitrenium ions following O-esterification, and 3. Peroxidative co-oxidation of arylamines catalysed by prostaglandin hydroperoxide synthetase (PHS).

1.   Acid-catalysed formation of carbonium—nitrenium ions from arylhydroxylamines

The concept of the electrophilicity of reactive forms of carcinogens, known from alkylating agents, was founded on observations made with N-aryl carcinogens. Thus, the activation of hydroxylamines by protonation, known from organic chemistry, was applied by Kriek (1965) to N-hydroxy-AF, which reacted in an acid-catalysed reaction with guanine in nucleic acids. Similar reactions were found later by others for the carcinogens N-hydroxy-N'-acetyl-benzidine (N-hydroxy ABZ) and N-hydroxy-NA (Beland *et al.*, 1983).

  The conversion of urinary hydroxylamines to carcinogenic electrophiles under mildly acidic conditions in the bladder lumen and the binding of these electrophiles to urothelial DNA have been proposed by Kadlubar *et al.* (1981) as a critical event in the initiation of bladder cancer. Production of the arylhydroxylamine is followed by formation of the N-glucuronide, a stable transportable form of the primary metabolite. At relatively acidic pH in the urinary bladder, the hydroxylamine is released, which in turn yields the carbonium–nitrenium ion. However, other activation mechanisms of aromatic amines and their N-hydroxy derivatives have been recently described, which do not require an acidic pH value. These mechanisms are discussed in the sections below.

2.   Formation of carbonium—nitrenium ions following O-esterification

The N-hydroxy derivatives may be metabolized further to reactive ester intermediates by hydroxamic acid-dependent transacetylase (King & Allaben, 1980), sulphotransferase (Sekura *et al.*, 1980), or by other cellular enzyme systems. The unstable O-acetyl esters are too reactive to be isolated, but their formation can be detected by trapping with nucleophiles (Met, Guo, DNA). Recently, we have found that hydroxylamines can be activated to

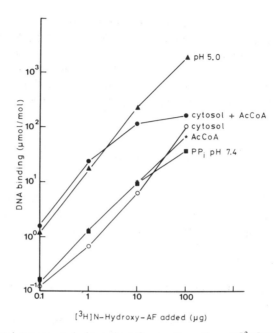

$[^3H]$N–Hydroxy–AF added (µg)

Fig. 1 – *S*-acetyl coenzyme A-dependent *O*-ester formation of $[^3H]$*N*-hydroxy-2-aminofluorene, catalysed by rat liver cytosol and subsequent binding to DNA. Experimental conditions are described in the text.

DNA-bound products by an *S*-acetyl coenzyme A-dependent *O*-ester formation catalysed by a cytosolic enzyme in rat liver (Fig. 1).

Assays were performed in helium-saturated 50 mM-sodium pyrophosphate (pH 7.4) containing 2.5 mg/ml calf thymus DNA, 1.0 mg/ml cytosol protein, 1 mM-AcCoA and 0.25–250 µM radiolabelled substrate (*N*-hydroxy-AF or *N*-hydroxy-AAF). The incubations were conducted for 30 min at 37 °C and the DNA was then isolated by extraction with phenol/cresol. The DNA preparations were purified by precipitation with ethanol followed by dialysis. The extent of covalent binding was determined by liquid scintillation counting after hydrolysis in 0.2 M-HCl at 37 °C for 18 h. Adducts were analysed following hydrolysis with TFA by h.p.l.c. as developed by Westra & Visser (1979). The major adduct in these experiments was found to be dGuo-8-AF, but some other unidentified products were also present. Further details of this study will be published elsewhere (Kriek *et al.*, in preparation).

A similar AcCoA-dependent mechanism has been proposed recently by Flammang & Kadlubar (1984) for the activation of *N*-hydroxy-DMABP, *N*-hydroxy-ABP, *N*-hydroxy-AF, *N*-hydroxy-ABZ and *N*-hydroxy-NA (in order of decreasing reactivity). It should also be noted that Kato *et al.* (1984) reported the AcCoA-dependent DNA binding of the heterocyclic hydroxylamines *N*-hydroxy-Trp-P-2 and *N*-hydroxy-Glu-P-1. A similar mechanism was proposed by Saito *et al.* (1983) for the activation of these compounds to mutagens in *Salmonella typhimurium* TA98/1,8-DNP$_6$.

DNA adducts of ionic intermediates of arylamines

The reactive sites in nucleic acids, *in vitro* as well as *in vivo*, have been summarized in Fig. 2.

Favoured sites of reaction are the C-8 of guanine and the exocyclic amino groups of the purines. Neither pyrimidines nor phosphate groups appear to be reactive. The specificity for guanine in DNA found *in vitro*, is even greater *in vivo* (Kriek, 1983). In all cases which have now been adequately examined, the major adduct *in vivo* was consistently an *N*-(deoxyguanosin-8-yl)-arylamine. Acetylated derivatives have been found only in a few cases and were always minor products. Benzidine adducts of this type were always present as dGuo-8-*N'*-ABZ. However, other unidentified products were present in DNA from dog bladder after treatment with benzidine (Beland *et al.*, 1983). Identification of these products will be of interest in view of the recently detected peroxidative activation of benzidine.

Mutagenicity

The mutagenic potential of arylamines and amides in the *S.typhimurium* tester strains is correlated with the formation of dGuo-8-arylamine products. In fact, Beranek *et al.* (1982) demonstrated that dGuo-8-AF is the only adduct formed in the microsome-activated reversion assay and the mutation frequency correlates with the fraction of DNA that is modified, when the *S.typhimurium* tester strain TA 1538 was exposed to *N*-hydroxy-AAF.

In a similar fashion, Saito *et al.* (1983) showed that an AcCoA-dependent enzyme in the *S.typhimurium* TA98/1,8-DNP$_6$ catalysed the DNA binding of the heterocyclic amines, *N*-hydroxy-Trp-P-2 and *N*-hydroxy-Glu-P-1 as the cause of mutagenicity. This tester strain is resistant to the mutagenicity of

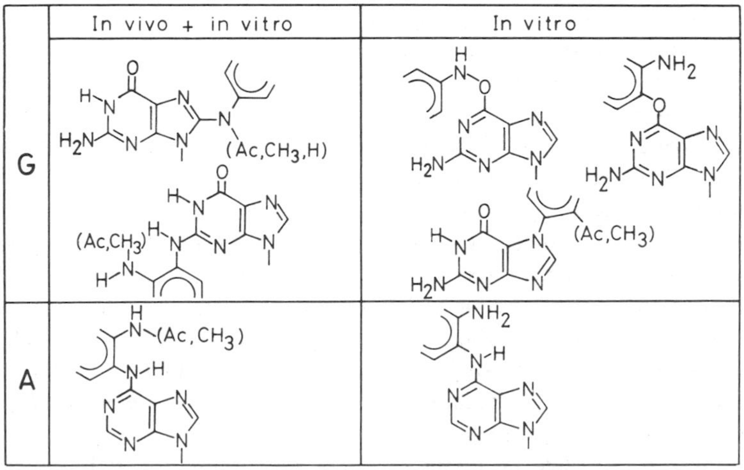

Fig. 2 – Identified sites of reaction in DNA with activated aromatic amines and amides *in vivo* and *in vitro*.

1,8-dinitropyrene and is also non-responsive to some nitroso- and
N-hydroxyarylamines.

Thus, the similarity of the adducts formed *in vitro*, in mutagenicity tests as
well as *in vivo*, combined with our recent observation that the removal of
dGuo-8-AF in rat liver *in vivo* is biphasic, suggests that dGuo-8-arylamines
may be critical lesions for tumor initiation (Westra *et al.*., 1983).

3. Peroxidative activation of arylamines

Haemprotein peroxidases are present in mammalian tissues and are capable
of catalysing the activation of aromatic amine carcinogens. The one-electron
non-enzymatic and enzymatic oxidation of *N*-hydroxy-AAF was first detected
by Forrester and Bartsch around 1970. Bartsch *et al.* (1972) demonstrated
that not only horseradish peroxidase, but also the mammalian enzymes
myeloperoxidase and lactoperoxidase catalyse the coversion of *N*-
hydroxy-AAF and *N*-acetoxy-AAF and 2-nitrosofluorene.

The one-electron or radical activation of arylamines has received little
attention for a number of years until recently when Zenser *et al.* (1983)
showed that prostaglandin hydroperoxide synthetase (PHS), which is widely
distributed in extrahepatic tissues, particularly in kidney and urinary tract,
mediated the arachidonic acid dependent co-oxidation of arylamines to DNA
bound products.

Purified PHS has been shown to consist of two separate activities: fatty acid
cyclooxygenase and prostaglandin hydroperoxidase. The first enzyme is
responsible for the initial bis-dioxygenation of the unsaturated fatty acid
(arachidonic acid). The hydroperoxidase is responsible for the cleavage of the
15-hydroperoxy group, giving prostaglandin H which is the common substrate
for the synthesis of prostaglandins and thromboxanes.

The hydroperoxidase activity of PHS is responsible for the co-oxidation of
carcinogens, which has been demonstrated with microsomes, tissue slices and
in vivo. Recent reports have appeared on the PHS-activated binding of AF,
benzidine and 2-NA to DNA *in vitro*. Krauss *et al.* (1984) reported that
PHS-activated AF was bound to poly G and poly C, but not to poly A or poly
U. When DNA was reacted in the same system, at least six adducts were
found, one of which was dGuo-8-AF.

The PHS-catalysed DNA binding of benzidine and 2-naphthylamine was
studied by Yamazoe *et al.* (1984). For 2-NA, both *N*-hydroxy-2-NA and
2-amino-1-naphthol (AN) have been implicated as reactive metabolites.
When these compounds were incubated with PHS, arachidonate and DNA at
pH 7.8 under aerobic conditions, the binding of AN to DNA was much higher
than that of *N*-hydroxy-2-NA. The major DNA adduct was identified as
4-(deoxyguanosin-N^2-yl)-1,2-naphthoquinoneimine. The second adduct was
suggested to be a N^6-deoxyadenosine derivative.

Similar products were detected in DNA modified by the PHS mediated
activation of 2-NA. In comparison, the peroxidative activation of benzidine
yielded only one DNA adduct, which was chromatographically different from
dGuo-8-N'-acetyl-BZ.

From the *in vitro* studies it is clear that the DNA adduct pattern of radical activation is quite different from that observed with ionic intermediates. Further studies are in progress to see whether the PHS-mediated activation of aromatic amines is significant *in vivo*.

The co-oxidative activation of benzidine (BZ) and 2-naphthylamine (2-NA) has also been shown to result in the formation of mutagens, but the DNA modifications responsible for mutagenicity still have to be identified.

METHODS OF DETECTION AND QUANTITATION (BIOMONITORING) OF ARYLAMINE-DNA ADDUCTS

The most important methods now available for the detection and quantitation of carcinogen-DNA adducts are: (1) h.p.l.c. of radiolabelled carcinogen-DNA adducts; (2) ^{32}P-post labelling of DNA hydrolysate (Randerath *et al.*, 1981); (3) Immunological procedures with specific antibodies raised against carcinogen-DNA or carcinogen–nucleoside adducts.

(1) H.p.l.c. of Radiolabelled Carcinogen—DNA adducts

Rapid and convenient identification and quantitation of arylamine–DNA adducts has been difficult, mainly because of problems encountered in the degradation of modified DNA and instability of certain adducts. Several groups of investigators, including ourselves, have used enzymatic digestion of modified DNA to the constituent deoxynucleosides. However, this method has its limitations because the digestion is not always complete and the relatively long digestion times required may alter arylamine-DNA adducts. Some years ago Westra & Visser (1979) in our laboratory introduced trifluoroacetic acid (TFA) for the degradation of AAF- and AF-modified DNA. TFA degrades DNA to bases by breaking the phosphodiester and glycosidic bonds. The base adducts can be separated by h.p.l.c.; Gu-8-AAF and Gu-8-AF are stable to TFA treatment and can be determined quantitatively. The utility of the method was demonstrated later by Tang & Lieberman (1983) who showed that Gu-N^2-AAF can also be determined when the TFA hydrolysis is performed under anhydrous conditions. The TFA/h.p.l.c. method is reliable and gives quantitative and reproducible results in our hands.

(2) ^{32}P-post-labelling of DNA Hydrolysate

Based on recent developments in the methodology of sequence analysis of nucleic acids, a new ^{32}P-post-labelling method for the analysis of carcinogen-DNA adducts was developed a few years ago by Randerath and his group in Texas. Following enzymatic hydrolysis of carcinogen- modified DNA, ^{32}P is incorporated into the non-radioactive nucleotides. The ^{32}P-nucleoside-diphosphate mixture is then chromatographed on thin layers of a strong anion exchanger. Evidence for the presence of chemically altered nucleotides is provided by the appearance of extra areas on the chromatograms, as detected by autoradiography. The method appeared to be particularly useful for aromatic carcinogens, including aromatic amines, and a detection limit of 1 adduct per 10^8 nucleotides was reported (Reddy *et al.*, 1984). The post-labelling method offers an interesting alternative for the immunological procedure discussed below. One of the major advantages is that adducts of unknown chemical structure can be detected, which is not possible by any other method, except radioactive labelling of the carcinogen.

(3) Immunological Quantification of AAF- and AF-DNA Adducts

In recent years highly sensitive immunological procedures have been developed which are capable of detecting femtomole amounts of carcinogen-modified DNA bases, employing antibodies raised against protein-bound nucleoside adducts (Poirier, 1981). These antibodies have been utilized in sensitive radioimmunoassays, enzyme immunoassays, in immunohistochemical and in electronmicroscopic investigations. This work was pioneered by Poirier in the USA and by Rajewski in Germany. Recent work on the immunological detection and quantitation of AAF- and AF-modified DNA in our laboratory is briefly summarized below.

Anti Guo-8-AAF was raised in rabbits with the BSA conjugate of Guo-8-AAF. The antiserum is highly specific for dGuo-8-AAF (affinity constant 6×10^9 l/mol) and dGuo-8-AF, but does not recognize dGuo-N^2-AAF, imidazole ring-opened dGuo-8-AF, dGuo or DNA (Van der Laken *et al.*, 1982). Because dGuo-8-AF is unstable under certain conditions, we decided not to raise antibodies against Guo-8-AF, but chose instead to follow a different route for the determination of dGuo-8-AF in DNA. Antibodies were raised against the guanine imidazole ring-opened form of dGuo-8-AF (rodGuo-8-AF) and employed in the enzyme immunoassay. As expected this antibody had high cross-reactivity with AF-DNA and AAF-DNA.

Analysis of mixtures consisting of two or three of the adducts could be accomplished by conversion of AF-DNA and AAF-DNA into the imidazole ring-opened (roAF-DNA) form (Kriek & Westra, 1980). The total amount of adducts can be measured after heating the mixture in 0.1 M-NaOH at 75 °C for 2 h, which converts AF-DNA and AAF-DNA into roAF-DNA. After heating the mixture at 100 °C for 30 min at pH 7, which converts AF-DNA into roAF-DNA, but leaves AAF-DNA unaffected, the content of AF- and roAF-DNA was measured. The difference between these values gives the amount of AAF-DNA. The successive reactions are illustrated in Fig. 3. In this manner mixtures of the three adducts in DNA can be analysed employing only one antibody (Kriek *et al.*, 1983).

The imidazole ring-opened form of dGuo-8-AF has not been detected in rat liver DNA following treatment with AAF, which is in disagreement with a previous report by Rio *et al.* (1982). In the course of our studies it appeared that formation of the ring-open form in AF-modified DNA is due to hydrolysis, dependent on the method of isolation of DNA, which may explain the observed difference.

Immunological methods have now been used in several laboratories for the cytochemical visualization of specific carcinogen-induced DNA modifications

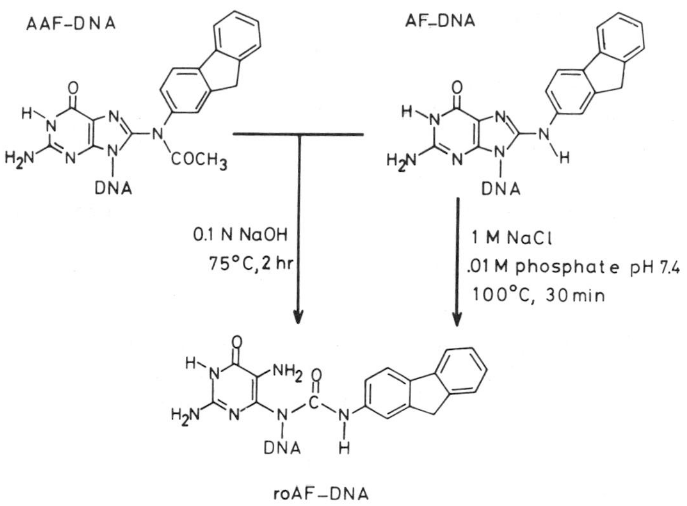

Fig. 3 – Conversion reactions of AAF- and AF-modified DNA into the guanine imidazole ring-opened form, roAF-DNA.

in cultured cells or in fixed animal tissue samples. Immunological detection of these DNA modifications in *in vivo* systems will be of great value because the biological activity of many carcinogens, mutagens and clastogens is highly specific with regard to cell type, site within an organ or tissue, and differentiation state of the cell. Our anti Guo-8-AAF antibody has also been employed to visualize the localization of dGuo-8-AAF/AF in DNA in liver sections from rats treated with AAF (Menkveld *et al.*, 1985). Positive staining was still detected after a single dose of 0.5 mg AAF/kg corresponding to a detection limit of 0.4 μmol/mole DNA-P, which is approximately 4000 carcinogen molecules per diploid genome.

REFERENCES

Bartsch, H., Miller, J. A. & Miller, E. C. (1972), *Biochim. Biophys. Acta*, **273**, 40.

Beland, F. A., Beranek, D. T., Dooley, K. L., Heflich, R. H. & Kadlubar, F. F. (1983), *Environm. Health Perspect.*, **49**, 125.

Beranek, D. T., White, G. L., Heflich, R. H. & Beland, F. A. (1982), *Proc. Natl. Acad. Sci. USA*, **79**, 5175.

Egan, H., Fishbein, L., Castegnaro, M., O'Neill, I. K. & Bartsch, H. (eds.) (1981), *IARC Scientific Publications*, **40**, International Agency for Research on Cancer, Lyon.

Flammang, T. J. & Kadlubar, F. F. (1984), *Proc. Amer. Assoc. Cancer Res.*, **25**, 474.

Kadlubar, F. F., Unruh, L. E., Flammang, T. J., Sparks, D., Mitchum, R. K. & Mulder, G. J. (1981), *Chem.-Biol. Interact.*, **33**, 129.

Kadlubar, F. F. & Beland, F. A. (eds.) (1983), *Environm. Health Perspect.*, **49**, 1.

Kato, R., Sairo, A., Shinohara, A., Yamazoe, Y. & Kamataki, T. (1984), *Proc. Amer. Assoc. Cancer Res.*, **25**, 475.

King, C. M. & Allaben, W. T. (1980), in *Enzymatic Basis of Detoxification*, (ed. Jakoby, W. B.) Vol. II, Academic Press, New York, p. 187.

Krauss, R. S., Reed, G. A. & Eling, T. E. (1984), *Proc. Amer. Assoc. Cancer Res.*, **25**, 331.

Kriek, E. (1965), *Biochem. Biophys. Res. Commun.*, **20**, 793.

Kriek, E. & Westra, J. G. (1980), *Carcinogenesis*, **1**, 459.

Kriek, E. (1983), in *Progress in Clinical and Biological Research*, Vol. 123B, 13th International Cancer Congress, Part B, Biology of Cancer (1) (eds. Mirand, E. A., Hutchinson, W. B. & Mihich, E.) Alan R. Liss, Inc., New York, p. 161.

Kriek, E., Welling, M. & Van Der Laken, C. J. (1983), *Proc. Amer. Assoc. Cancer Res.*, **24**, 247.

Menkveld, G. J., Van Der Laken, C. J., Hermsne, G., Kriek, E., Scherer, E. & Den Engelse, L. (1985), *Carcinogenesis*, **6**, 263.

Miller, J. A. & Miller, E. C. (1983). *Environm. Health Perspect.*, **49**, 3.

Poirier, M. C. (1981), *J. Natl. Cancer Inst.*, **67**, 515.

Randerath, K., Reddy, M. V. & Gupta, R. C. (1981), *Proc. Natl. Acad. Sci. USA*, **78**, 6126.

Reddy, M. V., Gupta, R. C., Randerath, E. & Randerath, K. (1984), *Carcinogenesis*, **5**, 231.

Rio, P., Bazgar, S. & Leng, M. (1982), *Carcinogenesis*, **3**, 225.

Saito, K., Yamazoe, Y., Kamataki, T. & Kato, R. (1983), *Biochem. Biophys Res. Commun.*, **116**, 141.

Sekura, R. D., Lyon, E. S., Marcus, C. J. & Wang, J.-L. (1980), in *Enzymatic Basis of Detoxification*, (ed. Jakoby, W. B.) Vol. II, Academic Press, New York, p. 199.

Tang, M. & Lieberman, M. W. (1983), *Carcinogenesis*, **4**, 1001.

Thorgeirsson, S. S., Weisburger, E. K., King, C. M. & Scribner, J. D. (eds.) (1981), National Cancer Institute Monograph 58, *Carcinogenic and Mutagenic N-Substituted Aryl Compounds*, US Government Printing Office, Washington, DC 20402.

Van Der Laken, C. J., Hagenaars, A. M., Hermsen, G., Kriek, E., Kuipers, A. J., Nagel, J., Scherer, E. & Welling, M. (1982), *Carcinogenesis*, **3**, 569.

Westra, J. G. & Visser, A. (1979), *Cancer Lett.*, **8**, 155.

Westra, J. G., Visser, A. & Tulp, A. (1983), *Environm. Health Perspect.*, **49**, 87.

Yamazoe, Y., Miller, D. W., Gupta, R. C., Zenser, T. V., Weis, C. C. & Kadlubar, F. F. (1984), *Proc. Amer. Assoc. Cancer Res.*, **25**, 356.

Zenser, T. V., Cohen, S. M., Mattamal, M. B., Wise, R. W., Rapp, N. S. & Davis, B. B. (1983), *Environm. Health Perspect.*, **49**, 33.

Chapter 42

NON-ENZYMATIC REARRANGEMENT AND COUPLING REACTIONS OF *N*-OXIDIZED COMPOUNDS

B. Lindeke, Department of Organic Pharmaceutical Chemistry, Biomedical Centre, University of Uppsala, Box 574, S-751 23 Uppsala, Sweden

1. In nitrogen as opposed to carbon oxidation an array of products must be considered due to the multiplicity of oxidation states of the nitrogen atom. Two fundamental processes can be discerned: those where electrons and protons are merely removed (*N*-oxidation) and those where oxygen is added (*N*-oxygenation).
2. Several *N*-oxidation products are prone to undergo condensation and coupling reactions, disproportionation and serve as oxidants or reductants.
3. Tertiary amine oxides are frequently susceptible to rearrangement and decomposition reactions.
4. In the adverse reactions of *N*-oxidized compounds radical mechanisms are far more important than previously realized.

INTRODUCTION

The interest in research on biological oxidation of amines is frequently due to the metabolic formation of secondary products which could cause toxicity, carcinogenicity or mutagenicity. Oxidation at nitrogen is now well established as an activating step for the carcinogenic aromatic amines, and for amine induced haemolysis (Miller, *et al*., 1964; Kiese, 1966; Weisburger & Weis-

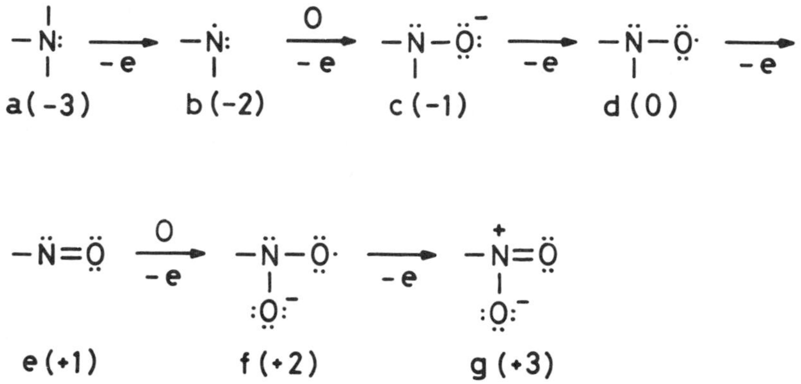

$$-\overset{|}{\underset{|}{\overset{\cdot}{N}}}: \xrightarrow{-e} -\overset{|}{\underset{|}{\overset{\cdot}{N}}}: \xrightarrow[-e]{O} -\overset{|}{N}-\ddot{\overset{..}{O}}:^{-} \xrightarrow{-e} -\overset{|}{N}-\ddot{\overset{..}{O}}\cdot \xrightarrow{-e}$$

a(−3)   b(−2)   c(−1)   d(0)

$$-\ddot{N}=\ddot{O} \xrightarrow[-e]{O} -\ddot{N}-\ddot{O}\cdot \xrightarrow{-e} -\overset{+}{N}=\ddot{O}$$
$$\qquad\qquad\quad \underset{:\ddot{O}:^{-}}{|} \qquad\qquad \underset{:\ddot{O}:^{-}}{|}$$

e(+1)   f(+2)   g(+3)

Fig. 1   Conventional oxidation states of nitrogen in various nitrogenous compounds. Elemental nitrogen is defined as zero oxidation state. Electropositive atoms bonded to the nitrogen decrease its oxidation state, while more electronegative atoms cause an increase. (a) Amines, amides, carbamates, ureas, nitriles and imines. (b) Radicals corresponding to a (c) Hydroxylamines, hydroxylamides, oximes, nitrones, *N*-oxides. (d) Radicals corresponding to c (e). Nitroso compounds (f) Radicals corresponding to f. (g) Nitro compounds.

burger, 1973; Uehleke, 1971; Miller, 1978; O'Brien, 1984). However, reductive activation of nitrogenous compounds can also occur as in nitrofurans and nitroimidazoles, which generate reactive products that can form covalent bonds with proteins and other macromolecules (Peterson *et al.*, 1979; Sasame & Boyd, 1979). Also, hydrolysis and subsequent oxidation of nitrate esters generate nitrogenous species, probably, *S*-nitrosothiols, which although not directly toxic, exert pharmacological effects (Ignarro *et al.*, 1981). Accordingly, in the toxicity of nitrogenous compounds, species generated by oxidation as well as reduction of a precursor have to be considered.

Focusing on the fundamental chemistry of *N*-oxidation, as opposed to that of carbon, an array of products must be dealt with, this due to the multiplicity of oxidation states of the nitrogen atom, ranging from −3 to +3 (Fig. 1). The elementary process of hydrogen abstraction from amines (a) produces reactive nitrogen radicals (b). Dependent on structure, the unpaired electron can be located on the nitrogen, as in aliphatic amines, or delocalized over the carbon skeleton, as in aromatic amines. Further oxidation generates hydroxylamines (c) and nitroxide radicals (d). Two electron oxidation of tertiary amines results in the formation of *N*-oxides with a formal valency of the nitrogen of −1. When applicable, electron abstraction from nitroxide radicals generates *C*-nitroso compounds (e), the subsequent stepwise oxidation of which yields the nitro anion radical (f) and the nitro compound (g).

Although each step in Fig. 1 represents an oxidation at the nitrogen atom, two fundamental processes can be discerned; those where electrons and protons are merely removed (formation of b, d, e and g) and those where oxygen is added (formation of c and f). Thus, it has to be kept in mind that

N-oxidation does not necessarily comprise an oxygenation of the nitrogen atom.

Several of the above nitrogen oxidation products are prone to undergo non-enzymatic condensation and coupling reactions or to decompose. The topic of the present contribution will be to discuss some of these reactions, and to consider when and under which conditions they occur.

AMINE RADICALS

In the last few years it has become clear that enzymes other than the cytochrome P-450s can be quite important for the bioactivation of xenobiotics. Thus, various peroxidases and prostaglandin synthetase (PGS) could catalyse the covalent binding of potential toxicants to the target tissues (Marnett, 1981; O'Brien, 1984). With substrates such as certain aromatic amines, peroxidase-type reactions will result in the initial formation of amine radicals. Strong evidence for the intermediary formation of amine radicals have in fact recently been shown also for cytochrome P-450 catalysed reactions. Miwa *et al.* (1983) presented evidence that the *N*-demethylation of *N,N*-dimethylaniline proceeds via α-carbon deprotonation from an initially formed aminium cation radical. By using spin-trapping techniques Kubow *et al.* (1984) were able to show the initial formation of a nitrogen centred radical from 3-methylindole in goat lung microsomes.

With respect to drug metabolism, aromatic amine radicals are the most intriguing. This is because the radical chemistry of aromatic amines is characterized by various coupling reactions through the aromatic rings, whereas aliphatic amine radicals are decomposed merely by fragmentations and hydrolytic reactions. In a recent review, O'Brien (1984) has described some of the products formed in the metabolic activation of aromatic amines such as benzidines, phenylenediamines and aminophenols. The products preferentially constituted dimers and trimers, formed through coupling reactions. Similar structures (Fig. 2) have recently been encountered in studies on the oxidation of *p*-phenetidine by horse radish peroxidase or prostaglandin synthetase (Ross *et al.*, 1985; Larsson *et al.*, 1984; cf. also Lindquist *et al.*, this volume).

The products formed in peroxidase catalysed reactions appear to be quite similar to those seen in anodic oxidation of aromatic amines, (Nelson, 1974; Bacon & Adams, 1968). The latter reaction is characterized by head-to-tail, tail-to-tail or head-to-head coupling of initially formed amine radicals. The product composition is highly dependent on the milieu in which the oxidation takes place. The coupled products can, when applicable, undergo subsequent hydrolysis. Compound **2** in Fig. 2 is thus formed by hydrolysis of the corresponding diimine.

The initially formed aromatic amine radicals are by nature very reactive, and are prone to interact with for example DNA (O'Brien, 1984; Larsson *et al.*, 1984). They oxidize glutathione, resulting in the formation of thiyl radicals (Ross *et al.*, 1985). Also some of the coupling products are reactive—e.g.

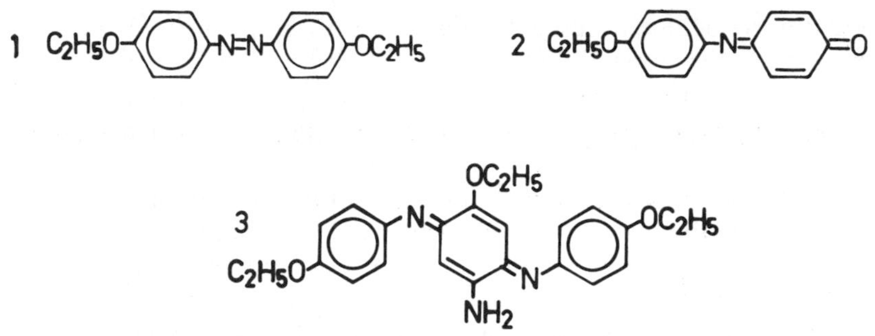

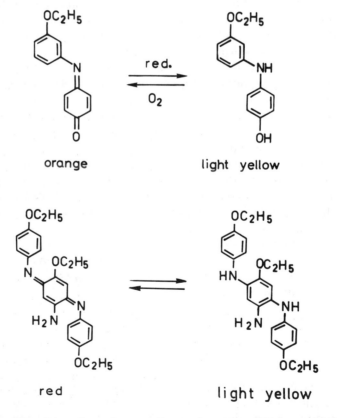

Fig. 2   Oxidation products identified in HRP- and PGS-catalyzed oxidations of
p-phentidine. **1**. 4,4′-diethoxyazobenzene, **2**. 4-(4-ethoxyphenylimino)-2,5-cyclo-
hexadien-1-on and **3**. 3,6-bis(4-ethoxyphenylimino)-4-ethoxy-1,4-cyclohexadienyl-
amine.

orange   light yellow

red   light yellow

Fig. 3.   Examples   of   redox   cycling   of   *p*-phenetidine   metabolites.
4-(4-ethoxyphenylimino)-2,5-cyclohexadien-1-one   (top)   and   3,6-bis(4-ethoxy
phenylimino)-4-ethoxy-1,4-cyclohexadienylamine (bottom). Reductions were
achieved with, dithionite, ascorbate, dithiotreitol or NADPH cytochorme P-450
reductase.

compound **2** in Fig. 2—and can bind to proteins or undergo nucleophilic attack from thiols. They can also undergo redox-cycling (Fig. 3).

HYDROXYLAMINES, NITROXIDES, *C*-NITROSO COMPOUNDS

The metabolic formation of primary and secondary *N*-hydroxylamines are most frequently the result of two electron oxidations of primary or secondary amines, catalysed by monooxygenases such as cytochrome P-450 or the microsomal NADPH-dependent flavin-containing monooxygenase (Wislocki *et al.*, 1980; Ziegler, 1980; Damani, 1982). The hydroxylamines are readily further oxidized to nitroxide radicals and, depending on structure, to nitroso compounds, oximes or nitrones. Nitroxide radicals are prone to disproportionation or like the amine radicals, they sometimes undergo coupling reactions.

The non- and postenzymatic chemistry of these nitrogenous species has been dealt with in detail in a previous review (Lindeke, 1982) and therefore the present review will only deal with one recent example; the chemistry of the oxidation of *N*-hydroxyphentermine (2-methyl-1-phenyl-2-propyl-*N*-hydroxylamine) by cytochrome P-450 systems.

N-Hydroxyphentermine is a primary hydroxylamine with the nitrogen attached to a tertiary carbon. This prevents tautomerization to an oxime subsequent to the formation of the corresponding nitroso compound. The NADPH-dependent oxidation of *N*-hydroxyphentermine, which is clearly catalysed by cytochrome P-450 enzymes (Duncan & Cho, 1982), is associated with a marked increase in NADPH-dependent hydrogen peroxide production (Cho *et al.*, 1982). Recent studies (Cho *et al.*, 1984) have shown that oxidation of the hydroxylamine to a nitroxide radical occurs concomitantly to the formation of a superoxide anion radical. The nitroxide radical then disproportionates to yield the hydroxylamine and the nitroso compound, the latter being oxidized to the corresonding nitro compound by hydrogen peroxide generated from the superoxide anion, or by molecular oxygen. The last step is believed to be a non-enzyme catalysed chemical oxidation. It should be noted that cytochrome P-450 in this sequence (Fig. 4) acts as a hydroxylamine oxidase rather than a monooxygenase (Equation 1).

$$R{-}NHOH + O_2 \longrightarrow R{-}N{=}O + H_2O \qquad (1)$$

N-OXIDES

Tertiary amine *N*-oxide formation results in a marked decrease in the pK_a of the amine. The *N*-oxides have considerable dipole moments and display high polarity and water solubility. In contrast to the *N*-hydroxylamines, they are generally less chemically reactive than the corresponding amines. However, certain *N*-oxides are indeed very unstable inasmuch as they can undergo various rearrangements (Lindeke, 1982). Work in our laboratory has been

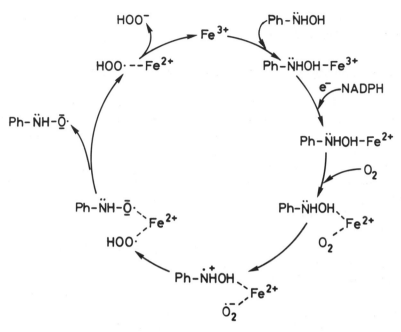

Fig. 4.   Proposed reaction sequence accounting for the generation of H_2O_2 and
nitroxide radicals in the interaction of *N*-hydroxyphentermine with cytochrome
P-450. Subsequent to its formation the nitroxide radical disproportionates to the
hydroxylamine and the nitroso compound. The latter is then oxidized to the nitro
compound, either by the concomitantly generated peroxide or by oxygen. Ph =
$C_2H_5CH_2C(CH_3)_2-$.

devoted to the investigation of two types of *N*-oxide rearrangements—the
Cope elimination (Cope *et al.*, 1943) and the Meisenheimer rearrangement
(Meisenheimer, 1919).

   In the Cope elimination, tertiary amine *N*-oxides with hydrogen atoms β to
the nitrogen decompose to secondary hydroxylamines and an alkene. The
Meisenheimer rearrangement involves the migration of a group from nitrogen
to oxygen in tertiary *N*-oxides having no β-hydrogen atoms. The Meisenheimer rearrangement has been shown to occur in the *N*-oxides (Fig. 5) of
certain α-acetylenic amines such as pargyline (*N*-benzyl-*N*-methylpropargy-
lamine) (Hallström *et al.*, 1981), while the Cope elimination appears to be
very   important   in   the   *N*-oxides   of   pargyline   analogues   containing
β-hydrogens, e.g. *N*-methyl-*N*-(1-phenylethyl)propargylamine (Fig. 5). Of
the two *N*-oxides, that of pargyline is considerably more stable than that of
the α-methyl-substituted analogue, which would indicate the Cope elimination as being energetically more favoured than the Meisenheimer rearrangement. Both reactions are however, hampered by hydration of the *N*-oxides,
and if they occur during the course of metabolism of the corresponding
amines, the rearrangements are likely to occur in the biological matrix prior
to hydration of the *N*-oxides. Interestingly, while pargyline *N*-oxide is found

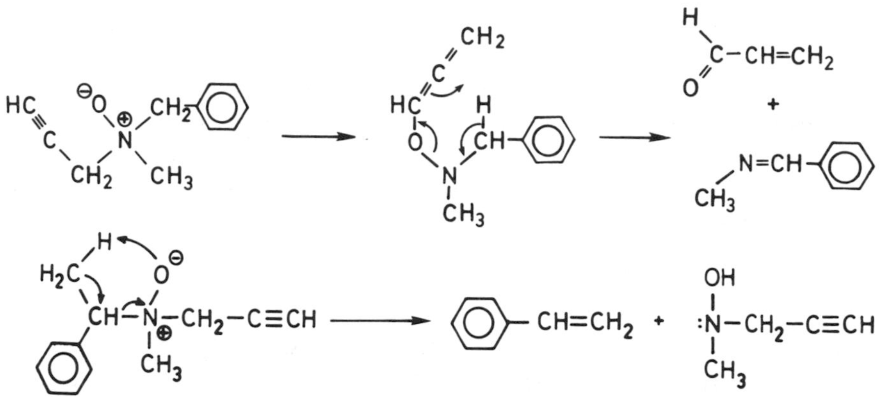

Fig. 5.   Meisenheimer rearrangement (top) and Cope elimination (bottom) in Pargyline   (N-benzyl-N-methylpropargylamine)   N-oxide   and   N-methyl-N-(1-
phenylethyl)propargylamine N-oxide, respectively.

to be a major metabolite of pargyline, only minute amounts of N-oxides could
be retrieved from incubations of the α-methyl-substituted pargyline congeners
(cf. Weli & Lindeke this volume). Whether this is due solely to the greater
instabilities of the latter N-oxides remains to be investigated.

The decomposition of N-oxides can result in the formation of products of
potential toxicity. Thus, Meisenheimer rearrangement of pargyline N-oxide
generates acrolein, a well known hepatotoxin. Also the Cope elimination
depicted in Fig. 5 will, apart from the formation of a styrene, concomitantly
generate a very reactive secondary N-hydroxylamine.

NITRO COMPOUNDS

The toxicity of organic nitro compounds in most cases seems to be associated
with their reductive metabolism. The crucial event appears to be the initial
formation of an anion radical which could decay by non-enzymatic processes.
In the reduction of nitrofurans (Peterson et al., 1979; Sasame & Boyd, 1979),
such a radical could either, concomitant to superoxide production, be oxidized back to the nitro stage, or undergo disproportionation to the nitro and
nitroso compounds. However, further reduction to the N-hydroxylamine
stage is also possible and depending on structure, subsequent hydrolysis of the
hydroxylamine could yield fragments with toxic properties. This could be a
sequence of events which is responsible for the toxicity of metronidazole
(Knox et al., 1983) and misonidazole (Raleigh & Liu, 1983). In the latter
case glyoxal was indicated as a fragment with toxic properties.

CONCLUSIONS

Further to my previous review on 'The Non- and Postenzymatic Chemistry of
N-oxygenated compounds' (Lindeke, 1982), the few examples from recent

investigations, discussed in this contribution, clearly show that N-oxidized compounds are prone to undergo, what from a toxicological point of view must be regarded as, adverse reactions. The results also indicate that in such reactions radical mechanisms are far more important than previously realized. Thus, one would predict that investigations of peroxidative mechanisms are bound to increase in the future at the expense of those involving the cytochrome P-450 enzymes.

REFERENCES

Bacon, J. & Adams, R. N. (1968), *J. Am. Chem. Soc.*, **90**, 6596.

Cho, A. K., Maynard, M. S., Matsumoto, R. M., Lindeke, B., Paulsen, U. & Miwa, G. T. (1982), *Mol. Pharmacol.*, **22**, 465.

Cho, A. K., Duncan, J. D., Fukoto, J. M. & Lindeke, B. (1984), in *Foreign Compound Metabolism*, (eds. Caldwell, J. & Paulson, G. D.) Taylor & Francis London, p. 79.

Cope, A. C., Pike, R. A. & Spencer, C. F. (1953), *J. Am. Chem. Soc.*, **75**, 3212.

Damani, L. A. (1982), in *Metabolic Basis of Detoxication*, (eds. Jacoby, W. B., Bend, J. R. & Caldwell, J.) Academic Press, New York, p. 127.

Duncan, J. D. & Cho, A. K. (1982), *Mol. Pharmacol.*, **22**, 235.

Duncan, J. D., Di Stefano, E. W., Miwa, G. T. & Cho, A. K. (1984), *Biochemistry* (submitted for publication).

Hallström, G., Lindeke, B., Khutier, A.-H. & Al Iraqi, M. A. (1981), *Chem.-Biol. Interact.*, **34**, 185.

Ignarro, L. J., Lippton, H., Edwards, J. C., Baricos, W. H., Hyman, A. L., Kadowitz, P. J. & Guetter, C. A. (1981), *J. Pharmacol. Exp. Therap.*, **218**, 739.

Kiese, M. (1966), *Pharmacol. Rev.*, **18**, 1091.

Knox, R. J., Knight, R. C. & Edwards, D. J. (1983), *Biochem. Pharmacol.*, **32**, 2149.

Kubow, S., Jansen, E. G. & Bray, T. M. (1984), *J. Biol. Chem.*, **259**, 4447.

Larsson, R., Ross, D., Nordenskjöld, M., Lindeke, B., Olsson, L.-I. & Moldéus, P. (1984), *Chem.-Biol. Interact.*, **52**, 1.

Lindeke, B. (1982), *Drug Metab. Rev.*, **13**, 71.

Marnett, L. J. (1981), *Life Sci.*, **29**, 531.

Meisenheimer, J. (1919), *Ber.*, **52**, 1667.

Miller, E. C. (1978), *Cancer Res.*, **38**, 1479.

Miller, E. C., Miller, J. A. & Enomoto, M. (1964), *Cancer Res.*, **24**, 2018.

Miwa, G. T., Walsh, J. S., Kedderis, G. L. & Hollenberg, P. F. (1983), *J. Biol. Chem.*, **258**, 14445.

Nelson, R. F. (1974), in *Techniques of Chemistry*, Vol. V, Part I, (ed. Weinberg, N. L.) John Wiley & Sons, New York, p. 535.

O'Brien, P. J. (1984), in *Free Radicals in Biology*, Vol. VI, (ed. Pryor, W. A.) Academic Press, New York, p. 289.

Peterson, F. J., Mason, R. P., Hovsepian, J. & Holtzman, J. (1979), *J. Biol. Chem.*, **254**, 4009.

Raleigh, J. A. & Liu, S. F. (1983), *Biochem. Pharmacol.*, **32**, 1444.

Ross, D., Larsson, R., Andersson, B., Nilsson, H., Lindquist, T., Lindeke, B. & Moldéus, P. (1984), *Biochem. Pharmacol.*, **34**, 343.

Sasame, H. A. & Boyd, M. R. (1979), *Life Sci.*, **24**, 1091.

Uehleke, H. (1971), *Xenobiotica*, **1**, 327.

Weisburger, J. H. & Weisburger, E. K. (1973), *Pharmacol. Rev.*, **25**, 1.

Wislocki, P. G., Miwa, G. T. and Lu, A. Y. H. (1980), in *Enzymatic Basis of Detoxication*, (ed. Jakoby, W. B.) Academic Press, New York, p. 135.

Ziegler, D. M. (1980), in *Enzymatic Basis of Detoxication* (ed. Jakoby, W. B.) Academic Press, New York, p. 201.

Chapter 43

REACTIONS OF NITROSOARENES WITH SULPHYDRYL GROUPS: REACTION MECHANISM AND BIOLOGICAL SIGNIFICANCE

P. Eyer, Institut für Pharmakologie und Toxikologie der Medizinischen Fakultät der Ludwig-Maximilians-Universität München, Nussbaumstrasse 26, D-8000 München 2, FRG

1. An overview is presented of a new type of reaction, which may modify the biological effects of nitrosoarenes.
2. Nitrosoarenes and thiols form an intermediate semimercaptal which (a) rearranges to a sulphinamide, (b) is thiolytically cleaved by a second thiol yielding a hydroxylamine, (c) is reduced by another thiol leading to a sulphenamide, or is reduced by thiols—possibly via a mercaptal—giving the amine.
3. Reaction pathways and rates are markedly influenced by substituents, type and concentration of the thiol, and by pH.
4. Examples of the biological significance of these reactions are given for nitrosobenzene, nitrosophenetol, and nitrosochloramphenicol. Reactions of glutathione are responsible for the rapid extraction of the nitrosoarenes by the liver and lead to glutathione depletion and alteration of the bile flow. In blood, these reactions sequester nitrosoarenes in red cells and markedly influence the kinetics of ferrihaemoglobin formation.

INTRODUCTION

If one screens chemical textbooks for reactions of nitrosoarenes with thiols, one notices a paucity of information on these reactions. Instead, authors of such textbooks give warnings on the potential hazards of C-nitroso compounds, like haemotoxicity, carcinogenicity and the explosive nature of such compounds. Therefore, understandably anxiety may have prevented chemists from working with nitrosoarenes, and the foul odour of thiols may have added to this dislike. Hence, reactions of nitrosoarenes with thiols, which provide interesting new synthetic pathways, have been largely overlooked by chemists. Indeed, to my knowledge, toxicologists are apparently the only scientists who have dealt with such reactions.

Since the discovery of metabolic N-oxygenation *in vivo* by Kiese (1959) and Cramer *et al.* (1960), this pathway is generally accepted as being one of the first activation steps in the toxication of arylamines, leading to methaemoglobinaemia and carcinogenesis. Despite the wealth of literature available on the further activation and detoxication of N-hydroxy compounds, detailed reports on the influence of thiols on those reactions are scanty (Eyer, 1979). The first report on the reaction of a nitrosoarene with glutathione originates from the Miller's' laboratory (Lotlikar *et al.*, 1965), where a water-soluble product was obtained upon incubation of nitrosofluorene with glutathione. This derivative hydrolysed under acidic or alkaline conditions to yield aminofluorene. No further attempts were made to identify this compound. In 1976, Kiese and Taeger reported on an analogous compound which they had isolated from red cells incubated with phenylhydroxylamine. This review describes recent studies carried out in this and other laboratories.

REACTIONS OF NITROSOBENZENE AND PHENYLHYDROXYLAMINE WITH GLUTATHIONE

When we studied reactions of phenylhydroxylamine with glutathione we found that no reaction took place under strict anaerobic conditions at pH 7.4 and 37 °C. In contrast, nitrosobenzene reacted rapidly with glutathione, with formation of phenylhydroxylamine, stoichiometric amounts of glutathione disulphide, and a water-soluble product which hydrolysed under acidic or alkaline conditions into aniline and a stoichiometric amount of glutathione sulphinic acid. The identity of the isolated compound as glutathione sulphinanilide was established by ^{1}H-n.m.r. and i.r. spectroscopy. The i.r. spectrum showed an intense absorption band at 1055 cm^{-1}, as one would expect for a sulphoxide structure. Formation of phenylhydroxylamine and the sulphinamide implies complex chemical reactions. When we followed these reactions spectroscopically by repetitive scanning, we observed changes in the u.v. spectrum of the reaction mixture without formation of an isosbestic point. At 254 nm there was a rapid increase in absorbance which gradually declined giving the final spectrum of glutathione sulphinanilide. This apparently indicated formation of a transient intermediate during sulphinamide formation.

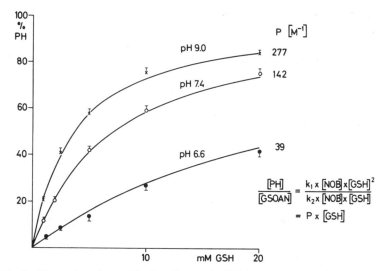

Fig. 1 – Formation of phenylhydroxylamine (% PH) in the reaction of nitrosobenzene (1 mM) with GSH at various pH. (37 °C, N$_2$, 5 min reaction; means of 3 exp. ± SD). (PH = phenylhydroxylamine; NOB = nitrosobenzene; GSOAN = glutathione sulphinanilide).

Phenylhydroxylamine formation was negligible in the above experiment since we used rather low glutathione concentrations. This is important, because the yield in phenylhydroxylamine formation depends on the glutathione concentration as shown in Fig. 1. At low glutathione concentrations, phenylhydroxylamine formation is poor and sulphinamide formation prevails; at increasing glutathione concentration, phenylhydroxylamine formation increases. The experimental data fit slopes calculated by the inset equation, which was derived by assuming that the proportion of both products depends on their individual formation rates. Apparently, formation of phenylhydroxylamine and glutathione disulphide needs two equivalents of glutathione, whereas sulphinamide formation needs only one. By cancelling that fraction, we derive the lower equation (see Fig. 1) from which it is evident that the proportion of phenylhydroxylamine formed at any nitrosobenzene concentration depends on the glutathione concentration only. The quotient P of both rate constants increases with increasing pH value.

When we followed the rapid initial reaction of nitrosobenzene with glutathione spectroscopically with a stop-flow technique we became aware of another phenomenon. The absorbance of nitrosobenzene at 306 nm decreases biphasically, and the proportion of the rapid *vs* the slow phase depends on the glutathione concentration. Analysis of the data obtained for various nitrosobenzene and glutathione concentrations clearly indicated that the rapid phase is due to *reversible* adduct formation leading to an equilibrium. This equilibrium is disturbed by irreversible reactions like sulphinamide and phenylhydroxylamine formation, both of which proceed at much slower

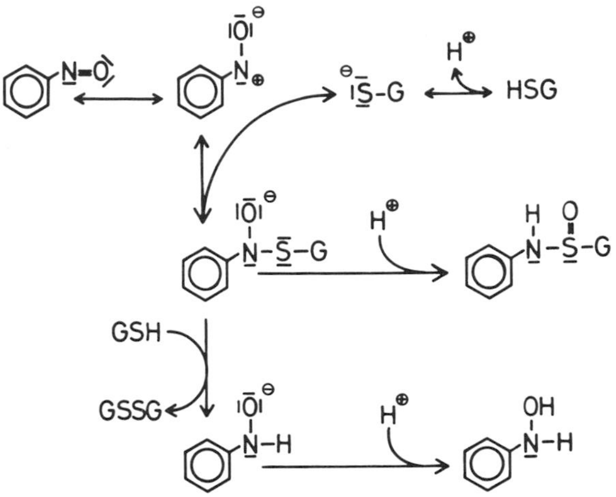

Fig. 2 – Tentative scheme of the proposed reactions of nitrosobenzene with glutathione.

rates. The assumption of a reversible adduct formation was also tested experimentally. At the very early stages of the reaction, nearly all nitrosobenzene could be extracted with ether from the mixture, although it did not exhibit its typical absorption at 306 nm. With these additional data we derived our reaction scheme (Fig. 2). In this tentative scheme nitrosobenzene and glutathione form a transient semimercaptal which either rearranges to the sulphinamide, a reaction which is catalysed by protons, or is thiolytically cleaved by a second glutathione molecule yielding phenylhydroxylamine and glutathione disulphide. This latter reaction resembles the diazene reduction with glutathione as described by Kosower *et al.* (1969) for several azoesters. As may be anticipated from this scheme, thiolytic cleavage of the semimercaptal is favoured by increase in the thiol concentration at the expense of the sulphinamide rearrangement.

REACTIVITIES OF *PARA*-SUBSTITUTED NITROSOARENES WITH GLUTATHIONE

We have since extended our investigations to other nitrosoarenes and found a good Hammett correlation for different *p*-substituted monocyclic nitrosoarenes (Fig. 3). The second order rate constant for the reacton with glutathione leading to the assumed semimercaptal is $5 \times 10^3 \, M^{-1} \, s^{-1}$ at pH 7.4 and 37 °C for nitrosobenzene, and differs by about five orders of magnitude between the compounds tested. In addition to the different reactivities, the pattern of reaction products after completion of the reaction varied considerably. As shown in Fig. 4, the yield of hydroxylamine at a given glutathione concentration varies greatly, and increases with increasing electron-withdrawing substituents. Hence, at low glutathione concentration, such as 0.1 mM, we hardly detected any hydroxylamine with nitrosobenzene,

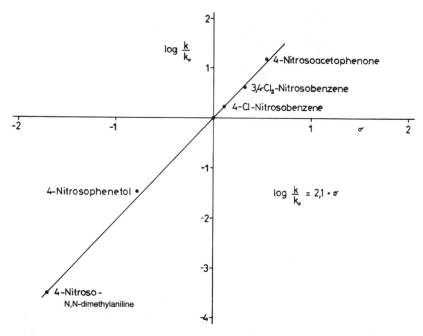

Fig. 3 – Relative reactivities of various *para*-substituted nitrosoarenes in the reaction with glutathione *vs* Hammett σ constants (pH 7.4, 37 °C, N₂).

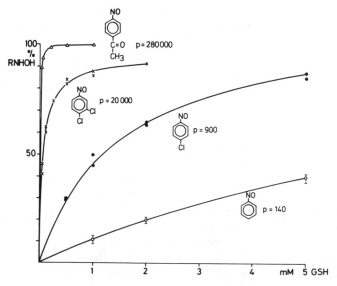

Fig. 4 – Yields of hydroxylamines in the reaction of various nitrosoarenes (0.01 mM) with GSH (pH 7.4, 37 °C, N₂, 10 min reaction).

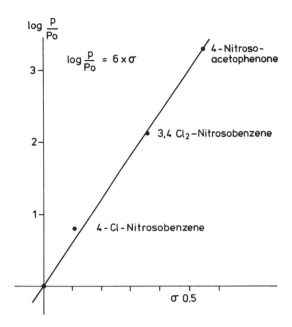

Fig. 5 – Relative formation rates of hydroxylamines in the reaction of various nitrosoarenes with GHS *vs* Hammett σ constants (pH 7.4, 37 °C, N_2). For *P* see Figs 1 and 4.

whereas with nitrosoacetophenone, hydroxylamine formation prevailed and we detected only traces of the sulphinamide. A Hammett plot using the constant *P* as a measure of hydroxylamine formation *vs* Hammett σ constant is shown in Fig. 5. The reaction constant ρ, which has a value of 6, is remarkably high. As expected from these data, nitrosoarenes with highly negative σ constants, like 4-nitrosophenetol and 4-nitroso-*N,N*-dimethylaniline, do not form any detectable amounts of the hydroxylamine. Interestingly, sulphinamide formation was also poor with 4-nitrosophenetol, and was not detected at all with 4-nitroso-*N,N*-dimethylaniline (Diepold *et al.*, 1982). These latter compounds are mainly reduced to the amine with formation of two equivalents of glutathione disulphide. Similar observations have also been reported by Neumann and coworkers with 4-nitrosotoluene, a compound with a Hammett σ constant somewhere between that of nitrosobenzene and 4-nitrosophenetol. From their results (Dölle *et al.*, 1980) they proposed a pathway (Fig. 6) by which the semimercaptal may add yet another thiol, leading to an intermediate complete mercaptal, which might be successively cleaved by further thiols leading to an intermediate sulphenamide, and finally to the amine.

Mulder *et al.* (1982) have published results of reactions of nitrosofluorene with glutathione. These authors have succeeded in isolating and identifying a glutathione sulphinamide and a glutathione sulphenamide. The latter compound decomposed readily to aminofluorene, a process that was accelerated

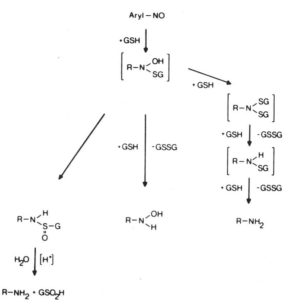

Fig. 6 – Tentative scheme of reactions of nitrosoarenes with mercaptans (from Dölle *et al.*, 1980).

by high concentrations of glutathione. These results support the reaction pathway previously proposed by Neumann's group (see above). Nevertheless, the existence of a semimercaptal and a complete mercaptal is still hypothetical and awaits further elucidation. To this end we tried another experimental approach (see next section).

REACTIONS OF NITROSOBENZENE WITH SELECTED THIOLS TO STABLE SEMIMERCAPTAL INTERMEDIATES

In the last few months, we have studied reactions of nitrosobenzene with various selected thiols in order to get more stable intermediates (Klehr & Eyer, unpublished work†). For steric reasons, tert-butylmercaptan was thought to be a particularly appropriate thiol to give a more stable semimercaptal because of the shielding methyl groups, which should hinder the sulphinamide rearrangement and the access of another thiol. Unfortunately, semimercaptal formation with this thiol is also a very slow process and has a second order rate constant of only 4 M^{-1} s^{-1} at pH 7.4 and 37 °C, compared with 5×10^3 obtained with glutathione. The only reaction product detected was the sulphinamide. Because of the low water-solubility of tert-butylmercaptan, adduct formation could not be sufficiently accelerated by increasing the concentration of both reactants.

---

†Part of H. Klehr's PhD Thesis.

We had more success with the water-soluble 1-thioglycerol. Since phenylhydroxylamine formation is significant only at rather high thiol concentrations (above 10 mM), we were able to work with larger quantitites and succeded in isolating two crystalline products: a sulphinamide and a sulphenamide, both of which were characterised by ^{1}H-n.m.r. and ^{13}C-n.m.r., mass and i.r. spectroscopy. Since we have no indication of an intermediate complete mercaptal (by means of h.p.l.c. analysis), the reaction mechanism of sulphenamide formation still remains obscure. In contrast to Mulder *et al.* (1982), our data do not indicate that the sulphenamide may be formed by a direct reaction of phenylhydroxylamine with a thiol. The apparent second order rate constants of semimercaptal formation in the reactions of nitrosobenzene with various thiols are presented in Table 1. The marked differ-

Table 1

Apparent second order rate constants of semimercaptal formation with nitrosobenzene at pH 7.4, 37 °C (M^{-1} s^{-1})

RSH	k_{obs}	pK_{SH}[a]	k'
HS-CH$_2$-CH$_2$OH	1 300	9.6	200 000
HS-CH$_2$-CH$_2$NH$_2$	12 000	8.35	120 000
HS-CH$_2$-CHOH-CH$_2$OH	1 700		
HS-CH$_2$-CH(NH$_2$)-COOH	12 000	8.6	200 000
HS-C-(CH$_3$)$_3$	4	10.7	8 000
HS-C$_6$H$_5$	32	6.5	40
Glutathione	5 000	8.75	120 000

[a]Data from Jocelyn (1972).
k_{obs} = -d(nitrosobenzene)/dt × (nitrosobenzene)$^{-1}$ × (RSH)$^{-1}$.
k' = -d(nitrosobenzene)/dt × (nitrosobenzene)$^{-1}$ × (RS$^-$)$^{-1}$.

ences in reactivities observed at pH 7.4 are partly due to the different pK values of the SH groups (Jocelyn, 1972), since the thiolate anion reacts in such nucleophilic addition reactions. Hence, the difference of k' is much smaller with the exception of tert-butylmercaptan (possibly a steric effect) and thiophenol (possibly an electronic effect).

Taken together, it is apparent that all nitrosoarenes and thiols investigated to date form the assumed semimercaptal. Because this reaction is reversible, isolation of the semimercaptal may be difficult. From this intermediate, sulphinamide rearrangement, hydroxylamine and sulphenamide formation lead to secondary, rather unstable products. It can be anticipated that these novel reaction pathways might be interesting for the synthetic chemist, since they readily lead to sulphinamides and sulphenamides.

NITROSOARENE—THIOL REACTIONS: BIOLOGICAL SIGNIFICANCE

What may be especially interesting, however, is the biological significance of the reactions outlined in the sections above. A few examples of our studies with red cells (Eyer *et al.*, 1980; Diepold *et al.*, 1982; Eyer *et al.*, 1984), and with the isolated perfused rat liver (Eyer *et al.*, 1980; Eyer & Kampffmeyer, 1982; Eckert *et al.*, 1983) will be discussed in order to indicate the potential biological role of such reactions.

In red cells, hydroxylamines and oxyhaemoglobin are co-oxidized to nitro-soarenes and ferrihaemoglobin. Because of an enzymatic cycle, first described by Kiese *et al.* (1950), in which the nitrosoarene is reduced back to the hydroxylamine, severe methaemoglobinaemia can occur even with small, catalytic amounts of hydroxylamines. This cycle, however, does not go on indefinately, since nitrosoarenes are disposed of in red cells by side reactions, e.g. reduction to the amine and reactions with thiols. The marked influence of glutathione can be well demonstrated for 4-nitrosophenetol, which is thought to be the main cause of methaemoglobinaemia in infants after administration of phenacetin. In the experiment depicted in Fig. 7, we incubated freshly dialysed human haemolysate with 0.1 mM-4-nitrosophenetol in the absence or presence of 1 mM-glutathione. As seen in the left panel, ferrihaemoglobin formation proceeded as long as nitrosophenetol was present, and only small amounts of phenetidine were formed. As shown in the right panel, glutathione quickly removed nitrosophenetol, terminated ferrihaemoglobin

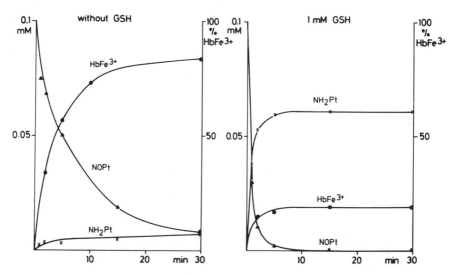

Fig. 7 – Influence of GSH on the decrease in nitrosophenetol (NOPt), ferrihaemoglobin (HbFe^{3+}) and phenetidine (NH$_2$Pt) formation in the reaction of nitrosophenetol with haemolysate. (10 g Hb/100 ml of freshly dialysed human haemolysate was reacted with 0.1 mM-nitrosophenetol ±1 mM-GSH in the presence of 0.02 mM-NADPH in a regenerating system; pH 7.4, 37 °C, air; means of 3–4 experiments).

formation at a much lower level, and caused a remarkable increase in phenetidine formation. In addition, glutathione diminished covalent binding of nitrosophenetol to haemoglobin, which is responsible for the lack of soluble metabolites.

Recently we extended our studies to reactions of nitrosochloramphenicol with red blood cells. The nitroso analogue of chloramphenicol, which was first synthesized by Corbett & Chippko (1978), was shown by Yunis *et al.* (1980) to be considerably more toxic to bone marrow cells and mitochondria than chloramphenicol itself. This compound may be the toxic metabolite responsible for the rare aplastic anaemia after chloramphenicol administration. If the *para* nitro group of chloramphenicol is reduced by intestinal bacteria or human liver, reaction products will at first be in contact with the blood before reaching the critical target (i.e. bone marrow). Hence, it seemed rational to study the fate of nitrosochloramphenicol in human blood *in vitro*.

We studied the reaction of nitrosochloramphenicol with glutathione (Eyer & Schneller, 1983), purified human haemoglobin, washed red cells, plasma, and whole blood (Eyer *et al.*, 1984). The results are summarized in Fig. 8. More than 90% of the nitrosochloramphenicol was eliminated from red cell suspensions within 15 seconds. The most important reaction in red cells is the rapid semimercaptal formation with glutathione (second order rate constant 5.5×10^3 $\text{M}^{-1}\text{s}^{-1}$). As in the case of nitrosobenzene, the semimercaptal rearranges to the sulphinamide ($k = 0.05$ s^{-1}), or is thiolytically cleaved by another glutathione with formation of the hydroxylamine ($k = 7.1$ $\text{M}^{-1}\text{s}^{-1}$). The reaction of nitrosochloramphenicol with the reactive sulphydryl groups in haemoglobin (the cysteine residues in the β-chains, position 93) also yields a

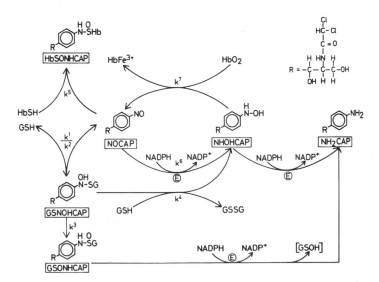

Fig. 8 – Reactions of nitrosochloramphenicol in red cells. Ⓔ indicates enzymatic reactions.

sulphinamide. This reaction appears to be monophasic because adduct formation is probably the rate-limiting step ($k = 5$ M^{-1} s^{-1}).

In addition to the above reactions, nitrosochloramphenicol is reduced enzymatically to the hydroxylamine in the presence of NADPH. This reaction, with a k_M of 10^{-5} M for nitrosochloramphenicol and an apparent V_{max} of 2×10^{-6} mole min^{-1} ml^{-1}, is only effective at rather low nitrosochloramphenicol concentrations. This is because of the limited amount of NADPH-regeneration, which normally does not exceed 80 nmole min^{-1} ml^{-1} blood. With the various reaction constants obtained by the *in vitro* experiments, it was possible to calculate reaction curves for various nitrosochloramphenicol concentrations. Such a computerized simulation indicates the importance of semimercaptal formation of nitrosochloramphenicol with glutathione, which allows the very rapid sequestration of nitrosochloramphenicol from the blood within seconds. From these data it seems very unlikely that nitrosochloramphenicol itself can be transported unchanged by the blood to bone marrow to any significant extent.

We were interested in ascertaining whether these fast reactions of nitrosoarenes with thiols also occur in the liver, where the concentration of glutathione is about 10 mM, and where *N*-oxygenation products should be highest after ingestion of an arylamine. We perfused isolated rat livers (haemoglobin-free media in a single-pass mode) with various nitrosoarenes and determined biotransformation rates and the glutathione status of the liver. By one single passage, nearly all prehepatic nitrosoarenes were reduced to the hydroxylamines and amines. With nitrosobenzene only 5% of the dose was converted into glutathione sulphinanilide. Phenylhydroxylamine formation by the glutathione pathway resulted in increased glutathione disulphide, which is usually excreted by the bile and released in the venous effluent. Such a glutathione release is also observed when glutathione disulphide is generated by other reactions, e.g. by organic hydroperoxides in the glutathione peroxidase reaction. Formation of glutathione sulphinanilide and glutathione disulphide release diminished hepatic glutathione. We were surprised, however, that only 10% of the infused nitrosobenzene had reacted with glutathione. The major part of nitrosobenzene was reduced by enzymes with high specific activity and a low k_M for nitrosoarenes, like alcohol dehydrogenase, which has a k_M for most monocyclic nitrosoarenes of around 10^{-6} M.

In contrast to nitrosobenzene and nitrosochloramphenicol, 4-nitrosophenetol reacted more intensely with liver glutathione, and about 0.2 equivalents of hepatic glutathione was lost per equivalent of nitrosophenetol infused. Moreover, nitrosophenetol markedly inhibited the bile flow of the perfused liver. The most exciting result with nitrosophenetol, however, was its inhibitory effect on the excretion of glutathione disulphide. In order to decide whether the lowered glutathione release was due to decreased glutathione disulphide formation or inhibition of the excretory mechanism, we stimulated at first the glutathione disulphide release by continuous infusion with tert-butylhydroperoxide. Upon a short pulse with

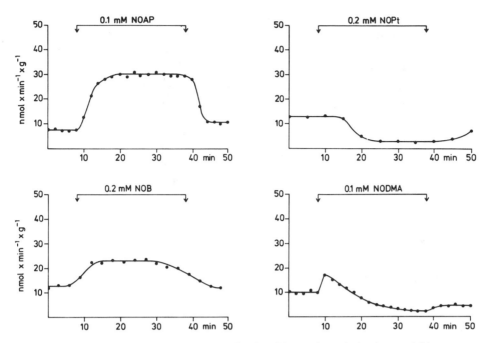

Fig. 9 – Influence of nitrosoarenes on the glutathione release during haemoglobin-free single-pass perfusion of isolated rat livers. (NOAP = 4-nitrosoacetophenone; NOB = nitrosobenzene; NOPt = 4-nitrosophenetol; NODMA = 4-nitroso-N,N-dimethylaniline). Glutathione release: nmol per min per g liver wet wt.

4-nitrosophenetol, this stimulated glutathione release was irreversibly inhibited (Eyer & Kampffmeyer, 1982).

This strange behaviour of nitrosophenetol is also shared by 4-nitroso-N,N-dimethylaniline. In Fig. 9 four different nitrosoarenes are compared with respect to their effects on the glutathione release from the isolated perfused rat liver. As expected, nitrosoacetophenone, which reacts with glutathione predominantly yielding glutathione disulphide formation, behaves like a hydroperoxide. The effect of the twofold dose of nitrosobenzene on glutathione excretion is smaller. The same load with nitrosophenetol inhibits glutathione excretion, and the same effect is observed with 4-nitroso-N,N-dimethylaniline, even at half that dose. Whether nitrosoarenes with electron-donating substituents form transient glutathionyl conjugates which compete for the glutathione disulphide excretion is under current investigation.

CONCLUSIONS

In conclusion, it is apparent that reactions of nitrosoarenes with thiols may be of biological significance. Such reactions are responsible for the limitation of an otherwise fatal methaemoglobinaemia, as exemplified for nitrosophenetol. However, these reactions also cause covalent binding to protein thiol groups, deplete cellular glutathione and enhance cellular levels of glutathione disulphide, which may disturb enzyme regulation by mixed disulphide formation or impair membrane function. Finally, we do not know whether the various glutathionyl derivatives can be regarded as real detoxication products. As pointed out by Mulder *et al.* (1982), sulphenamides may undergo protonation of the sulphur atom, resulting in loss of glutathione and formation of an electrophilic nitrenium ion. Hence, reactions of glutathione with nitrosoarenes in the liver may also provide an appropriate transport form of a reactive intermediate which exerts its toxic or carcinogenic effect on a distant target.

ACKNOWLEDGEMENT

This study was supported by Deutsche Forschungsgemeinschaft, Schwerpunktprogramm: *Mechanismen toxischer Wirkungen von Fremdstoffen*.

REFERENCES

Corbett, M. D. & Chippko, B. R. (1978), *Antimicrob. Agents Chemother.*, **13**, 193.

Cramer, J. W., Miller, J. A. & Miller, E. C. (1960), *J. biol. Chem.*, **235**, 885.

Diepold, C., Eyer, P., Kampffmeyer, H. & Reinhardt, K. (1982), in *Biological Reactive Intermediates: 2. Chemical Mechanisms and Biological Effects* (eds. Snyder, R., Parke, D. V., Kocsis, J. J., Jollow, D. J., Gibson, G. G. & Witmer, C. M.) Plenum Press, New York and London, p. 1173.

Dölle, B., Töpner, W. & Neumann, H.-G. (1980), *Xenobiotica*, **10**, 527.

Eckert, K.-G., Eyer, P. & Kampffmeyer, H. (1983), *Arch. Pharmac. Suppl.*, **322**, R 107.

Eyer, P. (1979), *Chem.-Biol. Interact.*, **24**, 227.

Eyer, P., Kampffmeyer, H., Maister, H. & Rösch-Oehme, E. (1980), *Xenobiotica*, **10**, 499.

Eyer, P. & Lierheimer, E. (1980), *Xenobiotica*, **10**, 517.

Eyer, P. & Kampffmeyer, H. (1982), *Chem.-Biol. Interact.*, **42**, 209.

Eyer, P. & Schneller, M. (1983), *Biochem. Pharmac.*, **32**, 1029.

Eyer, P., Lierheimer, E. & Schneller, M. (1984), *Biochem. Pharmac.*, **33**, 2299.

Jocelyn, P. C. (1972), *Biochemistry of the SH Group*, Academic Press, London and New York.

Kiese, M., Reinwein, D. & Waller, H. (1950), *Arch. exp. Path. Pharmak.*, **210**, 393.

Kiese, M. (1959), *Arch. exp. Path. Pharmak.*, **235**, 354.

Kiese, M. & Taeger, K. (1976), *Arch. Pharmac.*, **292**, 59.

Klehr, H. (1985), Ph.D. Thesis, Ludwig-Maximilians-Universität München.

Kosower, N. S., Kosower, E. M. & Wertheim, B. (1969), *Biochem. Biophys. Res. Commun.*, **37**, 593.

Lotlikar, P. D., Miller, E. C., Miller, J. A. & Margreth, A. (1965), *Cancer Res.*, **25**, 1743.

Mulder, G. J., Unruh, L. E., Evans, F. E., Ketterer, B. & Kadlubar, F. F. (1982), *Chem.-Biol. Interact*, **39**, 111.

Yunis, A. A., Miller, A. M., Salem, Z., Corbett, M. D. & Akimura, G. K. (1980), *J. Lab. Clin. Med.*, **96**, 36.

Chapter 44

THE REACTIONS OF *C*-NITROSO AROMATICS WITH α-OXO ACIDS

M. D. Corbett and **Bernadette R. Corbett**, Centre for Environmental Toxicology, The University of Florida, Gainesville, Florida 32611, USA

1. The adventitious reaction of *C*-nitroso aromatic compounds with catalytic intermediates of thiamine-containing enzymes results in the conversion of the nitroso group to one of three possible types of hydroxamic acids.
2. Evidence is summarized which suggests that some thiamine-dependent enzymes might contribute to the overall metabolism of arylamines or nitroaromatic xenobiotics in mammalian species.
3. The unusual chemical reaction between *C*-nitroso aromatic compounds and glyoxylic acid in aqueous solution results in the production of hydroxamic acids possessing the *N*-formyl group.
4. The proposed chemical mechanism for the nitroso–glyoxylate reaction is contrasted to the mechanism which explains the action of thiamine-dependent enzymes on *C*-nitroso aromatic compounds.

INTRODUCTION

Two types of aromatic amine metabolites of particular interest to biochemical toxicologists are the aromatic *C*-nitroso compounds and the arylhydroxyl-amines along with their acyl derivatives, the *N*-aryl hydroxamic acids. These functional groups are intermediate in oxidation state between the much less reactive aromatic amine and aromatic nitro compounds. Extensive research from a number of laboratories over the past two decades has established the toxicological significance of such intermediary oxidation states (Weisburger

& Weisburger, 1973; Hlavica, 1982). In general, the genotoxicity of aromatic amines is the result of the metabolic oxidation to the hydroxylamine state, which is at the same oxidation state as is the hydroxamic acid group. Further oxidation of the hydroxylamine can occur chemically or enzymatically to give the C-nitroso metabolite (Corbett *et al.*, 1980; Corbett & Corbett, 1983). Some evidence suggests that the hydroxylamine metabolite can react directly with DNA to give a covalent adduct, which can be the initial insult that eventually leads to a mutagenic or carcinogenic response. The facile redox interconversion of the hydroxylamine and nitroso oxidation states would allow for an identical pathway of nitroso aromatic binding to nucleic acid. On the other hand, it may be necessary in some cases for further bioactivation of the hydroxylamine (e.g. O-esterification) before a sufficiently electrophilic metabolite is produced. Hydroxamic acids are proximate genotoxic metabolites that require further bioactivation (e.g. O-sulphation, N–O acyltransfer) in order to react with DNA. Obviously the metabolic generation of hydroxylamines, hydroxamic acids and nitroso compounds in humans is an undesirable process. A complete knowledge of the production, reactions and ultimate fate of such reactive metabolites might provide us with the ability to prevent the genotoxic effects of those chemicals that are metabolized to such species. Much of the research from our laboratory over the past 10 years has been directed to a better understanding of the biochemical reactions of aromatic nitroso compounds.

Two new and quite general reactions of aromatic nitroso compounds have been elucidated in our laboratory. The two reactions initially appear to be quite similar to one another since both reactions involve the condensation of a nitroso compound with an α-oxo-acid, and the products of both reactions are hydroxamic acids. The similarity of the two reactions does not go beyond these common features, since the mechanisms of the two reactions are probably quite different. The nature of the N-acyl group in the product hydroxamic acids is also unique for each of the basic reactions, and for this reason the two reactions account for the production of some new and unusual types of hydroxamic acids.

One of the general reactions requires a thiamine-containing enzyme to bring about the reaction of aromatic nitroso compounds with either certain α-oxo-acids or keto sugars, while the other reaction is a purely chemical reaction that is independent of thiamine. Both of the reactions can occur under physiological conditions.

THIAMINE ENZYME MEDIATED CONVERSION OF NITROSO AROMATICS TO HYDROXAMIC ACIDS

On the basis of the accepted chemical mechanism by which thiamine pyrophosphate (TPP) functions as an enzyme cofactor, it was postulated that the 'active aldehyde' forms of TPP should react with C-nitroso functional groups by transfer of a C_2 unit (Corbett, 1974). The reality of this proposed

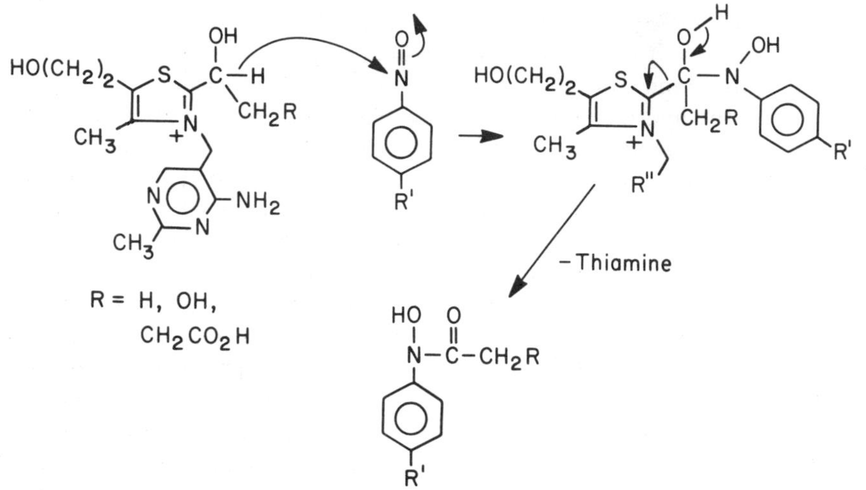

Fig. 1 – Nucleophilic mechanism for thiamine-catalysed conversion of nitroso aromatics to hydroxamic acids.

reaction was first demonstrated by the reaction of α-hydroxyethylthiamine (also known commonly as 'active acetaldehyde') with nitrosobenzene at 70 °C for 3 hours in dimethylformamide (Fig. 1; R,R′ = H). A low yield (13%) of *N*-hydroxyacetanilide was isolated from the reaction mixture (Corbett, 1974). Figure 1 depicts the mechanism originally proposed for such thiamine-catalysed reactions. The mechanism is now referred to as the nucleophilic mechanism for hydroxamic acid production to distinguish it from another potential mechanism that was subsequently proposed (Corbett & Chipko, 1980).

Following the discovery that the proposed reaction between nitrosobenzene and 'active acetaldehyde' did, in fact, result in the production of *N*-hydroxyacetanilide, we turned our attention to enzymes known to employ thiamine (as TPP) as their cofactor. These studies were greatly facilitated by our development of a special solvent that enabled us to analyse hydroxamic acids by high performance liquid chromatography (h.p.l.c.). The addition of a small amount of desferal mesylate to buffered aqueous methanolic solvents eliminated the pronounced tailing of hydroxamic acids that occurred during reverse-phase partition chromatography on C-18 bonded h.p.l.c. packings. Desferal mesylate is a trihydroxamic acid with an extremely high affinity for ferric ion. We proposed (Corbett & Chipko, 1979) that the desired effect of desferal mesylate was due to its ability to strongly chelate traces of ferric ion that are present in the matrix of silica gel packings. With the iron so sequestered, it became possible to chromatograph aromatic hydroxamic acids efficiently without the chemisorption effects that arise from hydroxamate–ferric chelation. With this h.p.l.c. technique, we have been able to analyse directly and quantitatively aliquots of enzymatic reactions or cell culture media for the presence of hydroxamic acid products.

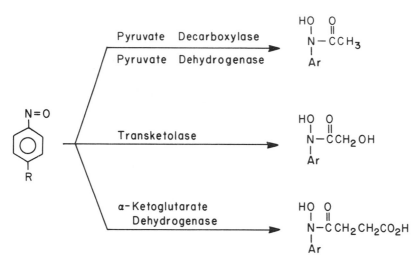

Fig. 2 – Thiamine-catalysed enzymatic conversion of nitroso aromatics to hydroxamic acids.

There are four common enzymes that utilize TPP as a cofactor, and which we have employed in our research. These are pyruvate decarboxylase, pyruvate dehydrogenase, α-ketoglutarate dehydrogenase and transketolase. The chemistry at the active sites of these distinct enzymes is quite similar in that the bound TPP cofactor acts to stabilize an acyl carbanion, and then either to pass it on to an acceptor carbonyl group or to undergo oxidation and hydrolysis to a carboxylic acid (Breslow, 1958; Ullrich *et al.*, 1970). The ability of a *C*-nitroso xenobiotic to substitute for the carbonyl acceptor or otherwise to interact with an enzyme bound acyl carbanion formed the basis for our original hypothesis on hydroxamic acid production. As expected on the basis of the generally accepted mechanisms for TPP-dependent enzymes, the nature of the *N*-acyl group in the hydroxamic acid product depends totally on the type of enzyme under investigation. Figure 2 illustrates this relationship. The *N*-acetyl type of hydroxamic acid produced by the pyruvate-metabolizing enzymes is a well known type of hydroxamic acid; however, the *N*-glycolyl and *N*-succinyl types of hydroxamic acids are essentially newly discovered functionalities. In addition to the expected differences of the *N*-acyl group in hydroxamic acid products, these enzymes have displayed certain unique features during their action on *C*-nitroso aromatic substrates.

Pyruvate decarboxylase was found to convert 4-chloronitrosobenzene to the corresponding *N*-acetyl hydroxamic acid; however, this was not the major product. H.p.l.c. analysis revealed that 4-chlorophenylhydroxylamine was the major product (Corbett & Chipko, 1980). This hydroxylamine was not produced by chemical or enzymatic hydrolysis of the corresponding hydroxamic acid. An alternative mechanism was proposed that more readily explains the production of these two products via a redox process, and is shown in Fig. 3. The reaction of pyruvate decarboxylase with nitroso aromatics is best

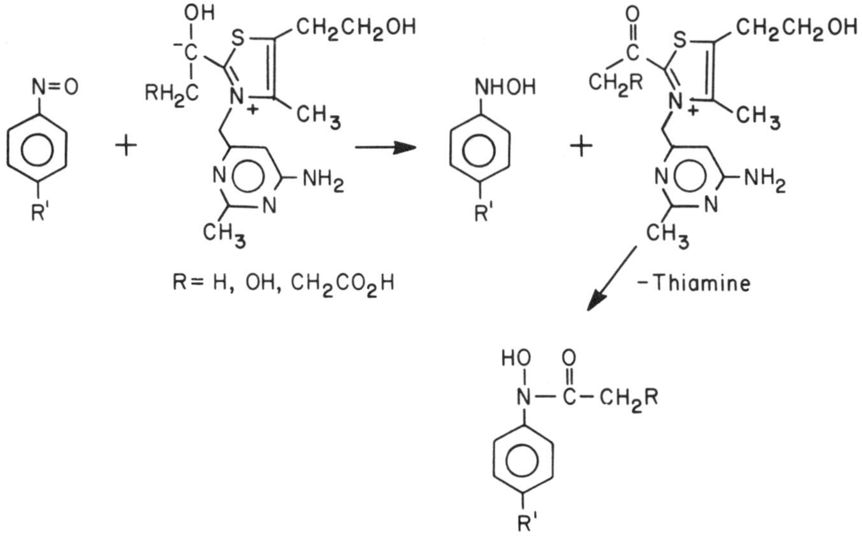

Fig. 3 – Redox mechanism for thiamine-catalysed conversion of nitroso aromatics to hydroxamic acids.

explained by the redox mechanism, while the action of other thiamine-dependent enzymes can be readily explained by the nucleophilic mechanism (Fig. 1).

The products resulting from the action of α-ketoglutarate dehydrogenase on 4-chloronitrosobenzene suggested that the active site of this enzyme catalyses a Bamberger-like rearrangement on an enzyme-bound arylhydroxylamine or hydroxamic acid (Corbett *et al.*, 1982; 1983a). The hydroxamic acid product from this enzymatic rection is the *N*-succinyl type (Fig. 2). The ring hydroxylated isomer of this hydroxamic acid is also a major product, and has been proposed to result from an active-site directed rearrangement. The enzyme does not catalyse the rearrangement of exogenous hydroxamic acid to the *o*-hydroxylated succinanilide (Corbett *et al.*, 1983a). These general results have been found for three different sources of the enzyme (Corbett & Doerge, in press).

Transketolase was found to convert nitrosobenzene to *N*-hydroxyglycolanilide when either D-fructose-6-phosphate or D-xylulose-5-phosphate was used as the C$_2$ donor (Corbett & Chipko, 1977). As expected, the reaction appears to be general for other nitroso aromatics (Corbett & Chipko, 1980; Corbett & Corbett, unpublished results). To date only the hydroxamic acid product has been detected in transketolase reactions, although we have evidence of a second minor metabolite of unknown structure. Most recently research with the known carcinogen, 4-nitrosobiphenyl, has shown that transketolase from *Escherichia coli* will convert this nitroso aromatic to the

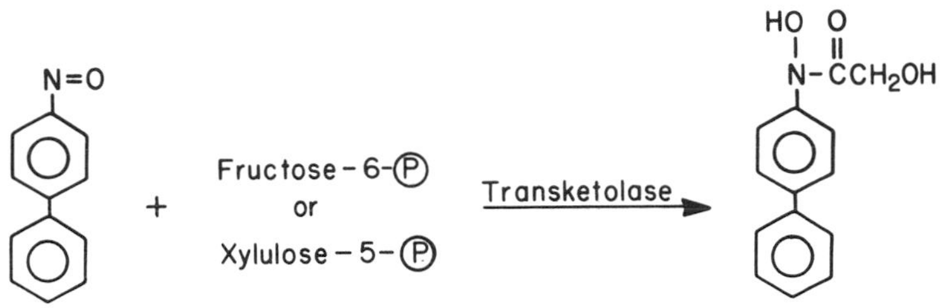

Fig. 4 – Transketolase-catalysed conversion of 4-nitrosobiphenyl to the glycol-derived hydroxamic acid.

glycol-derived hydroxamic acid (Fig. 4). Similar results were obtained with rat liver transketolase.

The transketolase-catalysed conversion of nitroso aromatics to the glycol-derived hydroxamic acids is currently receiving the greatest attention in this laboratory. The basis for this interest comes from our observation that the most frequently observed hydroxamic acid product in *in vitro* and *in vivo* systems is the glycol type of hydroxamic acid. The unique structure of this type of hydroxamic acid is strong evidence that in complex biochemical systems it is transketolase alone that accounts for the production of this unique metabolite. This transketolase-type of metabolite from a nitroso aromatic has been observed in a unicellular alga, *Chlorella pyrenoidosa* (Corbett *et al.*, 1978), rat liver homogenate (Corbett *et al.*, 1979), primary rat hepatocytes and the leukocyte fraction of human blood (Corbett & Corbett, unpublished results). The pathway from a nitroso aromatic metabolite to the glycol-type of hydroxamic acid is probably not a major metabolic pathway in mammals, since this type of metabolite has never been reported for a xenobiotic outside of our own studies. Nevertheless, the toxicological significance of even a minor metabolic pathway could be of considerable overall importance.

THE NITROSO GLYOXYLATE REACTION

The direct chemical reaction between the nitroso functional group and glyoxylic acid to produce a hydroxamic acid was discovered quite by accident. During a study of the action of pyruvate decarboxylase on 4-chloronitrosobenzene in the presence of unusual α-oxo-acid substrates, the rather rapid reaction of these two chemicals was found to proceed in the absence of enzyme (Fig. 5). The resulting product was the N-formyl hydroxamic acid (N-hydroxyformanilide), and the reaction was found to be a general property of probably all C-nitroso aromatics (Corbett & Corbett, 1980). Numerous other α-oxo-acids and simple aldehydes have been investigated for the ability to convert a nitroso aromatic to a hydroxamic acid under modest conditions in aqueous solvents. These studies have shown that this reaction is unique to glyoxylic acid.

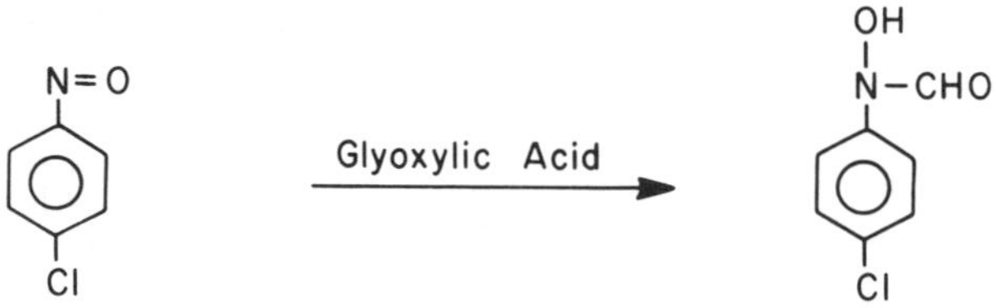

Fig. 5 – The reaction of 4-chloronitrosobenzene with glyoxylic acid.

There was no precedence for this organic reaction in the chemical literature, so the general features of the novel reaction were elucidated. Most
notable was the observation that the condensation of glyoxylic acid with a
nitroso aromatic proceeded in aqueous solutions, and only very slowly, if at
all, in organic solvents. The reaction was first-order with respect to each of the
other reactants, and the rate of the reaction was increased by the presence of
electron-donating groups in the *para* position. Radiotracer experiments with
$[1-^{14}C]$glyoxylate demonstrated the release of $^{14}CO_2$ during the reaction, and
also showed that the *N*-formyl group of the hydroxamic acid product arose
from the C-2 of glyoxylic acid. A possible mechanism for the nitroso-
glyoxylate reaction is presented in Fig. 6 (Corbett & Corbett, 1980). Most
noteworthy about this proposed mechanism is that the electron-pair on the
nitrogen atom of the nitroso functional group serves as a nucleophile. This
contrasts to most reactions of *C*-nitroso compounds, which are proposed to
proceed via an initial nucleophilic attack on the nitroso group. In fact, the
thiamine-enzyme catalysed reactions provide a very good example of a reaction that proceeds by nucleophilic attack on the nitroso functional group.

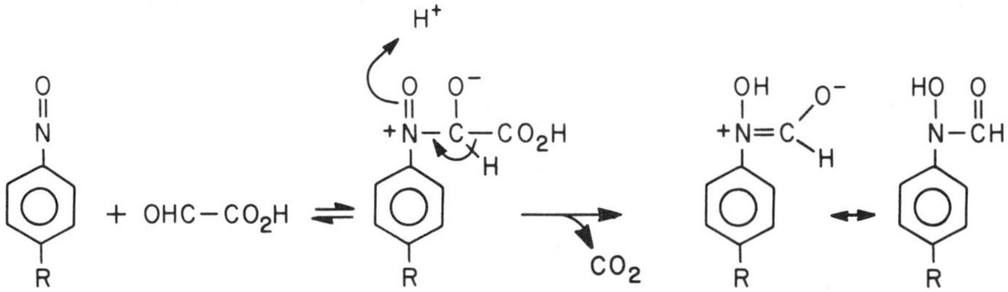

Fig. 6 – Proposed mechanism of the nitroso–glyoxylate reaction.

The nitroso-glyoxylate reaction has proved to be a very convenient
synthetic route to *N*-hydroxyformanilides (Corbett & Corbett, 1980; Corbett
et al., 1983b). Perhaps of even greater importance is the likely possibility that
this reaction could occur *in vivo* in certain situations, and therefore might be
another metabolic reaction to which *C*-nitroso aromatics are subjected.

CONCLUSION

Basic chemical principles allowed for the prediction of the reactions of nitroso aromatics with certain intermediates of thiamine-dependent enzymes. In contrast, it was a degree of serendipity that led to the discovery of the nitroso–glyoxylate reaction. Both types of reaction are certainly compatible with physiological conditions; thus, it remains to be shown how important these processes are to the overall metabolic fate of nitroso aromatics. In general, the conversion of a *C*-nitroso metabolite to a hydroxamic acid would be considered as a detoxification process, even though the product hydroxamic acids are themselves considered to be proximate genotoxicants. Since hydroxamic acids are much more stable than *C*-nitroso analogues (in a typical biochemical system), it is possible that such conversions of nitroso compounds serve to latentiate the potential toxicity of the nitroso group. This possibility and the toxicity of the more unusual types of hydroxamic acids are currently under investigation.

ACKNOWLEDGEMENTS

This research has been supported by grants from the National Cancer Institute and the National Institute of Occupational Safety and Health and by a Research Career Development Award from the National Institute of Environmental Health Sciences of the United States Department of Health and Human Services.

REFERENCES

Breslow, R. (1958), *J. Amer. Chem. Soc.*, **80**, 3719.
Corbett, M. D. (1974), *Bioorg. Chem.*, **3**, 361.
Corbett, M. D. & Chipko, B. R. (1977), *Biochem. J.*, **165**, 263.
Corbett, M. D., Chipko, B. R. & Paul, J. H. (1978), *J. Envir. Pathol. Tox.*, **1**, 259.
Corbett, M. D. & Chipko, B. R. (1979), *Anal. Biochem.*, **98**, 169.
Corbett, M. D., Baden, D. G. & Chipko, B. R. (1979), *Bioorg. Chem.*, **8**, 227.
Corbett, M. D. & Chipko, B. R. (1980), *Bioorg. Chem.*, **9**, 273.
Corbett, M. D. & Corbett, B. R. (1980), *J. Org. Chem.*, **45**, 2834.
Corbett, M. D., Chipko, B. R. & Batchelor, A. O. (1980), *Biochem. J.*, **187**, 893.
Corbett, M. D., Corbett, B. R. & Doerge, D. R. (1982), *J. Chem. Soc. Perkin Trans. I*, 345.
Corbett, M. D. & Corbett, B. R. (1983), *J. Agric. Food Chem.*, **31**, 1276.
Corbett, M. D., Doerge, D. R. & Corbett, B. R. (1983a), *J. Chem. Soc. Perkin Trans. I*, 765.

Corbett, M. D., Wei, C., Fernando, S. Y., Doerge, D. R. & Corbett, B. R. (1983b), *Carcinogenesis*, **4**, 1615.
Hlavica, P. (1982), *CRC Crit. Rev. Biochem.*, **12**, 39.
Ullrich, J., Ostrovsky, Y. M., Eyzaguirre, J. & Holzer, H. (1970), *Vitam. Horm.*, **17**, 365.
Weisburger, J. H. & Weisburger, E. K. (1973), *Pharmacol. Revs.*, **25**, 1.

Chapter 45

MECHANISM OF HAEMOGLOBIN OXIDATION BY N-HYDROXY-N-ARYLACETAMIDES IN VIVO AND IN VITRO*

W. Lenk and Heidrun Sterzl, Institut für Pharmakologie und Toxikologie der LM-Universität München, D-8000 München 2, Nussbaumstr. 26, FRG.

1. In rats in vivo, N-hydroxy-4-chloroacetanilide did not itself produce ferri-haemoglobin, but by conversion to N-hydroxy-4-chloroaniline.
2. In erythrocytes, N-hydroxy-4-chloroacetanilide itself produced ferri-haemoglobin and was co-oxidized to 4-chloroacetanilide and 4-chloro-nitrosobenzene.
3. Purified haemoglobin was co-oxidized with N-hydroxy-4-chloroacetanilide to yield 4-chloroacetanilide, 4-chloronitrosobenzene, and 4-chloronitro-benzene.

INTRODUCTION

Aromatic amines and their N-acetyl derivatives can be N-oxygenated both in vivo and in vitro to yield arylhydroxylamines and N-hydroxy-N-aryl-acetamides (i.e. N-arylacetohydroxamic acids), respectively. These products have in common the ability to oxidize haemoglobin as an expression of their acute toxicity, and to initiate tumour growth as an expression of their chronic toxicity. The ferrihaemoglobin-forming activity, as well as the carcinogenic activity of arylhydroxylamines and N-hydroxy-N-arylacetamides, depends on the aryl residue attached to the nitrogen atom. We have compared

---

*Part of the PhD thesis of H. Sterzl, LM-Universität München, 1984.

the mechanism of haemoglobin oxidation by *N*-hydroxy-*N*-arylacetamides with that of arylhydroxylamines at three different levels of molecular haemoglobin organization, i.e. with oxyhaemoglobin in erythrocytes *in vivo*, with oxyhaemoglobin in erythrocytes *in vitro*, and with purified human haemoglobin.

The abbreviations used in this text are as follows:

$HbFe^{2+}$	ferrohaemoglobin; desoxyhaemoglobin
$HbFe^{3+}$	ferrihaemoglobin
HbO_2	oxyhaemoglobin
CO-Hb	carbonylhaemoglobin
NOH-A	*N*-hydroxyaniline, phenylhydroxylamine
NOH-4EA	*N*-hydroxy-4-ethoxyaniline, *N*-hydroxy-4-phenetidine
NOH-4ClA	*N*-hydroxy-4-chloroaniline
NOH-3.4Cl$_2$A	*N*-hydroxy-3,4-dichloroaniline
NOH-4AB	*N*-hydroxy-4-aminobiphenyl
NOH-2AF	*N*-hydroxy-2-aminofluorene
NOH-AA	*N*-hydroxyacetanilide, *N*-acetylphenylhydroxylamine
NOH-4EAA	*N*-hydroxy-4-ethoxyacetanilide, *N*-hydroxyphenacetin
NOH-4ClAA	*N*-hydroxy-4-chloroacetanilide
NOH-3.4Cl$_2$AA	*N*-hydroxy-3,4-dichloroacetanilide
NOH-4AAB	*N*-hydroxy-4-acetylaminobiphenyl
NOH-2AAF	*N*-hydroxy-2-acetylaminofluorene
NO-4ClB	4-chloronitrosobenzene
NO$_2$-4ClB	4-chloronitrobenzene
4-ClAA	4-chloroacetanilide
4-ClA	4-chloroaniline

EXPERIMENTAL

Chemicals and methods

Synthesis and molecular properties of arylhydroxylamines, *N*-hydroxy-*N*-arylacetamides, of 4-chloroacetanilide, 4-chloronitroso-, and 4-chloronitrobenzene have been described recently (Lenk *et al.*, in press).

Female Sprague–Dawley rats of 230–360 g weight were used for the *in vivo* and *in vitro* experiments; the animals were housed in stainless steel cages and fed Alma diet for rats and mice and water *ad libitum*. Bovine erythrocytes were obtained by centrifugation of bovine blood from the slaughterhouse, which contained 20 g sodium citrate and 2 g glucose per litre. They were washed three times with 0.9% saline. Purified human haemoglobin was prepared from human erythrocytes as described by Eyer *et al.* (1975). Carbonylhaemoglobin was prepared by passing CO at room temperature through a solution of purified human haemoglobin. After addition of a small amount of $Na_2S_2O_4$, the solution was applied to a column of Sephadex^R G 25; CO-Hb was eluted by using CO-saturated 0.2 M-phosphate buffer pH 7.4.

HbFe^{3+} concentration was determined at 540 or 550 nm by measuring the increase in absorbance caused by addition of CN$^-$ to the sample cuvette as described by Kiese (1974).

Determination of N-hydroxy-4-chloroacetanilide, 4-chloroacetanilide, 4-chloroaniline, 4-chloronitroso-, and 4-chloronitrobenzene

Blood of rats injected i.p. with NOH-4ClAA and bovine erythrocyte suspensions incubated with NOH-4ClAA were haemolysed by addition of an equal volume of water. Haemolysates or solutions of human haemoglobin incubated with NOH-4ClAA were shaken twice with distilled diethyl ether; the combined ether extracts were shaken twice with 2 M-NaOH to remove NOH-4ClAA from the neutral and basic metabolites in the ether extract (=*ether 1*). The alkaline phases were acidified and NOH-4ClAA was re-extracted into ether (=*ether 2*). The residual amount of NO-4ClB bound to HbFe^{2+} was extracted into ether after oxidation of the haemolysate with 10% K$_3$Fe(CN)$_6$ (=*ether 3*).

The components of *ether 1* and *ether 3* extracts were identified and determined by h.p.l.c. using a μ-Bondapak C$_{18}$-column and methanol–water (70:30 v/v) as solvent. The following retention times were found for the synthetic compounds: 4-ClAA: 2.7 min, 4-ClA: 2.4 min, NO-4ClB: 4.1 min, and NO$_2$-4ClB: 3.6 min. Known amounts of these compounds either injected as single compounds or as a mixture facilitated the determination of the metabolites.

NOH-4ClAA present in *ether 2* extracts was determined either directly by reading the absorbance at 257 nm of its methanol solution or as NO-4ClB according to Herr & Kiese (1959) after oxidation by K$_3$Fe(CN)$_6$. Blood of rats injected with NOH-4ClA and bovine erythrocyte suspensions as well as solutions of human haemoglobin incubated with NOH-4ClA were analysed for NO-4ClB, NO$_2$-4ClB, and 4-ClA as described above by determining the metabolites in *ether 1* and *ether 3* extracts, except that no *ether 2* extract was prepared, because NOH-4ClAA was not present.

RESULTS

Haemoglobin oxidation by various arylhydroxylamines and N-hydroxy-N-arylacetamides in the rat

In 1959, Hustedt and Kiese described experiments on haemoglobin oxidation in cats and dogs by N-hydroxyacetanilide. A few minutes after i.v. injection of NOH-AA, high HbFe^{3+} and nitrosobenzene concentrations were determined in the blood of cats, whereas after i.v. injection of higher doses of NOH-AA into dogs, lower HbFe^{3+} and nitrosobenzene concentrations were determined in the blood, indicating that the hydroxamic acid itself did not produce HbFe^{3+}, but after N-deacetylation to yield the active phenylhydroxyl-

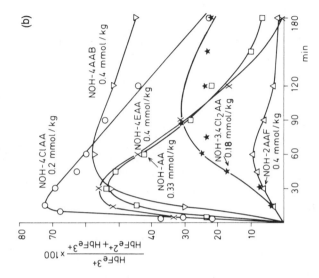

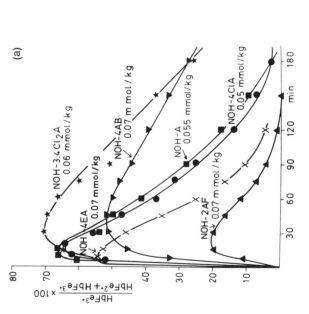

Fig. 1 – Kinetics of ferrihaemoglobin-formation *in vivo* by arylhydroxylamines or their *N*-acetyl derivatives.
(a) HbFe^{3+}-formation in female rats after i.p. injection of the doses of six arylhydroxylamines indicated in the graph. Symbols indicate means from experiments on six rats each and NOH-4EA, NOH-4AB, NOH-2AF and on three rats each and NOH-A, NOH-4ClA, and NOH-3.4Cl$_2$A. For i.p. injection, the pure compounds were suspended in 0.25% agar (prepared from 0.9% saline) and injected as quickly as possible.
(b) HbFe^{3+}-formation in male and female rats after i.p. injection of the doses of six *N*-hydroxy-*N*-arylacetamides indicated in the graph. Symbols indicate means from experiments on nine female rats and NOH-2AAF, six female and three male rats and NOH-4AAB, four female rats and NOH-AA, and three female rats each and NOH-4ClAA and NOH-3.4Cl$_2$AA. For i.p. injection, the pure compounds were finely ground and suspended in 0.25% agar (prepared from 0.9% saline).

amine. Hustedt and Kiese (1959) implicated enzymatic N-deacetylation in the liver as a prerequisite for the $HbFe^{3+}$-forming activity of this compound.

We have tested the 4 monocyclic arylhydroxylamines NOH-A, NOH-4EA, NOH-4ClA, and NOH-3.4Cl$_2$A and the 2 polycyclic analogs NOH-4AB and NOH-AF for their $HbFe^{3+}$-forming capability in the rat. The results, illustrated in Fig. 1(a) showed, that NOH-4ClA exceeded the other arylhydroxylamines tested in their activity and that NOH-2AF was the least active compound. We have also tested the corresponding N-acetyl derivatives in the rat. The results, illustrated in Fig. 1(b) showed that (i) NOH-4ClAA was the most active and NOH-2AAF was the least active hydroxamic acid tested, and (ii) synthetic N-acetylation decreased the $HbFe^{3+}$-forming activity of arylhydroxylamines, 4- to 14-fold higher molar doses of the N-arylacetohydroxamic acids being required for the production of the same maximal $HbFe^{3+}$ concentrations in the rat. If it is true that N-arylacetohydroxamic acids are active *in vivo* only after N-deacetylation, then the different molar ratios of arylhydroxylamine to N-arylacetohydroxamic acid necessary to produce the same maximal $HbFe^{3+}$ concentration, would reflect differences in the efficiency of N-deacetylation and(or) in the availability of the corresponding arylhydroxylamine (or nitrosoarene) for oxyhaemoglobin in erythrocytes. Therefore we have carried out the following experiments with NOH-4ClAA and NOH-4ClA to demonstrate the differences in the mechanism of haemoglobin oxidation, if there are any, between N-arylacetohydroxamic acids and arylhydroxylamines. Three minutes after i.p. injection of 0.27 mmol/kg NOH-4ClAA in the rat, similar $HbFe^{3+}$ and NO-4ClB concentrations were determined in the blood as 3 min after i.p. injection of 0.06 mmol/kg NOH-4ClA. As is shown in Fig. 2, a

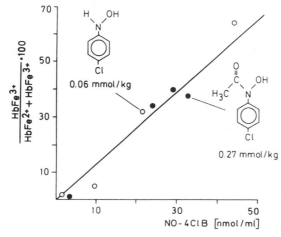

Fig. 2 – Ferrihaemoglobin and 4-chloronitrosobenzene in the blood after 0.27 mmol/kg N-hydroxy-4-chloroacetanilide (●) or 0.06 mmol/kg N-hydroxy-4-chloroaniline (○). Symbols indicate data from single experiments with four female rats.

linear relationship was established between $HbFe^{3+}$ and NO-4ClB concentration up to 60% $HbFe^{3+}$, whether NOH-4ClAA or NOH-4ClA was injected. This finding proved that NO-4ClB solely accounted for the $HbFe^{3+}$ concentrations determined and that indeed *N*-arylacetohydroxamic acids oxidize haemoglobin *in vivo* after *N*-deacetylation.

When we determined the kinetics of $HbFe^{3+}$-formation in the rat after i.p. injection of 10 mg/kg NOH-4ClAA both in the absence and presence of phosphoric acid bis(4-nitrophenyl ester) (50 mg/kg, i.p. injected 30 min before), the inhibitor of microsomal carboxylesterases *in vivo* (Heymann *et-al.*, 1969) did not affect the kinetics of NOH-4ClAA-induced haemoglobin oxidation. This finding is an indication that *N*-deacetylation of NOH-4ClAA *in vivo* may be prediated less by enzymic than by oxidative mechanisms.

Oxidation of haemoglobin in bovine erythrocytes *in vitro* by *N*-hydroxy-4-chloroacetanilide and *N*-hydroxy-4-chloroaniline

On incubation of bovine erythrocytes (9 mM-$HbFe^{2+}$) with NOH-4ClA (10 μM), 730 mol $HbFe^{3+}$ were produced per mol NOH-4ClA within 3 h in the presence of 11 mM-glucose, but 140 mol in the absence of glucose, and 90 mol $HbFe^{3+}$ per mol NOH-4ClA in the presence of 18 mM-lactate, see Fig. 3(b). The observation that glucose increased the apparent catalytic activity of NOH-4ClA 5.2-fold, is explained with the metabolic regeneration of NOH-4ClA from the oxidation product NO-4ClB by NADPH-dependent diaphorases of the erythrocyte cytosol. Lactate apparently decreased the catalytic activity of NOH-4ClA by 40%, because (i) metabolic regeneration of NOH-4ClA by NADH-dependent diaphorases apparently was not stimulated and (ii) NADH-dependent diaphorases can accelerate $HbFe^{3+}$-reduction.

In contrast, on incubation of bovine erythrocytes (9 mM-$HbFe^{2+}$) with NOH-4ClAA (0.3 mM), only 25 mol $HbFe^{3+}$ were produced per mol NOH-4ClAA within 3 h in the presence or absence of glucose or lactate, see Fig. 3(a). This is an indication that NOH-4ClAA did not require metabolic regeneration for its activity, and that the hydroxamic acid itself was capable of oxidizing haemoglobin without prior *N*-deacetylation, although being 90-fold less active than the corresponding arylhydroxylamine.

In order to establish a relationship between haemoglobin oxidation and product formation, which might help elucidate the mechanism of haemoglobin oxidation by *N*-arylacetohydroxamic acids, we have analysed the reaction mixture of bovine erythrocytes and NOH-4ClAA at 2, 30, and 67 min of incubation at 37 °C by t.l.c. and h.p.l.c. The results shown in Table 1 showed that with increasing $HbFe^{3+}$ concentration, the concentration of 4-ClAA and 4-ClA increased, whereas the concentration of NO-4ClB and NO_2-4ClB remained constant after an initial increase. This is an indication that the chemically inert and non-volatile 4-ClAA and 4-ClA accumulated during $HbFe^{2+}$ oxidation, but that NO-4ClB escaped quantitative determination by its (i) volatility, (ii) reaction with the SH-groups of glutathione and(or) cysteine moieties of haemoglobin to yield 4-ClA, and (iii) partial autoxidation to NO_2-4ClB.

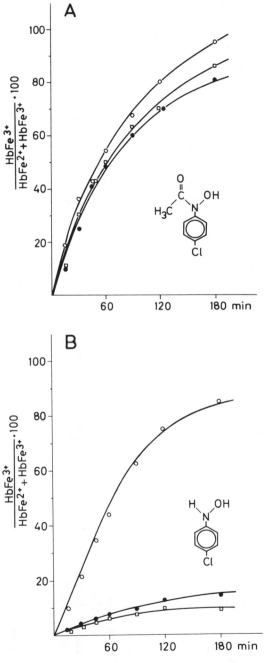

Fig. 3 – Kinetics of ferrihaemoglobin formation in bovine erythrocytes *in vitro* by
N-hydroxy-4-chloroacetanilide (a) or *N*-hydroxy-4-chloroaniline (b). Symbols represent data from single experiments with bovine erythrocytes (9 mM-HbFe^{2+}) and
0.3 mM-NOH-4ClAA (a) or 10 μM-NOH-4ClA (b) in Krebs–Ringer phosphate
buffer pH 7.4 at 37 °C containing 11 mM-glucose (—O—O—), 18 mM-sodium lactate (—□—□—) or no additive (—●—●—).

Table 1

Product pattern of the reaction of *N*-hydroxy-4-chloroacetanilide and 4-chloronitrosobenzene with oxyhaemoglobin in bovine erythrocytes *in vitro*

t (min) substrate	1.05 mM *N*-hydroxy-4-chloro-acetanilide			1.0 mM 4-chloronitrosobenzene		
	2	30	67	2	30	67
HbFe^{3+} (%)	3.6	57.1	87.3	22.3	64.4	83.8
N-Hydroxy-4-chloroacetanilide recovered (μM)	934.3	769.7	630	—	—	—
4-Chloroacetanilide (μM)	0.5	12.6	24.2	—	—	—
4-Chloronitrosobenzene (μM)	0.1	5.3	5.3	401.2	318.9	216.7
4-Chloronitrobenzene (μM)	0.1	1.1	1.1	7.9	8.2	7.2
4-Chloroaniline (μM)	0.3	1.1	4.2	12.9	25.8	36.4
data from *n* exp.	1	1	1	2	2	2

HbFe^{2+} concn.: 10 mM and 9 mM, respectively.
Krebs–Ringer phosphate buffer pH 7.4, containing 11 mM-glucose; incubation temperature 37 °C. concns. were determined by h.p.l.c.

NO-4ClB produced from NOH-4ClAA during haemoglobin oxidation obviously did not participate in the oxidation of $HbFe^{2+}$, because the major fraction was rapidly inactivated by reduction to 4-ClA. Whereas NOH-4ClA was completely (95%) oxidized to NO-4ClB during haemoglobin oxidation ('co-oxidation'), only 4–7% of the employed NOH-4ClAA was accounted for by the four metabolites. These results have provided support for a mechanism of haemoglobin oxidation by N-arylacetohydroxamic acids, by which only a small fraction of NOH-4ClAA was co-oxidized with haemoglobin.

Oxidation of purified human haemoglobin by N-hydroxy-4-chloroacetanilide and N-hydroxy-4-chloroaniline

In order to prevent metabolic regeneration of NOH-4ClA, $HbFe^{3+}$-reduction, and modification of the product pattern derived from the co-oxidation of NOH-4ClAA, erythrocyte membrane and cytosolic enzymes were excluded by using purified human haemoglobin.

On addition of 2.7 mM-NOH-4ClA to haemoglobin (2.7 mM-$HbFe^{2+}$), at once 0.7 mol $HbFe^{3+}$ was formed per mol NOH-4ClA; lower NOH-4ClA concentrations caused the oxidation of 1 mol $HbFe^{3+}$ per mol NOH-4ClA, see Fig. 4(b). This is an indication that NOH-4ClA lacked catalytic activity in the absence of reducing equivalents. As can also be seen from Fig. 4(b), a second slow phase of haemoglobin oxidation was observed, which we interpret as the transfer of electrons from $HbFe^{2+}$ complexed with NO-4ClB to molecular oxygen, by which also part of the ligand NO-4ClB was reduced to 4-ClA.

In contrast, 2.7 mM-NOH-4ClAA oxidized haemoglobin (2.7 mM-$HbFe^{2+}$) much slower, but 1 mol NOH-4ClAA oxidized up to 45 mol $HbFe^{2+}$ depending on NOH-4ClAA concentration, see Fig. 4(a). This is an indication for the limited catalytic activity of NOH-4ClAA. Since NOH-4ClAA is co-oxidized during haemoglobin oxidation, we implicate acetyl 4-chlorophenyl nitroxide as the catalytically active molecule.

Haemoglobin oxidation by NOH-4ClA as well as NOH-4ClAA both followed second order kinetics, rate constants being determined to 7840 ± 300 litres mol^{-1} sec^{-1} for NOH-4ClA and 3.4 ± 0.2 litres mol^{-1} sec^{-1} for NOH-4ClAA. We have also determined the activation energy for haemoglobin oxidation by NOH-4ClA as 3.5 kcal mol^{-1} and by NOH-4ClAA as 12.7 kcal mol^{-1}, indicating that the two-electron oxidation of NOH-4ClA to NO-4ClB was energetically more favoured than the postulated one-electron oxidation of NOH-4ClAA to the secondary aromatic nitroxide. Relevant half-wave potentials are given in the discussion section.

In order to establish a relationship between haemoglobin oxidation and product formation, we have analysed the reaction mixture of haemoglobin (3.13 mM-$HbFe^{2+}$) and NOH-4ClAA (1.05 mM) at 2 and 60 min incubation at 37 °C by h.p.l.c. As can be seen from Table 2, the concentration of 4-ClAA increased with increasing $HbFe^{3+}$ concentration, indicating that it was formed by oxidation. But the concentration of NO-4ClB and NO_2-4ClB decreased after an initial increase, indicating losses due to their volatility and

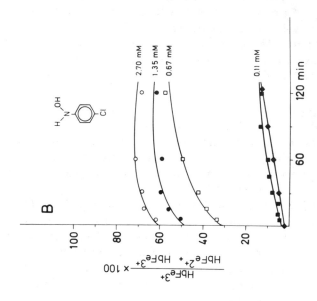

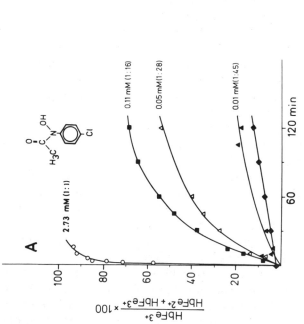

Fig. 4 – Kinetics of ferrihaemoglobin-formation in solutions of purified human haemoglobin by *N*-hydroxy-4-chloroacetanilide (a) or *N*-hydroxy-4-chloroaniline (b).
A solution of isolated haemoglobin (2.73 mM-HbFe²⁺) in 0.2 M-phosphate buffer pH 7.4 was incubated at 37 °C with various concns. of NOH-4ClAA (a) or NOH-4ClA (b) indicated in the graph. Symbols mean data from single experiments and are corrected on the basis of haemoglobin autoxidation (––◆––).

Table 2

Product pattern of the reaction of N-hydroxy-4-chloroacetanilide with oxyhaemoglobin and carbonylhaemoglobin and of 4-chloronitrosobenzene with oxyhaemoglobin *in vitro*

	3.13 mM HbFe²⁺		1.0 mM CO-Hb	Control		3.3 mM HbFe²⁺ 4-chloronitroso-benzene (mM) 3.0	
N-Hydroxy-4-chloroacetanilide (mM)	1.05		1.0	1.0			
t (min)	2	60	60	2	60	2	60
HbFe³⁺ (%)	42	96	—	—	—	8.5	68.2
N-Hydroxy-4-chloroacetanilide recovered (μM)	794	776	872	898	936	—	—
4-Chloroacetanilide (μM)	19	53	1	1	1	—	—
4-Chloronitrosobenzene (μM)	9	5	3	1	1	912	601
4-Chloronitrobenzene (μM)	7	4	1	1	1	14	23
4-Chloroaniline (μM)	*	*	*	*	*	1	15
means from n exp.	3	3	4	4	4	1	1

*traces, not unambiguously identified.
0.2 M-phosphate buffer pH 7.4, incubation temperature 37 °C.
concs. were determined by h.p.l.c.
enzyme source: purified human haemoglobin.

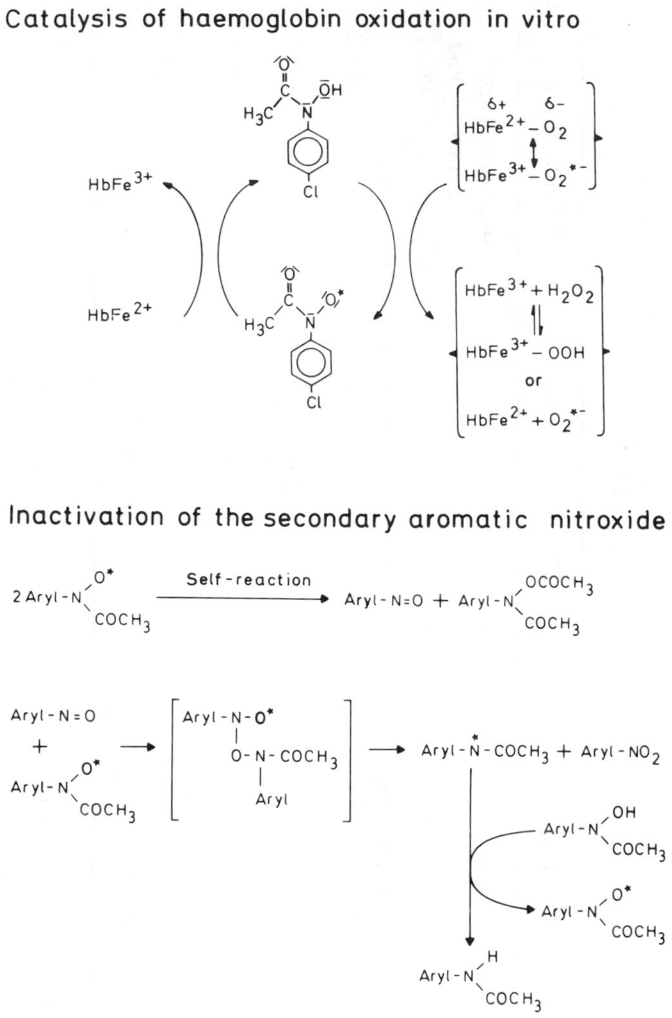

Catalysis of haemoglobin oxidation in vitro

Inactivation of the secondary aromatic nitroxide

Scheme 1 – Catalysis   of   haemoglobin   oxidation   *in vitro*   by   *N*-hydroxy-4-chloroacetanilide and inactivation of the secondary aromatic nitroxide according to Forrester *et al.* (1970).

In the first section the mechanism of haemoglobin oxidation by NOH-4ClAA is depicted; we consider the formation of acetyl-4-chlorophenyl nitroxide the rate limiting step in the coupled oxidation; since superoxide dismutase and catalase did not affect $HbFe^{2+}$-oxidation, we assume that the dioxygen species of HbO_2 is not released as O_2^* or H_2O_2, but is fully reduced *in situ*.

In the second section reactions are outlined, by which acetyl-4-chlorophenyl nitroxide, produced by oxidation of 0.1 M NOH-4ClAA in benzene with PbO_2 or $KMnO_4$ undergoes self-reaction. Experimental conditions and product pattern (two pairs of products) were as described by Forrester *et al.* (1970) for NOH-AA and NOH-4AAB.

reactivity of NO-4ClB with the cysteine moieties of haemoglobin. Therefore, 4-ClAA, NO-4ClB, and NO_2-4ClB are considered oxidation products, which probably have arisen from acetyl 4-chlorophenyl nitroxide by self-reaction. These products account for approximately 7% of NOH-4ClAA. As can also be seen from Table 2, these products were not formed in the absence of molecular oxygen or oxyhaemoglobin. This is proof, that $HbFe^{2+}$ and NOH-4ClAA are co-oxidized only in the presence of oxyhaemoglobin.

DISCUSSION

If NOH-4ClAA is allowed to react with HbO_2 *in vitro*, $HbFe^{2+}$ is completely oxidized. $HbFe^{3+}$ formation is accompanied by the release of molecular oxygen. But since less oxygen is released than was present as HbO_2, the dioxygen species of the remaining HbO_2 was the acceptor for electrons from $HbFe^{2+}$ and the external donors NOH-4ClAA or NOH-4ClA, see Scheme 1. We consider the formation of the catalytically active radical as the rate-limiting step of the coupled oxidation and assume that it catalyses the transfer of electrons from $HbFe^{2+}$ to the dioxygen species of HbO_2, and that its (rare) self-reaction yielded 4-ClAA, NO-4ClB, and NO_2-4ClB according to a mechanism postulated by Forrester *et al.* (1970) for the chemical oxidation of *N*-arylacethydroxamic acids. Such an assumption is in agreement with the laws of thermodynamics, since electrons can flow from $HbFe^{2+}$ ($E_{1/2} = +125$ mV, Antonioni *et al.*, 1964) and NOH-4ClAA ($E_{1/2} = +610$ mV, Riedl, 1983) or NOH-4ClA ($E_{1/2} = +80$ mV, Riedl, 1983) to the dioxygen species of HbO_2, which has superoxide character ($O_2^* + 2H^+ + e \rightleftharpoons H_2O_2$, $E_{1/2} = +900$ mV; $H_2O_2 + 2H^+ + 2e \rightleftharpoons 2H_2O$, $E_{1/2} = +1350$ mV, James, 1978). HbO_2 therefore acts as a superoxidase.

In vivo, however, NOH-4ClA formed from NOH-4ClAA by enzymatic and(or) oxidative *N*-deacetylation, is metabolically regenerated from NO-4ClB several times and produces $HbFe^{3+}$ much faster than its *N*-acetyl derivative. It remains to be demonstrated which portion of NO-4ClB and 4-ClAA is produced in the liver by enzymatic *N*-deacetylation and reduction, respectively, and which portion in erythrocytes by oxidation.

REFERENCES

Antonioni, E., Wyman, J., Brunori, M., Taylor, J. F., Rossi-Fanelli, A. & Caputo, A. (1964), *J. Biol. Chem.*, **239**, 907.

Eyer, P., Hertle, H., Kiese, M. & Klein, G. (1975), *Molec. Pharmacol.*, **11**, 326.

Forrester, A. R., Ogilvy, M. M. & Thomson, R. H. (1970), *J. Chem. Soc. (C) (London)*, 1081.

Herr, F. & Kiese, M. (1959), *Naunyn-Schmiedeberg's Arch. exp. Path. Pharmacol.*, **235**, 351.

Heymann, E., Krisch, K., Büch, H. & Buzello, W. (1969), *Biochem. Pharmacol.*, **18**, 801.

Hustedt, G. & Kiese, M. (1959), *Naunyn-Schmiedeberg's Arch. exp. Path. Pharmacol.*, **236**, 435.

James, B. R. (1978), in *The Porphyrins*, Vol. 5 (ed. Dolphin, D.) Academic Press, London, p. 209.

Kiese, M. (1974), *Ferrihaemoglobinemia, a comprehensive treatise*. CRC press Inc., Cleveland, Ohio, pp. 3–5.

Lenk, W., Riedl, M. & Sterzl, H., submitted for publication to *Xenobiotica*.

Riedl, M. (1983), Diplomarbeit LM-Universität München.

Chapter 46

CYTOTOXIC ACTIVITY AND MODE OF ACTION OF SOME NEW *N*-OXIDES OF *N,N*-DIALKYLAMINOALKYLESTERS OF DODECANOIC ACID

M. Miko, Department of Microbiology and Biochemistry, Slovak Polytechnic University, 812 37 Bratislava, Czechoslovakia

F. Devinsky[†], Department of Inorganic and Organic Chemistry, Faculty of Pharmacy, Comenius University, 832 32 Bratislava, Czechoslovakia

1. Cytotoxicity and mode of action of some *N*-oxides of *N,N*-dialkylaminoalkylesters of dodecanoic acid have been studied.
2. On the basis of primary screening, one of the most active compounds, namely the *N*-oxide of 3-(*N,N*-dimethylaminopropyl)ester of dodecanoic acid has been chosen for detailed biochemical study.
3. The drug inhibited the incorporation rate of ^{14}C-precursors (adenine, valine, thymidine, uridine) into appropriate macromolecules of leukemia cells P 388 and Ehrlich carcinoma.
4. The amine oxide also interferes with energy-yielding processes (aerobic glycolysis, endogenous respiration).
5. Cytotoxicity is a consequence of the cytolytic activity of the compounds mentioned above. Membranous effects were demonstrated by measuring marker enzyme activities (LDH, MDH), protein concentration in cells and in the culture medium, as well as by morphological examination.

[†]To whom correspondence should be addressed.

INTRODUCTION

Amine oxides represent a large group of compounds derived from tertiary amines containing a strongly polarized N–O bond (Linton, 1940; Culvenor, 1953; French & Gens, 1973). A great number of amine oxides occurring in nature, or prepared synthetically, are biologically active compounds (anti-metabolites and chemotherapeutics, psychotropic and cancerostatic compounds, etc.). Though some non-aromatic amine oxides have found wide industrial utilization due to their good surface active properties (Lindner, 1964; Nowak, 1970), relatively little attention has been paid to their biological activity, in contrast to aromatic amine oxides (Ochiai, 1967; Bickel, 1969; Katritzky & Lagowski, 1971).

In view of the interesting biological activities shown by these compounds (Hlavica, 1982), in the present study we have investigated the cytotoxic activities and mode of action of the compounds mentioned above. The chemical structure of the substances studied has shown in Fig. 1 ($X = O, n = 2, 3$). Synthesis, properties and antimicrobial activity of the compounds have recently been described (Devínsky *et al.*, 1983). The compounds of this type belong to the so-called 'soft' antimicrobially active compounds (Bodor, 1980). The *N*-oxide of 3-(*N,N*-dimethylaminopropyl)ester of dodecanoic acid was used for the isolation of ADP and ATP from heart mitochondria (Kraemer *et al.*, 1977).

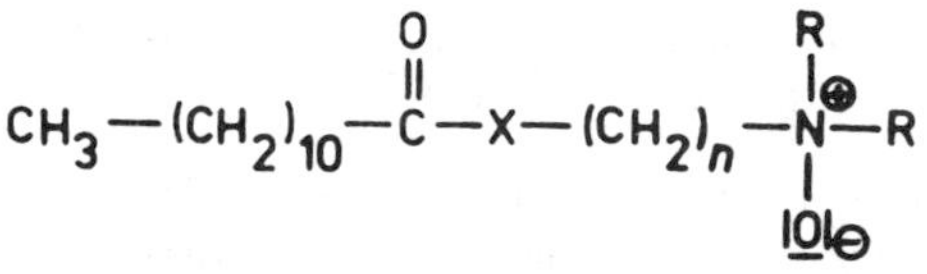

$$CH_3 - (CH_2)_{10} - \overset{O}{\overset{\|}{C}} - X - (CH_2)_n - \overset{R}{\underset{\underset{\ominus}{|O|}}{\overset{|\oplus}{N}}} - R$$

Fig. 1 – The chemical structure of *N*-oxides of *N,N*-dialkylaminoalkylesters of dodecanoic acid ($X = O, n = 2, 3$).

MATERIALS AND METHODS

The procedure used in evaluating the cytotoxic effect of the compounds was similar to that used when testing other metabolic inhibitors (Miko *et al.*, 1979a,b). The substances were dissolved in Krebs–Ringer phosphate medium shortly before experiments.

RESULTS AND DISCUSSION

Biochemical screening

A new system has been developed and is being used routinely for mass screening of candidate compounds for anticancer activity. In P 388 and Ehrlich ascites carcinoma (EAC) cells, the influence on metabolic activity is studied at selected concentrations of the substance, under defined conditions *in vitro*, by following the active synthesis of proteins and nucleic acids. Each substance was evaluated at at least four concentrations. The results from primary screening of the cytotoxic activity at EAC cells and leukemic cells P 388 are summarized in Table 1. The numbers represent c.p.m. or percentage of inhibition (or stimulation), respectively, in parentheses. The inhibitory effect was characterized by ID_{50} values (molar concentration of compound required for 50% reduction of the incorporation rate). As is seen from the results in Table 1, all three substances affected the incorporation of both precursors in both types of cells; this has been confirmed not only by percentage inhibition (in parentheses) but also by ID_{50} values. By comparing the ID_{50} values for both types of cells we can observe that for a 50% incorporation rate reduction of [^{14}C]adenine and/or [^{14}C]valine, substantially higher concentrations in case of P 388 are required than in the case of EAC cells. That means, EAC cells are 'more sensitive' to this group of compounds than P 388 cells.

On the basis of our previous results (Miko et al., 1979a,b), it is convenient to use an ID_{50} adenine : ID_{50} valine ratio, which is a suitable parameter indicating the possible primary mode of action of the substance investigated. All three ratios, as demonstrated in Table 1, are in the range 1.27 to 1.95. Such ratios are typical also for other biologically active compounds which interfere with energy-generating systems in the cells. Inhibition of energy metabolism may be due to direct interaction or through the disorganization of the membrane struture. The lengthening of the joining chain in N-oxides of N,N-dialkylaminoalkylesters of dodecanoic acid positively affected their cytotoxic activity. Maximum activity was achieved with the compound No. 3, namely the N-oxide of 3-(N,N-dimethylaminopropyl)ester of dodecanoic acid. Therefore, it was chosen for detailed biochemical studies.

Effect on macromolecule biosynthesis
The values from biochemical screening represent the first fundamental information about cytotoxic activity of new derivatives. The data obtained in a relatively short time exactly inform whether the tested substance shows cytotoxic activity at all, and perhaps also indicate the possible mechanism of action. To be able to pinpoint the primary effect of the compounds investigated, it is necessary to follow the inhibitory action in relation to both time and concentration dependence. Only when the time course is known is it possible to state at what time and concentration the inhibitory effect appears. Figure 2 demonstrates the effect of compound No. 3 (Table 1) upon the biosynthesis of macromolecules indicated by incorporation of [^{14}C]adenine and [^{14}C]valine into TCA-insoluble material of Ehrlich ascites cells. As can be seen from Fig. 2, [^{14}C]valine incorporation is inhibited more than [^{14}C]adenine incorporation. At the highest concentration, nearly complete inhibition of macromolecular biosynthesis takes place; that is true for adenine as well for valine. Some differences in adenine and valine incorporation are observed at lower concentrations. The lowest concentrations of the test substance do not significantly affect the [^{14}C]adenine incorporation rate, while valine incorporation is

Table 1

Effects of the *N*-oxides of *N,N*-dialkylaminoalkylesters of dodecanoic acid on [^{14}C]adenine (a) and [^{14}C]valine (b) incorporation into whole Ehrlich ascites carcinoma and P 388 cells

No.	n	R	Formula (m.w.)			μmol/l					ID$_{50}$	R
						0	75	150	300	600		
						Inhibition of incorporation in c.p.m. or % (in parentheses)						
1	2	CH$_3$	C$_{16}$H$_{33}$NO$_3$ (287.44)	EAC	(a)	1433(0)	1215(15.22)	970(32.31)	480(66.51)	63(95.61)	230	1.88
					(b)	1909(0)	1303(31.85)	713(62.66)	131(93.14)	64(96.65)	122	
				P388	(a)	1767(0)	1981(+12.11)	1223(30.79)	1198(32.21)	288(83.7)	405	1.95
					(b)	2303(0)	1920(16.63)	1560(32.26)	468(79.68)	132(94.27)	208	
2	2	C$_2$H$_5$	C$_{18}$H$_{37}$NO$_3$ (315.49)	EAC	(a)	1433(0)	1393(2.8)	1314(8.31)	499(65.18)	96(93.31)	260	1.42
					(b)	1909(0)	1178(38.3)	1196(37.35)	—	55(97.12)	182	
				P388	(a)	1767(0)	2085(+17.99)	1949(+10.33)	—	259(85.34)	<600	?
					(b)	2303(0)	2152(6.56)	1554(32.52)	—	171(92.57)	290	
3	3	CH$_3$	C$_{17}$H$_{35}$NO$_3$ (301.47)	EAC	(a)	1433(0)	1361(5.03)	605(57.79)	100(93.03)	124(91.35)	140	1.57
					(b)	1909(0)	1076(43.64)	362(81.04)	57(97.02)	77(95.97)	89	
				P388	(a)	1767(0)	1676(5.15)	1234(30.16)	511(71.08)	237(86.59)	225	1.27
					(b)	2303(0)	1979(14.07)	1359(40.99)	177(92.31)	166(92.79)	177	

$R = \dfrac{\text{ID}_{50}\ \text{adenine}}{\text{ID}_{50}\ \text{valine}}$

+stimulation over 100% against control sample.
Compounds were dissolved in Krebs–Ringer phosphate medium.

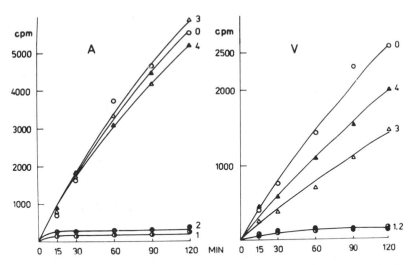

Fig. 2 – The effect of compound No. 3 (Table 1) on macromolecule synthesis of Ehrlich ascites cells. Incorporation of radioactive precursors into acid-insoluble fractions was determined by incubating EAC cells with [8-^{14}C]adenine sulphate (A), final concentration 0.312 μCi/ml and L-[U-^{14}C]valine (V), final concentration 0.275 μCi/ml, in the presence of compound No. 3 at various concentrations. Compound No. 3 and precursors were added to the cells at the same time. The test tubes were incubated at 37 °C, and 1-ml samples of each suspension were analysed for radioactivity in acid-insoluble material. The results are expressed as c.p.m./5 × 10^6 cells. Compound No. 3 concentrations: O = none, 1 = 600, 2 = 300, 3 = 150, 4 = 75 (μmol/l).

inhibited proportionally to the concentration tested. At the same time, the results appear to indicate that the inhibition takes place on addition of the substance to EAC cells suspension i.e. without a lag phase. Moreover, this substance affects protein synthesis to a higher degree than it does nucleic acid biosynthesis.

Effects on energy-yielding processes

As macromolecule biosynthesis is an energy requiring process, we investigated further the action of substance No. 3 on some bioenergetic functions. Figure 3 demonstrates the effect of compound No. 3 (Table 1) on aerobic glucose utilization and lactic acid formation by Ehrlich ascites cells. The compound, at the highest concentrations, causes a rapid and practically complete inhibition of glycolysis, as judged from the cessation of glucose consumption or lactate formation. However, at the lowest concentrations glycolysis is stimulated. The conversion of glucose to lactate in control cells is in the range 75 to 83% (Miko *et al.*, 1979a,b) of that in cells containing 75–150 μmol/l of the test substance.

Such a stimulation of glycolysis at low concentrations pointed to the potential interference with respiratory processes in EAC cells. In order to verify this, experiments were carried out; results of which are presented in Fig. 4. Sub-

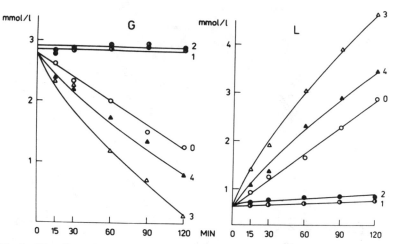

Fig. 3 – The effect of compound No. 3 (Table 1) on the kinetics of aerobic glucose utilization (G) and lactic acid formation (L) by Ehrlich ascites tumour cells. The cells were incubated at 37 °C in the presence of different concentrations of compound No. 3. The initial glucose concentration was 3 mmol/l. At various times, 1-ml samples of suspension were analysed for glucose and lactate. Other experimental conditions and symbols are the same as for Fig. 2.

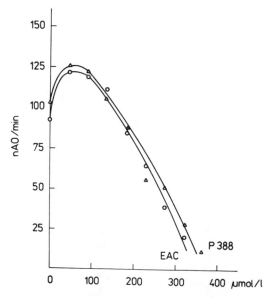

Fig. 4 – The effects of compound No. 3 (Table 1) on endogenous oxygen utilization by EAC and P 388 cells. Different concentrations of compound No. 3 were added to the respiring tumour cells. The ratio of oxygen consumption was determined immediately after the addition of compound No. 3 to the cells. Cell suspension (0.2 ml) containing 13.3 mg (EAC) and 12.44 mg (P 388) dry weight, was added to 2.0 ml of 0.9% NaCl solution–phosphate medium, pH 7.4. Oxygen uptake was measured at 30 °C.

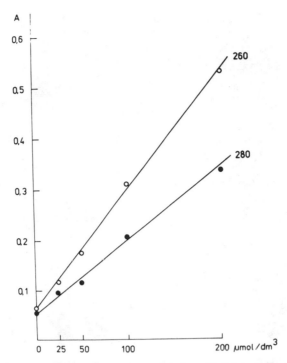

Fig. 5 – Release of ultraviolet-absorbing materials from Ehrlich ascites cells treated with compound No. 3 (Table 1) after 2 h incubation at 37 °C.

stance No. 3, at the lowest concentrations, first stimulates the endogenous respiration of tumour cells, followed by linear inhibition of oxygen consumption in both types of cells. The results from endogenous respiration indicate that the respiration is affected faster than glycolysis. In other words, the substance shows a higher affinity for mitochondria than for the cytosol, where all the glycolytic enzymes are found. Both stimulation of respiration at low concentrations and inhibition of respiration at higher concentrations are typical for uncouplers of oxidative phosphorylation. Though these results do not enable us to reach a definite conclusion about the exact mode of action, it is possible to state that the substance affects the respiratory processes of EAC cells and thereby also the production and/or utilization of ATP.

Biochemical evidence of cytolytic activity

The substance investigated showed a considerable inhibitory effect on all the metabolic processes examined, specially at the highest concentrations utilized. We assumed therefore, that the cytotoxic effect could be the consequence of cytolytic activity of the substance investigated. Therefore, we carried out experiments, results of which are presented in Fig. 5. Membranous effects were demonstrated by several methods. The results of such experiments indicate that the cells incubated with the amine oxide (25 to 200 μmol/l) for a longer

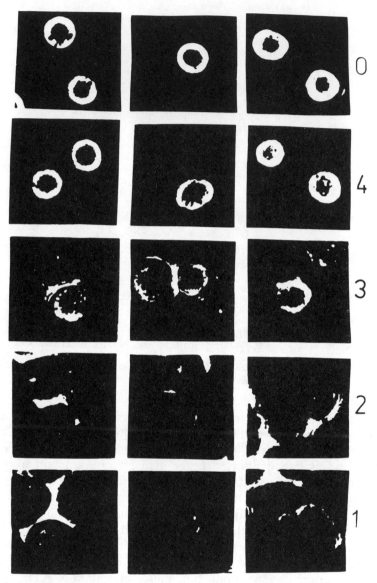

Fig. 6 – Morphological changes in Ehrlich ascites cells induced by different concentrations of compound No. 3 (Table 1) after 2 h incubation. O = none, 1 = 600, 2 = 300, 3 = 150, 4 = 75 (μmol/l).

time (2 h) released 260- and 280-nm absorbing materials, indicating damage of the treated cells (Fig. 5). Direct evidence comes from the measurement of both lactate (EC 1.1.27) and malate dehydrogenases (EC 1.1.37) activities, in the culture medium (Table 2). Lactate dehydrogenase (LDH) activity is a typical marker of cytoplasmic (soluble) proteins. Table 2 indicates a significant increase in activity of both enzymes (five times), especially at the highest

Table 2

Effect of the *N*-oxide of 3(*N*,*N*-dimethylaminopropyl)ester of dodecanoic acid (substance No. 3 Table 1) on the activities of lactate and malate dehydrogenases, respectively in the culture medium of Ehrlich ascites cells after incubatin for 2 hours

Enzyme	μmol/l of the inhibitor				
	0	75	150	300	600
	μkat/l				
LDH	1.86	2.53	2.70	9.71	9.80
MDH	2.18	2.26	2.35	10.02	10.88

Activity of LDH before experiment was 0.422 and MDH 0.436 μkat/l, respectively.

concentrations. Morphological changes in Ehrlich ascites cells induced after a 2 h incubation in the presence of different concentrations of compound No. 3 are shown in Fig. 6.

It is evident that biological membranes, which after interaction with amine oxide undergo changes in molecular organization, osmotic and permeability properties, are the site of action.

REFERENCES

Bickel, M. H. (1969), *Pharmacol. Rev.*, **21**, 325.

Bodor, N. (1980), *J. Med. Chem.*, **23**, 469.

Culvenor, C. C. J. (1953), *Rev. Pure Appl. Chem. (Australia)*, **3**, 83.

Devínsky, F., Lacko, I., Mlynarčik, D. & Krasnec, L. (1983), *Chem. Zvesti*, **37**, 263.

French, H. S. & Gens, C. M. (1973), *J. Amer. Chem. Soc.*, **59**, 2600.

Hlavica, P. (1982), *CRC Crit. Rev. Biochem.*, **12**, 39.

Katritzky, A. R. & Lagowski, J. M. (1971), *Chemistry of the Heterocyclic N-oxides*, Academic Press, London.

Kraemer, R., Aguila, H. & Klingenberg, M. (1977), *Biochemistry*, **16**, 4949.

Lindner, K. (1964), *Tenside*, **1**, 112.

Linton, E. P. (1940), *J. Amer. Chem. Soc.*, **62**, 1945.

Miko, M., Drobnica, L. & Chance, B. (1979a), *Cancer Res.*, **39**, 4242.

Miko, M., Drobnica, L., Jindra, A. & Semonsky, M. (1979b), *Neoplasma*, **26**, 449.

Nowak, G. A. (1970), *Kosmetik*, **43**, 951.

Ochiai, E. (1967), *Aromatic Amine Oxides*, Elsevier, Amsterdam.

Chapter 47

STUDIES ON THE MUTAGENIC METABOLITES OF *N*-NITROSOPYRROLIDINE USING *ESCHERICHIA COLI* DNA-REPAIR MUTANTS

A. J. Alldrick, I. R. Rowland, Theresia M. Coutts and **S. D. Gangolli**, The British Industrial Biological Research Association, Woodmansterne Road, Carshalton, Surrey SM5 4DS, UK

1. DNA repair-defective strains of *Escherichia coli* were exposed to increasing concentrations of *N*-nitrosopyrrolidine (NPYR) in the presence of an Aroclor-treated rat liver S9-preparation.
2. None of these strains exhibited increased sensitivity to the cytotoxic effects of NPYR compared with the repair-proficient parent. However strains deficient in excision-repair processes ($uvrA^-$) exhibited increased sensitivity to the mutagenic effects of NPYR.
3. Addition of SKF-525A, an inhibitor of cytochrome P-450, reduced but did not abolish the mutagenic effects of NPYR.
4. NPYR achieves its mutagenic effects by generating bulky DNA-lesions.
5. Cytochrome P-450 mediated α-hydroxylation is only one of a number of metabolic processes which can convert NPYR to a mutagenic moiety.

INTRODUCTION

All organisms possess DNA repair systems which are to a certain extent lesion-specific (reviewed by Lindahl, 1982). The availability of isogenic series of the bacterium *Escherichia coli* defective in a particular DNA repair function makes it possible to tain insight into the lesion(s) produced by a particular mutagen by comparing its cytotoxic and mutagenic effects on a number of repair-defective strains with their wild-type parent. Such a system was used by Green & Muriel (1975) in their studies with pyrrolizidine alkaloids. We have used the approach of Green and Muriel to gain further knowledge of the type of lesions produced by the cyclic nitrosamine *N*-nitrosopyrrolidine (NPYR).

This nitrosamine is found in both processed meats (Birdsall, 1977) and tobacco smoke (Hecht *et al.*, 1979). Although the pathways of NPYR metabolism have been extensively studied, little is known of the metabolites which interact with DNA or of the lesions they produce. The studies described here, therefore, were designed to gain insight into the DNA lesions produced by NPYR and by inference the metabolic routes which lead to the active mutagen.

EXPERIMENTAL

Bacteria are detailed in Table 1. Two isogenic series of *E. coli* B/r were used: WP2 and its DNA repair-defective derivatives and F26 and its derivatives showing altered sensitivity to simple alkylating agents. The mutagenic and cytotoxic potentials of known concentrations of NPYR were determined using a modification of the method of Alldrick *et al.* (1984). Exponential phase cultures of *E. coli* ($A_{580} = 0.1$) were exposed for 30 minutes to known concentrations of NPYR (challenge dose) in the presence of a liver S9 preparation derived from Aroclor 1254-treated rats (Ames *et al.*, 1975). After washing the cells, the viable count of organisms in the culture was deter-

Table 1

Bacterial strains of *E. coli* used

Strain	Genotype	Comments and Reference
WP2	*trpE*	Parent (Witkin & George, 1973)
WP2s	WP2*uvrA*	Lacks *uvr* error-free excision repair system (Hill, 1965)
CM571	WP2*recA*	Deficient in *recA* function—no *rec* dependent error-prone repair (Bridges *et al.*, 1972)
WP6	WP2*polA*	Reduced DNA polymerase I activity
F26	*his*⁻	Parent (Sedgwick & Robbins, 1980)
BS21	*adc-1*	Constitutive for repair of DNA alkylation (Sedgwick & Robbins, 1980)
BS23	*ada*⁻	Deficient in ability to repair alkylated DNA (Sedgwick, 1982)

mined. Mutagenicity was measured by determining the number of amino-acid prototroph revertants in the population.

RESULTS AND DISCUSSION

The cytotoxic effects of increasing challenge doses of NPYR (up to 460 μM) on both isogenic series were studied. NPYR was observed to exert a cytotoxic effect at concentrations above 300 mM (data not shown). However all strains irrespective of their DNA repair potential showed a similar response indicating that altered DNA repair capacity has no effect on the cytotoxic consequences of NPYR mutagenesis. Mutagenicity studies using challenge doses of 0, 46 and 96 mM-NPYR (Fig. 1) revealed that although NPYR exerted a mutagenic effect in the F26 series, the effect was the same irrespective of the DNA repair capacity of the cell. However, consideration of the results obtained with the WP2 series revealed that not only was the *uvrA*⁻ derivative the only strain to show increased NPYR mutagenicity compared with its parent, it was the only strain of this series to be mutated by NPYR.

It has been believed for some time that NPYR is converted to its mutagenic moiety by cytochrome P-450 α-hydroxylation (Hecker *et al.*, 1979). However recent studies by Cottrell *et al.* (1983) have disputed this finding since they observed that addition of SKF 525A, a mixed function oxidase inhibitor, to the S9 mix, resulted in only a slight reduction in NPYR mutagenesis. We have studied the effect of increasing concentrations of SKF 525A on the mutagenic effects of 21 and 48 mM-NPYR using *E. coli* WP2s

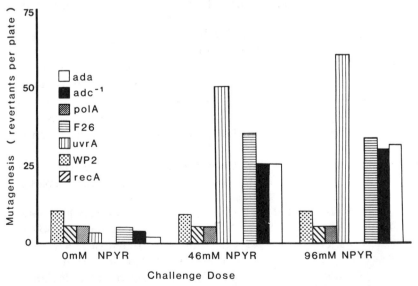

Fig. 1 – Mutagenicity of two different concentrations of NPYR on *E. coli* repair defective strains.

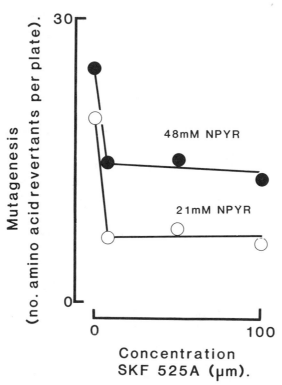

Fig. 2 – Effect of increasing concentrations of SKF 525A on NPYR mutagenesis.

(*uvrA*⁻). SKF 525A was observed to reduce (but not abolish) the mutagenic consequences of NPYR challenge (Fig. 2). This effect was greatest at a SKF 525A concentration of 10 μM and did not increase when the inhibitor concentration was increased to 100 μM.

These studies, using DNA repair-defective mutants of *E. coli*, demonstrate that NPYR exerts its mutagenic and cytotoxic effects through different mechanisms. Although possession of an altered DNA repair capacity did not alter the cytotoxic response of the cell, deficiency in the *uvr* excision repair pathway led to increased sensitivity to NPYR mutagenesis compared with its parent. This result coupled with the observation that mutagenesis was not observed in other members of the WP2 series strongly suggests that NPYR generates bulky DNA-lesions. This result is supported by the radiochemical studies of Hunt & Shank (1982) who concluded that NPYR generated a N^7-alkyl guanine derivative. The ability of NPYR to generate O^6-methyl guanine, like its aliphatic analogue *N*-nitrosodimethylamine (Pegg & Hui, 1978) seems unlikely since no increased mutagenicity was seen in BS23 (*ada*⁻), the strain defective in its ability to repair simple DNA alkylation damage. We have also observed that treatment of the S9 activation system with SKF 525A significantly inhibits but does not entirely abolish NYPR mutagenesis. The results suggests that cytochrome P-450 participates in the activation of NPYR to a mutagenic entity but that other metabolic pathways are also involved.

ACKNOWLEDGEMENTS

We thank Prof. B. Bridges and Drs B. Sedgwick and E. Witkin for bacterial strains. This work was supported by funds provided by the U.K. Ministry of Agriculture, Fisheries and Food. The results presented here are the property of the Ministry of Agriculture Fisheries and Food and are Crown Copyright.

REFERENCES

Alldrick, A. J., Rowland, I. R. & Gangolli, S. D. (1984), *Mutat. Res.*, **139**, 111.

Ames, B. N., McCann, J. & Yamasaki, E. (1975), *Mutat. Res.*, **31**, 347.

Birdsall, J. J. (1977), in *Nitrite in Food Products* (ed. Tinbergen, B. J. & Krol, B.) Proc. 2nd. Intl Symp. on Nitrite in Food Products, Zeist (1976), Wageningen Center for Agriculture Publishing and Documentation.

Bridges, B. A., Mottershead, R. P., Rothwell, A. M. & Green, M. H. L. (1972), *Chem.-Biol. Interact.*, **5**, 77.

Cottrell, R. C., Blowers, S. D., Walters, D. G., Lake, B. G., Purchase, R., Phillips, J. C. & Gangolli, S. D. (1983), *Carcinogenesis*, **4**, 311.

Green, M. H. L. & Muriel, M. J. (1975), *Mutat. Res.*, **28**, 331.

Hecht, S. S., Chen, C.-H. B. & Hoffman, D. (1979), *Acc. Chem. Research*, **12**, 92.

Hecker, L., Elespuru, R. K. & Farrelly, J. G. (1979), *Mutat. Res.*, **62**, 213.

Hill, R. F. (1965), *Photochem. Photobiol.*, **4**, 563.

Hunt, E. J. & Shank, R. C. (1982), *Biochem. Biophys. Res. Commun.*, **104**, 1343.

Lindahl, T. (1982), *Ann. Rev. Biochem.*, **51**, 61.

Pegg, A. E. & Hui, G. (1978), *Biochem. J.*, **173**, 739.

Rowland, I. R., Lake, B. G., Phillips, J. C. & Gangolli, S. D. (1980), *Mutat. Res.*, **72**, 63.

Sedgwick, B. (1982), *J. Bacteriol.*, **150**, 984.

Sedgwick, B. & Robbins, P. (1980), *Molec. gen. Genet.*, **180**, 85.

Witkin, E. M. & George, D. L. (1973), Genetics, **73** (suppl), 91.

INDEX